Laser Beam Propagation
in Nonlinear Optical
Media

Laser Beam Propagation in Nonlinear Optical Media

Shekhar Guha and Leonel P. Gonzalez

Air Force Research Laboratory

Wright-Patterson Air Force Base, OH, USA

CRC Press
Taylor & Francis Group
Boca Raton   London   New York

CRC Press is an imprint of the
Taylor & Francis Group, an **informa** business

MATLAB® is a trademark of The MathWorks, Inc. and is used with permission. The MathWorks does not warrant the accuracy of the text or exercises in this book. This book's use or discussion of MAT-LAB® software or related products does not constitute endorsement or sponsorship by The MathWorks of a particular pedagogical approach or particular use of the MATLAB® software.

The cover image shows the transverse spatial profile of a 4.6 mm wavelength beam recorded by a mid-wave infrared camera. The beam was created by second harmonic generation of a transverse excited atmospheric CO_2 laser propagated through two walk-off compensated $AgGaSe_2$ crystals.

CRC Press
Taylor & Francis Group
6000 Broken Sound Parkway NW, Suite 300
Boca Raton, FL 33487-2742

First issued in paperback 2017

© 2014 by Taylor & Francis Group, LLC
CRC Press is an imprint of Taylor & Francis Group, an Informa business

No claim to original U.S. Government works

Version Date: 20131119

ISBN 13: 978-1-4398-6638-2 (hbk)
ISBN 13: 978-1-138-07198-8 (pbk)

Visit the Taylor & Francis Web site at
http://www.taylorandfrancis.com

and the CRC Press Web site at
http://www.crcpress.com

To

Sudeshna, Rahul and Rakesh – from Shekhar

and

Trina, Noah, Ian, Connor and Joshua – from Leonel

Contents

List of Figures

List of Tables

Preface

This book arose out of a need felt by the authors, whose daily work is in experimental nonlinear optics, that existing books in their field, excellent as they are, and indeed from which they learned the subject, don't always have the right formulas or derivations needed for understanding or predicting experimental results. Some examples of topics in which the authors particularly felt this need are: walk-off angle expressions for light propagating in anisotropic crystals, detailed derivations of the effective d coefficients in uniaxial, biaxial and isotropic crystals, expressions for power and energy conversion efficiencies in various three-wave mixing processes in both homogenous or layered (quasi phase matched) materials, and the effects of beam focusing on power conversion efficiencies beyond the pioneering work by Boyd and Kleinman, for non-resonant and resonant processes.

This book is an attempt to be somewhat half-way between the monographs and textbooks by Bloembergen, Zernike and Midwinter, Yariv, Shen, Boyd, Powers etc. and the handbooks by Dmitriev, Gurzadyan and Nikogosyan, or by Sutherland. The derivations included here are often just the details left out as problems in textbooks. In the attempt to provide detailed derivations, the number of equations in this book has proliferated. Still the authors have kept the equations and derivations in the present format hoping that they may be as useful to some readers as they have been to them. The detailed derivations may also help the readers in catching any mistakes that may have crept into the formulas. The authors would appreciate it very much if the readers would let the publishers know of any mistakes they find in the book.

Many topics of current interest are not included in this book, such as ultrashort pulse effects and light propagation in nonlinear optical fibers and waveguides. The authors also didn't get a chance to delve into third-order or photorefractive nonlinearities. Some of these topics are covered in detail in books by Weiner, Agarwal and Banerjee.

The formulas and expressions in this book are presented in a manner suitable for implementation in computer programs. Some representative programs, written in MATLAB and Fortran, are included in the appendices as examples and guides. These, and programs based on these, can of course be written in other programming languages or on other platforms preferred by the reader. Numerical results for many cases of interest in nonlinear optics can also be obtained using the freely available software SNLO (http:/www.as-photonics.com).

Since a large number of symbols have been used in this book, some duplication is inevitable. An attempt was made to choose the symbols somewhat systematically, especially through the selection of subscripts and superscripts when dealing with similar parameters for different processes, such as sum-frequency generation, second harmonic generation, difference frequency generation and optical parametric oscillation.

MATLAB is a registered trademark of The MathWorks, Inc. For product information, please contact:
The MathWorks, Inc.
3 Apple Hill Drive
Natick, MA 01760-2098 USA
Tel:508 647 7000
Fax: 508 647 7001
E-mail: info@mathworks.com
Web: www.mathworks.com

Author Biographies

Dr. Shekhar Guha obtained his Ph.D. degree in Physics from University of Pittsburgh and did post doctoral work at University of Southern California. He has been working at the Air Force Research Laboratory since 1995. His research interests are in the field of nonlinear optical materials, especially in the infrared.

Dr. Leonel P. Gonzalez received his M.S. and Ph.D. degrees in Electro-Optics from the University of Dayton. He has worked in the commercial laser industry as well as in the telecommunications field. In 2002 he returned to the Air Force Research Laboratory and since then has been investigating nonlinear optical materials and their applications.

Acknowledgements

SG thanks his wife Sudeshna for all the support she provided during the writing of this book and without which it could not be written. It was a labor of love for him but meant considerable sacrifice on her part. SG also thanks his many teachers and colleagues who taught and helped him along in his career, especially Prof. Joel Falk at University of Pittsburgh, Prof. Martin Gundersen at the University of Southern California, and Profs. Eric W. Van Stryland and M.J. Soileau at the University of Central Florida. He also thanks Jonathan Slagle at Wright Patterson Air Force Base for helping him with MATLAB programming.

LG thanks his wife Trina and children Joshua, Connor, Ian and Noah for their love and support. LG also thanks Profs. Peter Powers and Joseph Haus from the University of Dayton and Dr. Shekhar Guha from the Air Force Research Laboratory for their teaching, advice and guidance provided over the years. He also thanks Dr. Phil Milsom from the Defense Science and Technology Laboratory for early discussions of FFT application to NLBP problems.

SG and LG jointly thank their everyday colleagues Jacob Barnes, Amelia Carpenter, Dr. Joel Murray, Derek Upchurch and Dr. Jean Wei for constant help, support and encouragement. They also thank their collaborators Prof. Qin Sheng at Baylor University, especially for help with the numerical methods, Dr. Zhi Gang Yu at SRI for help with Fortran programming, especially the FFT code and Peter Schunemann at BAE not just for the nonlinear optical materials including OPGaAs that he provided but also for his expert advice on general crystallographic matters. Finally, they thank the Air Force Research Laboratory for providing the stimulating research environment which made this work possible.

Acknowledgements

1

Light Propagation in Anisotropic Crystals

This chapter introduces the *principal axes* and *principal refractive indices* of a crystal, and addresses these questions: For light propagating in an anisotropic crystal, if the direction of the **k** vector is known with respect to the principal axes,

1. What are the directions along which the electric field vector **E** and the displacement vector **D** oscillate?

2. What is the angle between the **k** vector and the Poynting vector **S**?

3. What is the propagation equation describing the spatial change of the electric field amplitude for the propagation of an electromagnetic wave through an anisotropic medium?

Propagation in a biaxial crystal in an arbitrary direction will be considered first. Then, special directions of propagation in a biaxial crystal (along the principal dielectric axes and then along the principal planes) will be considered. Finally, propagation in an arbitrary direction in a uniaxial crystal will be described.

1.1 Introduction

Light propagates in vacuum with a constant speed c, which is given by $(\epsilon_0 \mu_0)^{-1/2}$, with ϵ_0, the vacuum permittivity, equal to $8.85 \times 10^{-12} \mathrm{CV^{-1}m^{-1}}$ and μ_0, the vacuum permeability, equal to $4\pi \times 10^{-7} \mathrm{Vs^2C^{-1}m^{-1}}$. In material media, the speed of light (v) depends on the properties of the medium, with v equal to c/n, where n, the refractive index of the medium, depends on the properties of the medium. Gases, liquids, most glasses and some crystalline solids are *optically isotropic*, and in these media the value of n is independent of the propagation direction of light. However, many other optically transparent solids are *optically anisotropic* because of their anisotropic crystalline structure, and for light propagation in them, the value of n is not the same for all directions of propagation. Moreover, the allowed directions of oscillation of the electric and the magnetic fields in anisotropic media also depend on the propagation direction.

The main aim of this chapter is to provide explicit expressions for the components of the electric field for a general propagation direction of light in a general optical medium in terms of the propagation angles and the principal refractive indices. The values of these components are needed for the calculation of 'effective nonlinear optical coefficients' in the next chapter. There are of course numerous textbooks describing light propagation in anisotropic media, particularly Ref. [1]. Reference [2] has an extensive collection of the needed equations and literature references relevant to nonlinear optics. But to our knowledge the 'electric field' components are not available (or at least not easily available) in past literature, and usually only the components of the 'displacement vector' are given. Explicit expressions for the walk-off angles between the propagation vector and the Poynting vector for different crystal classes are also presented here.

After the presentation of the general formalism, some special cases (such as, light traveling along principal axes and principal planes of biaxial crystals, and in uniaxial media) are discussed here. Since results for the special cases can be directly derived from the general equations, this discussion (which is lengthy because there are several cases to consider) is admittedly redundant. But this detailed discussion of the special cases is included here in the hope that it may add to the understanding of the topic and that the results, summarized in tables, may be useful as reference. A detailed derivation of the propagation equation for light traveling in an anisotropic medium is also provided.

1.2   Vectors Associated With Light Propagation

In a charge and current free, non-magnetic medium (i.e., with $\mu = \mu_0$) Maxwell's equations are written as

$$\nabla \cdot \widetilde{\mathbf{D}} = 0 \tag{1.1}$$

$$\nabla \cdot \widetilde{\mathbf{H}} = 0 \tag{1.2}$$

$$\nabla \times \widetilde{\mathbf{E}} = -\mu_0 \frac{\partial \widetilde{\mathbf{H}}}{\partial t} \tag{1.3}$$

$$\nabla \times \widetilde{\mathbf{H}} = \frac{\partial \widetilde{\mathbf{D}}}{\partial t}. \tag{1.4}$$

where $\widetilde{\mathbf{D}}$, $\widetilde{\mathbf{E}}$ and $\widetilde{\mathbf{H}}$ denote the electric displacement vector and the electric and magnetic fields respectively. The tilde symbol is used to indicate that these fields are oscillating in time and space.

For a propagating electromagnetic wave, the direction of energy flow is given by the Poynting vector $\widetilde{\mathbf{S}}$, where

$$\widetilde{\mathbf{S}} = \widetilde{\mathbf{E}} \times \widetilde{\mathbf{H}}. \tag{1.5}$$

Equations 1.3, 1.4 and 1.5 show that the vectors $\widetilde{\mathbf{E}}$, $\widetilde{\mathbf{D}}$ and $\widetilde{\mathbf{S}}$ are all perpendicular to $\widetilde{\mathbf{H}}$, which implies $\widetilde{\mathbf{E}}$, $\widetilde{\mathbf{D}}$ and $\widetilde{\mathbf{S}}$ must be co-planar, and also that $\widetilde{\mathbf{E}}$ and $\widetilde{\mathbf{S}}$ are perpendicular to each other. This is shown in Fig. 1.1.

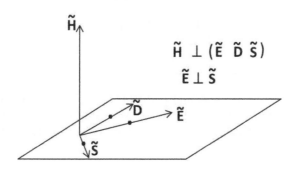

FIGURE 1.1: The vectors $\widetilde{\mathbf{E}}$, $\widetilde{\mathbf{D}}$ and $\widetilde{\mathbf{S}}$ are all perpendicular to $\widetilde{\mathbf{H}}$ and are coplanar. The vectors $\widetilde{\mathbf{E}}$ and $\widetilde{\mathbf{S}}$ are also perpendicular to each other. Vectors lying on the same plane are indicated by the dots.

1.2.1 Plane waves

Assuming monochromatic plane wave solutions for the fields $\widetilde{\mathbf{E}}$, $\widetilde{\mathbf{D}}$ and $\widetilde{\mathbf{H}}$, i.e., with

$$\widetilde{\mathbf{E}} = \mathbf{E_0}e^{i(\mathbf{k}\cdot\mathbf{r}-\omega t)} \tag{1.6}$$

$$\widetilde{\mathbf{D}} = \mathbf{D_0}e^{i(\mathbf{k}\cdot\mathbf{r}-\omega t)} \tag{1.7}$$

$$\widetilde{\mathbf{H}} = \mathbf{H_0}e^{i(\mathbf{k}\cdot\mathbf{r}-\omega t)} \tag{1.8}$$

it can be easily shown that

$$\nabla \times \widetilde{\mathbf{E}} = ie^{i(\mathbf{k}\cdot\mathbf{r}-\omega t)}(\mathbf{k} \times \mathbf{E_0}). \tag{1.9}$$

Using Eqns. 1.3, 1.6, 1.8 and 1.9, we obtain

$$(\mathbf{k} \times \mathbf{E_0}) = \mu_0 \omega \mathbf{H_0} \tag{1.10}$$

and also, similarly

$$(\mathbf{k} \times \mathbf{H_0}) = -\omega \mathbf{D_0}. \tag{1.11}$$

Thus $\mathbf{k}$ is also perpendicular to $\widetilde{\mathbf{H}}$ and must lie in the plane with $\widetilde{\mathbf{E}}$, $\widetilde{\mathbf{D}}$ and $\widetilde{\mathbf{S}}$. In addition, $\mathbf{k}$ is perpendicular to $\widetilde{\mathbf{D}}$. The vectors $\widetilde{\mathbf{E}}$, $\widetilde{\mathbf{D}}$, $\mathbf{k}$ and $\widetilde{\mathbf{S}}$ are all coplanar. Since $\widetilde{\mathbf{E}} \perp \widetilde{\mathbf{S}}$ and $\widetilde{\mathbf{D}} \perp \mathbf{k}$, the angle between $\widetilde{\mathbf{E}}$ and $\widetilde{\mathbf{D}}$ must be the same as the angle between $\mathbf{k}$ and $\widetilde{\mathbf{S}}$. This angle of deviation of the energy propagation

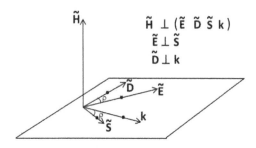

FIGURE 1.2: For plane waves and paraxial beams, the propagation vector **k** is coplanar with the vectors $\widetilde{\mathbf{E}}$, $\widetilde{\mathbf{D}}$ and $\widetilde{\mathbf{S}}$. Moreover, the vectors $\widetilde{\mathbf{D}}$ and **k** are perpendicular to each other as are the vectors the vectors $\widetilde{\mathbf{E}}$ and **S**. ρ denotes the 'walk-off' angle.

direction ($\widetilde{\mathbf{S}}$) from the phase propagation direction (**k**) is denoted by ρ and is called the 'walk-off angle'. In isotropic media, $\widetilde{\mathbf{E}}$ and $\widetilde{\mathbf{D}}$ are parallel, and ρ is zero. In anisotropic media, ρ can be zero only for certain special directions of **k**, but in general $\rho \neq 0$ for **k** in other directions. These results are shown in Fig. 1.2.

1.2.2 Non-plane waves

Laser beams of interest propagating in anisotropic media are often cylindrical (collimated) in shape, or they may be converging or diverging, i.e., in general they are not truly 'plane waves'. It is important to know if $\widetilde{\mathbf{D}}$ remains perpendicular to **k** for such beams because the expression for the refractive index of an anisotropic medium is usually deduced assuming $\mathbf{k} \cdot \widetilde{\mathbf{D}} = 0$, i.e., under the plane wave assumption.

Say we consider a light beam which is 'not' a plane wave, i.e., for which

$$\widetilde{\mathbf{E}} = \mathbf{A}e^{i(\mathbf{k} \cdot \mathbf{r} - \omega t)} \tag{1.12}$$

$$\widetilde{\mathbf{D}} = \mathfrak{D}e^{i(\mathbf{k} \cdot \mathbf{r} - \omega t)} \tag{1.13}$$

$$\widetilde{\mathbf{H}} = \mathcal{H}e^{i(\mathbf{k} \cdot \mathbf{r} - \omega t)} \tag{1.14}$$

where the amplitudes $\mathbf{A}$, $\mathfrak{D}$ and $\mathcal{H}$ are not necessarily constants but can be slowly varying functions of **r**. Defining

$$\psi \equiv e^{i\mathbf{k} \cdot \mathbf{r}}$$

so that $\nabla\psi = i\mathbf{k}\psi$, and using the vector identity

$$\nabla \cdot (\psi\mathfrak{D}) = \mathfrak{D} \cdot \nabla\psi + \psi\nabla \cdot \mathfrak{D} \tag{1.15}$$

in the Maxwell's equation $\nabla \cdot \widetilde{\mathbf{D}} = 0$, we obtain

$$\nabla \cdot \widetilde{\mathbf{D}} = e^{-i\omega t}\nabla \cdot (\psi\mathfrak{D}) = e^{-i\omega t}\psi(i\mathbf{k} \cdot \mathfrak{D} + \nabla \cdot \mathfrak{D}) = 0. \tag{1.16}$$

If θ denotes the angle between the $\mathbf{k}$ and $\widetilde{\mathbf{D}}$, Eqn. 1.16 shows that

$$|\cos\theta| = \left|\frac{\nabla \cdot \mathfrak{D}}{k\mathfrak{D}}\right|. \tag{1.17}$$

For a beam of light confined to a spatial region of the order of r_0 (the beam spot size), $\nabla \cdot \mathfrak{D}/\mathfrak{D}$ is approximately equal to $1/r_0$ so that

$$|\cos\theta| \approx \frac{1}{kr_0} = \frac{\lambda}{2\pi r_0}. \tag{1.18}$$

For most cases of interest here, $r_0 \geq \lambda$ so that $\cos\theta$ is small, implying $\theta \approx 90°$, i.e., for most practical purposes, $\mathbf{k}$ is perpendicular to $\widetilde{\mathbf{D}}$. However, this is only an approximate relationship, not as ironclad as the orthogonality of $\widetilde{\mathbf{E}}$ and $\widetilde{\mathbf{H}}$ and of $\widetilde{\mathbf{E}}$ and $\widetilde{\mathbf{S}}$ which follow directly from Maxwell's equations and the definition of $\widetilde{\mathbf{S}}$.

1.3 Anisotropic Media

The electric displacement vector $\widetilde{\mathbf{D}}$ is related to the electric field $\widetilde{\mathbf{E}}$ by the relation

$$\widetilde{\mathbf{D}} = \epsilon_0 \widetilde{\mathbf{E}} + \widetilde{\mathbf{P}} \tag{1.19}$$

where $\widetilde{\mathbf{P}}$ represents the macroscopically averaged electric dipole density of the material medium in the presence of the applied field [3]. If the medium is isotropic, the polarization $\widetilde{\mathbf{P}}$ induced by the electric field $\widetilde{\mathbf{E}}$ is parallel to $\widetilde{\mathbf{E}}$, That is, for isotropic media, we can write

$$\widetilde{\mathbf{P}} = \varepsilon_0 \chi \widetilde{\mathbf{E}}. \tag{1.20}$$

The parameter χ relating $\widetilde{\mathbf{P}}$ and $\widetilde{\mathbf{E}}$ is called the *electric susceptibility*. For isotropic media it is a scalar quantity independent of the relative direction between $\widetilde{\mathbf{P}}$ and $\widetilde{\mathbf{E}}$.

In an *anisotropic medium*, the electric field imposes a force on the microscopic charges in the direction of the field, but the net displacement of the charges can be in a different direction, imposed by the crystal structure. The macroscopically averaged electric dipole density, while being linearly proportional to the field, is in general in a different direction, i.e., $\widetilde{\mathbf{P}}$ is *not* parallel to $\widetilde{\mathbf{E}}$. In general, the Cartesian components of $\widetilde{\mathbf{P}}$ and $\widetilde{\mathbf{E}}$ can be related as

$$\widetilde{P}_x = \kappa_{xx}\widetilde{E}_x + \kappa_{xy}\widetilde{E}_y + \kappa_{xz}\widetilde{E}_z \tag{1.21}$$

$$\widetilde{P}_y = \kappa_{yx}\widetilde{E}_x + \kappa_{yy}\widetilde{E}_y + \kappa_{yz}\widetilde{E}_z \tag{1.22}$$

$$\widetilde{P}_z = \kappa_{zx}\widetilde{E}_x + \kappa_{zy}\widetilde{E}_z + \kappa_{zz}\widetilde{E}_z. \tag{1.23}$$

This can be written in tensor notation as

$$\widetilde{P}_i = \kappa_{ij}\widetilde{E}_j \tag{1.24}$$

where κ_{ij} is a tensor, the indices i and j of which run over the Cartesian coordinates x, y and z, and the Einstein summation convention has been assumed. The vector Eqn. 1.19 can then be rewritten in tensor form as

$$\widetilde{D}_i = \varepsilon_{ij}\widetilde{E}_j \tag{1.25}$$

where

$$\varepsilon_{ij} = \varepsilon_0(\delta_{ij} + \chi_{ij}). \tag{1.26}$$

ε_{ij} is called the *dielectric permittivity tensor*, δ_{ij} is the Kronecker delta symbol and

$$\chi_{ij} = \kappa_{ij}/\varepsilon_0. \tag{1.27}$$

Using the simple Lorentz model for electron oscillator motion under an applied electric field [4] or from energy flow consideration [1], it can be shown that the dielectric permittivity tensor is symmetric, i.e.,

$$\varepsilon_{ij} = \varepsilon_{ji}. \tag{1.28}$$

1.3.1   The principal coordinate axes

The coordinate system used to define the susceptibility and dielectric permittivity tensors has been left arbitrary so far. For any anisotropic crystal, a coordinate system (X, Y, Z) can always be chosen which diagonalizes the dielectric permittivity tensor, i.e., in which only the diagonal components of the tensor are non-zero. In such a coordinate system, the dielectric permittivity tensor takes the form

$$\varepsilon = \begin{pmatrix} \varepsilon_X & 0 & 0 \\ 0 & \varepsilon_Y & 0 \\ 0 & 0 & \varepsilon_Z \end{pmatrix} \tag{1.29}$$

X, Y, and Z are called the *principal coordinate axes* of the crystal. The electric displacement and the field are related in the principal coordinate axes system by the relations

$$\widetilde{D}_X = \varepsilon_X\widetilde{E}_{.X} \qquad \widetilde{D}_Y = \varepsilon_Y\widetilde{E}_Y \qquad \widetilde{D}_Z = \varepsilon_Z\widetilde{E}_Z. \tag{1.30}$$

ε_X, ε_Y and ε_Z are called the *principal dielectric permittivities* of the anisotropic crystal.

By convention, the principal coordinate axes are always chosen (in optics) such that the values of ε_X, ε_Y and ε_Z are either monotonically increasing or decreasing, i.e., either $\varepsilon_X > \varepsilon_Y > \varepsilon_Z$ or $\varepsilon_X < \varepsilon_Y < \varepsilon_Z$. We will also assume here that at the wavelengths of interest, the dielectric permittivities are dominantly real numbers. The principal coordinate axes system may not always coincide with the crystallographic coordinate axes (denoted by a, b, c in [6] or by x, y, z in [1]).

The three planes XY, YZ and ZX are called the 'principal planes'.

1.3.2 Three crystal classes

All crystals can be classified into three groups, isotropic, uniaxial and biaxial, depending on the relations between ε_X, ε_Y and ε_Z. In *isotropic* crystals all the principal dielectric permittivities are equal, i.e.,

$$\varepsilon_X = \varepsilon_Y = \varepsilon_Z.$$

In *biaxial* crystals all three principal dielectric permittivities are unequal, i.e.,

$$\varepsilon_X \neq \varepsilon_Y \neq \varepsilon_Z.$$

In *uniaxial* crystals only two principal dielectric permittivities are equal to each other. By convention, in a uniaxial crystal

$$\varepsilon_X = \varepsilon_Y \neq \varepsilon_Z.$$

The reason behind choosing the names *uniaxial* and *biaxial* will be discussed in the next section.

1.3.3 The principal refractive indices

We 'define' here three *principal refractive indices* n_X, n_Y and n_Z

$$n_X \equiv \sqrt{\varepsilon_X/\varepsilon_0} \qquad n_Y \equiv \sqrt{\varepsilon_Y/\varepsilon_0} \qquad n_Z \equiv \sqrt{\varepsilon_Z/\varepsilon_0} \qquad (1.31)$$

using which the Eqns. 1.30 take the forms

$$\widetilde{D}_X = \varepsilon_0 n_X^2 \widetilde{E}_X \qquad \widetilde{D}_Y = \varepsilon_0 n_Y^2 \widetilde{E}_Y \qquad \widetilde{D}_Z = \varepsilon_0 n_Z^2 \widetilde{E}_Z. \qquad (1.32)$$

In a biaxial crystal, n_X, n_Y and n_Z are all unequal. Some examples of biaxial crystals of importance in nonlinear optics are Lithium Triborate (LiB$_3$O$_5$ or LBO), Potassium Niobate (KNbO$_3$), Potassium Titanyl Phosphate (KTiOPO$_4$ or KTP), Potassium Titanyl Arsenate (KTiOAsO$_4$ or KTA), alpha Iodic acid (α HIO$_3$) etc.

In uniaxial crystals, two of the principal refractive indices are equal. By convention, the two equal indices are chosen to be the n_X and n_Y values and the indices are renamed as

$$n_o = n_X = n_Y \qquad \text{for uniaxial crystals.} \qquad (1.33)$$

If $n_Z > n_o$, the crystal is defined to be *positive uniaxial* and if $n_o > n_Z$, the crystal is defined to be *negative uniaxial*. Some examples of positive uniaxial crystals are Zinc Germanium Phosphide ($ZnGeP_2$ or ZGP), Cadmium Selenide (CdSe), Cadmium Germanium Arsenide ($CdGeAs_2$ or CGA), Cinnabar (HgS), Selenium (Se), Tellurium (Te) and Quartz (SiO_2). Some examples of negative uniaxial crystals are Potassium Dihydrogen Phosphate (KH_2PO_4 or KDP), Lithium Iodate ($LiIO_3$), Lithium Niobate ($LiNbO_3$), Silver Gallium Sulfide ($AgGaS_2$), Silver Gallium Selenide ($AgGaSe_2$), Gallium Selenide (GaSe), Rubidium Dihydrogen Phosphate (RbH_2PO_4 or RDP), Mercury Thiogallate ($HgGa_2S_4$), Pyrargyrite (Ag_3SbS_3) and Thallium Arsenide Sulfide (Tl_3AsS_3 or TAS).

1.4 Light Propagation In An Anisotropic Crystal

As shown in Section 1.2.1, for a plane wave (or a paraxial beam of light) propagating in any medium, the direction of oscillation of the electric displacement vector $\widetilde{\mathbf{D}}$ is perpendicular to the propagation direction, i.e., $\mathbf{k} \cdot \widetilde{\mathbf{D}} = 0$, where $\mathbf{k}$ is the propagation vector. In an isotropic medium, the electric field $\widetilde{\mathbf{E}}$ is parallel to $\widetilde{\mathbf{D}}$, and for a given direction of $\mathbf{k}$, the $\widetilde{\mathbf{D}}$ and $\widetilde{\mathbf{E}}$ vectors can be in any direction as long as they lie on a plane perpendicular to $\mathbf{k}$.

For light propagation in anisotropic media, for a *general* direction of $\mathbf{k}$ (with respect to the principal axes), $\widetilde{\mathbf{D}}$ and $\widetilde{\mathbf{E}}$ are *not* parallel. Moreover, the direction of $\widetilde{\mathbf{D}}$ is not only restricted to a plane perpendicular to $\mathbf{k}$, it is also further specified to be only in two allowed directions in that plane, as will be shown below. Similarly, for the given $\mathbf{k}$, the electric field $\widetilde{\mathbf{E}}$ is also restricted to two fixed directions, which are in general not parallel to the allowed directions $\widetilde{\mathbf{D}}$.

In *biaxial* crystals, (defined by the condition $\varepsilon_X \neq \varepsilon_Y \neq \varepsilon_Z$), there are two special directions called the directions of the *optic axes* such that if $\mathbf{k}$ is parallel to an optic axis, the $\widetilde{\mathbf{D}}$ and $\widetilde{\mathbf{E}}$ vectors are parallel and they can lie in *any* direction as long as they are restricted to the plane perpendicular to $\mathbf{k}$. Thus, for light propagation with $\mathbf{k}$ along the optic axes, the anisotropic medium behaves as an isotropic medium. It will be shown later that in biaxial crystals the two optic axes lie on the XZ principal plane, with the Z axis the bisector of the angle between the optic axes, and the tangent of the angle between the two axes is proportional to $\sqrt{n_X^2 - n_Y^2}$.

In uniaxial crystals $n_X = n_Y$, so the angle between the two axes goes to zero and the two axes coalesce into one (pointed along the Z direction), thereby justifying the name of this crystal class.

1.4.1 Allowed directions of $\widetilde{\mathbf{D}}$ and $\widetilde{\mathbf{E}}$ in an anisotropic medium

Here we determine the allowed oscillation directions of the propagating plane waves $\widetilde{\mathbf{D}}$ and $\widetilde{\mathbf{E}}$, for a general direction of $\mathbf{k}$ with respect to the principal axes. Say the unit vectors along the $\mathbf{k}$, $\widetilde{\mathbf{D}}$ and $\widetilde{\mathbf{E}}$ vectors are denoted by $\hat{m}$, $\hat{d}$ and $\hat{e}$, i.e.,

$$\mathbf{k} = \hat{m}k \quad \widetilde{\mathbf{D}} = \hat{d}\widetilde{D} \quad \widetilde{\mathbf{E}} = \hat{e}\widetilde{E}. \tag{1.34}$$

Maxwell's equations in a nonmagnetic medium given in Eqns. 1.1 to 1.4 can be rearranged in the form

$$\nabla \times (\nabla \times \widetilde{\mathbf{E}}) = -\mu_0 \frac{\partial^2 \widetilde{\mathbf{D}}}{\partial t^2}. \tag{1.35}$$

Rewriting Eqns. 1.12 and 1.13 as

$$\widetilde{\mathbf{E}} = \mathbf{E}e^{-i\omega t} \tag{1.36}$$

$$\widetilde{\mathbf{D}} = \mathbf{D}e^{-i\omega t} \tag{1.37}$$

where

$$\mathbf{E} = \mathbf{A}e^{i\mathbf{k} \cdot \mathbf{r}}, \quad \text{and} \quad \mathbf{D} = \mathfrak{D}e^{i\mathbf{k} \cdot \mathbf{r}}. \tag{1.38}$$

Equation 1.35 is thus rewritten as

$$\nabla \times (\nabla \times \mathbf{E}) = \mu_0 \omega^2 \mathbf{D}. \tag{1.39}$$

For monochromatic plane waves, i.e., for $\mathbf{E}$ in Eqn. 1.38 constant, the ∇ operator can be replaced by $i\mathbf{k}$, i.e., by $ik\hat{m}$, so that Eqn. 1.39 becomes

$$-k^2(\hat{m} \times \{\hat{m} \times \mathbf{E}\} = \mu_0 \omega^2 \mathbf{D}. \tag{1.40}$$

We will assume here that Eqn. 1.40 is valid even for non-planar waves, and that $\mathbf{k}$ and $\mathbf{D}$ are approximately perpendicular, as discussed before.

The waves propagate in the medium with speed ω/k. The refractive index n of the medium is the ratio of the wave speed to c, the speed of light in vacuum, i.e.,

$$n = \frac{c\,k}{\omega}. \tag{1.41}$$

From Eqns. 1.40 and 1.41 we obtain

$$-\hat{m} \times (\{\hat{m} \times \mathbf{E}\}) = \frac{\mathbf{D}}{\epsilon_0 n^2} \tag{1.42}$$

and using the vector identity $\mathbf{A} \times (\mathbf{B} \times \mathbf{C}) = \mathbf{B}(\mathbf{A} \cdot \mathbf{C}) - \mathbf{C}(\mathbf{A} \cdot \mathbf{B})$ we get

$$\mathbf{D} = \epsilon_0 n^2 \{\mathbf{E} - \hat{m}(\hat{m} \cdot \mathbf{E}).\} \tag{1.43}$$

For a given medium (i.e., one with known values of n_X, n_Y and n_Z), the allowed oscillation directions of $\mathbf{D}$ and $\mathbf{E}$ fields, for a specified direction of $\hat{m}$, can be determined using Eqn. 1.30 and Eqn. 1.43. The speeds of these waves propagating with the $\mathbf{k}$ vector in the $\hat{m}$ direction are also determined by Eqn. 1.43 through the solution for n.

Rewriting Eqn. 1.43 in terms of the vector components along the three principal coordinate axes, and using Eqns. 1.30 and 1.31, we obtain

$$\begin{aligned}
D_X &= \varepsilon_0 n^2 \{E_X - m_X(m_X E_X + m_Y E_Y + m_Z E_Z)\} &= \varepsilon_0 n_X^2 E_X \\
D_Y &= \varepsilon_0 n^2 \{E_Y - m_Y(m_X E_X + m_Y E_Y + m_Z E_Z)\} &= \varepsilon_0 n_Y^2 E_Y \\
D_Z &= \varepsilon_0 n^2 \{E_Z - m_Z(m_X E_X + m_Y E_Y + m_Z E_Z)\} &= \varepsilon_0 n_Z^2 E_Z
\end{aligned} \tag{1.44}$$

which can be solved to obtain the values of the components of $\mathbf{E}$:

$$E_X = \frac{n^2 m_X}{n^2 - n_X^2}(\hat{m} \cdot \mathbf{E})$$

$$E_Y = \frac{n^2 m_Y}{n^2 - n_Y^2}(\hat{m} \cdot \mathbf{E}) \tag{1.45}$$

$$E_Z = \frac{n^2 m_Z}{n^2 - n_Z^2}(\hat{m} \cdot \mathbf{E}). \tag{1.46}$$

Figure 1.2 shows that if ρ is the angle between $\mathbf{D}$ and $\mathbf{E}$, the angle between $\mathbf{E}$ and $\hat{m}$ is $90° - \rho$, i,e., $\hat{m} \cdot \mathbf{E} = E \sin \rho$, where E is the magnitude of the vector $\mathbf{E}$. The Cartesian components of $\hat{e}$, the unit vector in the direction of $\mathbf{E}$, are then given by

$$e_X = \frac{n^2 m_X}{n^2 - n_X^2} \sin \rho$$

$$e_Y = \frac{n^2 m_Y}{n^2 - n_Y^2} \sin \rho$$

$$e_Z = \frac{n^2 m_Z}{n^2 - n_Z^2} \sin \rho. \tag{1.47}$$

Since $\hat{e}_X^2 + \hat{e}_Y^2 + \hat{e}_Z^2 = 1$, the walk-off angle ρ is given by

$$\sin \rho = \frac{1}{n^2 \left[\left(\dfrac{m_X}{n^2 - n_X^2} \right)^2 + \left(\dfrac{m_Y}{n^2 - n_Y^2} \right)^2 + \left(\dfrac{m_Z}{n^2 - n_Z^2} \right)^2 \right]^{1/2}}. \tag{1.48}$$

The components of the unit vector $\hat{d}$ along the direction of $\mathbf{D}$ can similarly be determined. Taking the dot product of both sides of Eqn. 1.43 with themselves, we obtain

$$
\begin{aligned}
\mathbf{D} \cdot \mathbf{D} = D^2 &= \epsilon_0^2 n^4 [\mathbf{E} - \hat{m}(\hat{m} \cdot \mathbf{E})] \cdot [\mathbf{E} - \hat{m}(\hat{m} \cdot \mathbf{E})] \\
&= \epsilon_0^2 n^4 [\mathbf{E} \cdot \mathbf{E} - 2(\hat{m} \cdot \mathbf{E})(\hat{m} \cdot \mathbf{E}) + (\hat{m} \cdot \hat{m})(\hat{m} \cdot \mathbf{E})^2] \\
&= \epsilon_0^2 n^4 [\mathbf{E} \cdot \mathbf{E} - (\hat{m} \cdot \mathbf{E})^2] \\
&= \epsilon_0^2 n^4 [E^2 - E^2 \sin^2 \rho]
\end{aligned}
\tag{1.49}
$$

so that

$$
D = \epsilon_0 n^2 E \cos \rho.
\tag{1.50}
$$

Since $D_X = d_X D$, $E_X = e_X E$ and also $D_X = \varepsilon_0 n_X^2 E_X$, we obtain, using Eqns. 1.50 and 1.47

$$
\begin{aligned}
d_X = \frac{D_X}{D} &= \frac{\epsilon_0 n_X^2 E e_X}{D} \\
&= \frac{\epsilon_0 n_X^2 E}{\epsilon_0 n^2 E \cos \rho} e_X \\
&= \frac{n_X^2 m_X}{n^2 - n_X^2} \tan \rho.
\end{aligned}
\tag{1.51}
$$

Similarly,

$$
d_Y = \frac{n_Y^2 m_Y}{n^2 - n_Y^2} \tan \rho \quad \text{and} \quad d_Z = \frac{n_Z^2 m_Z}{n^2 - n_Z^2} \tan \rho.
\tag{1.52}
$$

1.4.2 Values of n for a given propagation direction

Since $\mathbf{D}$ and $\hat{m}$ are perpendicular to each other, i.e., $\hat{m} \cdot \hat{d} = 0$, we obtain from Eqns. 1.51 and 1.52

$$
\frac{n_X^2 m_X^2}{n^2 - n_X^2} + \frac{n_Y^2 m_Y^2}{n^2 - n_Y^2} + \frac{n_Z^2 m_Z^2}{n^2 - n_Z^2} = 0.
\tag{1.53}
$$

Equation 1.53 easily reduces to a quadratic equation in n^2, which can be expressed as

$$
\mathcal{A} n^4 - \mathcal{B} n^2 + \mathcal{C} = 0
\tag{1.54}
$$

where

$$
\begin{aligned}
\mathcal{A} &= n_X^2 \, m_X^2 + n_Y^2 \, m_Y^2 + n_Z^2 \, m_Z^2 \\
\mathcal{B} &= n_X^2 \, m_X^2 \, (n_Y^2 + n_Z^2) + n_Y^2 \, m_Y^2 \, (n_Z^2 + n_X^2) + n_Z^2 \, m_Z^2 \, (n_X^2 + n_Y^2) \\
\mathcal{C} &= n_X^2 \, n_Y^2 \, n_Z^2.
\end{aligned}
\tag{1.55}
$$

The discriminant of the Eqn. 1.54 is $\mathcal{D}^2$ where

$$\mathcal{D} \equiv (\mathcal{B}^2 - 4\mathcal{A}\mathcal{C})^{1/2}. \tag{1.56}$$

If the values of n_X, n_Y, n_Z and m_X, m_Y, m_Z are such that $\mathcal{B}^2$ is greater than $4\mathcal{A}\mathcal{C}$, then $\mathcal{D}$ is real and it is positive by definition. Eqn. 1.54 then has two roots for n^2, from which we get two possible values of n. Denoting these two values of n by n_s and n_f we have

$$n_s = \left(\frac{\mathcal{B} + \mathcal{D}}{2\mathcal{A}}\right)^{1/2} \quad \text{and} \quad n_f = \left(\frac{\mathcal{B} - \mathcal{D}}{2\mathcal{A}}\right)^{1/2}. \tag{1.57}$$

Since $\mathcal{B}$ is positive by its definition and $\mathcal{D}$ is less than or equal to $\mathcal{B}$ (from Eqn. 1.56) $n_s \geq n_f$. A wave propagating in the direction $\hat{m}$ in an anisotropic medium can have these two values for refractive index, corresponding to which there can be two waves traveling with speeds c/n_s and c/n_f for the given direction of propagation. These two waves are called the 'slow' and 'fast' waves respectively, since the speed c/n_s is less than the speed c/n_f. Equations 1.47, 1.51 and 1.52 show that the directions of the unit vectors $\hat{e}$ and $\hat{d}$ depend on n, so for each direction of $\hat{m}$ there are two unit vectors $\hat{e}$, say denoted by $\hat{e}_s$ and $\hat{e}_f$ for $n = n_s$ and n_f respectively and correspondingly two unit vectors $\hat{d}_s$ and $\hat{d}_f$.

1.4.3   Directions of D and E for the slow and fast waves

The components of $\hat{e}_s$, $\hat{e}_f$, $\hat{d}_s$ and $\hat{d}_f$ can be directly obtained from Eqns. 1.47, 1.51 and 1.52 by substituting n_s and n_f in place of n. From these equations we obtain

$$
\begin{aligned}
e_{sX} &= \frac{n_s^2 m_X}{n_s^2 - n_X^2} \sin \rho_s & e_{fX} &= \frac{n_f^2 m_X}{n_f^2 - n_X^2} \sin \rho_f \\[2mm]
e_{sY} &= \frac{n_s^2 m_Y}{n_s^2 - n_Y^2} \sin \rho_s & e_{fY} &= \frac{n_f^2 m_Y}{n_f^2 - n_Y^2} \sin \rho_f \\[2mm]
e_{sZ} &= \frac{n_s^2 m_Z}{n_s^2 - n_Z^2} \sin \rho_s & e_{fZ} &= \frac{n_f^2 m_Z}{n_f^2 - n_Z^2} \sin \rho_f
\end{aligned} \tag{1.58}
$$

and

$$
\begin{aligned}
d_{sX} &= \frac{n_X^2 m_X}{n_s^2 - n_X^2} \tan \rho_s & d_{fX} &= \frac{n_X^2 m_X}{n_f^2 - n_X^2} \tan \rho_f \\[2mm]
d_{sY} &= \frac{n_Y^2 m_Y}{n_s^2 - n_Y^2} \tan \rho_s & d_{fY} &= \frac{n_Y^2 m_Y}{n_f^2 - n_Y^2} \tan \rho_f \\[2mm]
d_{sZ} &= \frac{n_Z^2 m_Z}{n_s^2 - n_Z^2} \tan \rho_s & d_{fZ} &= \frac{n_Z^2 m_Z}{n_f^2 - n_Z^2} \tan \rho_f
\end{aligned} \tag{1.59}
$$

and from Eqn. 1.48 the angles ρ_s and ρ_f are given by

$$\sin \rho_s = \cfrac{1}{n_s^2 \left[\left(\cfrac{m_X}{n_s^2 - n_X^2} \right)^2 + \left(\cfrac{m_Y}{n_s^2 - n_Y^2} \right)^2 + \left(\cfrac{m_Z}{n_s^2 - n_Z^2} \right)^2 \right]^{1/2}}$$

$$\sin \rho_f = \cfrac{1}{n_f^2 \left[\left(\cfrac{m_X}{n_f^2 - n_X^2} \right)^2 + \left(\cfrac{m_Y}{n_f^2 - n_Y^2} \right)^2 + \left(\cfrac{m_Z}{n_f^2 - n_Z^2} \right)^2 \right]^{1/2}}.$$

$$(1.60)$$

Since Eqn. 1.48 was derived by taking a square root, the numerator on its right hand side could be -1 as well, so the angles ρ_s and ρ_f given in Eqn. 1.60 can be either positive or negative. We 'define' here the angles ρ_s and ρ_f to be both positive. Because of this choice made here, the direction (but not the magnitudes of the components) of $\hat{d}$ determined here contradict that 'chosen' in earlier work [7], [8]. We choose to maintain this definition of positive values for ρ_s and ρ_f for the sake of simplicity; the choice of negative values would of course be equally valid.

Rewriting Eqn. 1.53 as

$$\frac{m_X^2}{\dfrac{1}{n_X^2} - \dfrac{1}{n^2}} + \frac{m_Y^2}{\dfrac{1}{n_Y^2} - \dfrac{1}{n^2}} + \frac{m_Z^2}{\dfrac{1}{n_Z^2} - \dfrac{1}{n^2}} = 0 \qquad (1.61)$$

and substituting n_s and n_f for n in succession and subtracting we obtain

$$\frac{m_X^2}{\left(\dfrac{1}{n_X^2} - \dfrac{1}{n_s^2} \right) \left(\dfrac{1}{n_X^2} - \dfrac{1}{n_f^2} \right)} + \frac{m_Y^2}{\left(\dfrac{1}{n_Y^2} - \dfrac{1}{n_s^2} \right) \left(\dfrac{1}{n_Y^2} - \dfrac{1}{n_f^2} \right)}$$

$$+ \frac{m_Z^2}{\left(\dfrac{1}{n_Z^2} - \dfrac{1}{n_s^2} \right) \left(\dfrac{1}{n_Z^2} - \dfrac{1}{n_f^2} \right)}.$$

$$= 0 \qquad (1.62)$$

From Eqn. 1.59 the dot product of the unit vectors $\hat{d}_s$ and $\hat{d}_f$ is then equal to zero, since

$$\hat{d}_s \cdot \hat{d}_f = \tan \rho_s \tan \rho_f$$

$$\times \left[\frac{m_X^2}{\left(\frac{1}{n_X^2} - \frac{1}{n_s^2}\right)\left(\frac{1}{n_X^2} - \frac{1}{n_f^2}\right)} + \frac{m_Y^2}{\left(\frac{1}{n_Y^2} - \frac{1}{n_s^2}\right)\left(\frac{1}{n_Y^2} - \frac{1}{n_f^2}\right)} \right.$$

$$\left. + \frac{m_Z^2}{\left(\frac{1}{n_Z^2} - \frac{1}{n_s^2}\right)\left(\frac{1}{n_Z^2} - \frac{1}{n_f^2}\right)} \right] = 0. \tag{1.63}$$

Thus displacement vectors $\mathbf{D}_s$ and $\mathbf{D}_f$ of the two plane waves propagating in a given direction in an anisotropic crystal are always perpendicular to each other. For a general propagation direction, the electric field vectors $\mathbf{E}_s$ and $\mathbf{E}_f$ are not necessarily perpendicular to each other.

To summarize, in a given anisotropic crystal, if the directions of the dielectric principal coordinate axes (X, Y, Z) are known (say through x ray diffraction measurements), and the values of the principal refractive indices n_X, n_Y and n_Z are known (through independent optical measurements) then for a beam of light with the propagation vector $\mathbf{k}$ having direction cosines m_X, m_Y and m_Z with respect to the XYZ axes, there are two possible waves. These two waves, named 'slow' and 'fast', travel with speeds c/n_s and c/n_f where the values of n_s and n_f are given by Eqns. 1.57. The directions of the displacement vectors $\mathbf{D}_s$ and $\mathbf{D}_f$ of these two waves are given by Eqns. 1.59 and the directions of the electric fields $\mathbf{E}_s$ and $\mathbf{E}_f$ are given by Eqns. 1.58. The angles between the displacement vectors and the electric fields are denoted by ρ_s and ρ_f for the slow and the fast waves, respectively, and are determined by Eqns. 1.60.

These sets of equations provide almost all the information needed to describe light propagation in anisotropic materials for purpose of this book. We use these equations to determine the characteristics of the polarization components of the slow and fast waves.

Since it is hard to find in the literature a detailed description of the *field components* for light propagation in an arbitrary direction in a biaxial crystal, we provide some calculations in the next section for the cases of interest here, assuming some hypothetical values for n_X, n_Y and n_Z.

1.5 Characteristics Of The Slow And The Fast Waves In A Biaxial Crystal

The principal dielectric axes X, Y and Z are chosen in such a way that the value of n_Y is in between the values of n_X and n_Z. In some crystals, (such as cesium triborate, CBO or lithium triborate, LBO) n_X has the smallest value, and in some other crystals, (such as potassium niobate, KNbO$_3$, strontium formate dihydrate, Sr(COOH)$_2 \cdot$ 2H$_2$O or barium sodium niobate, Ba$_2$NaNb$_5$O$_{15}$), n_X has the largest value of the three principal refractive indices. The five crystals mentioned here all 'negative' biaxial crystals, and were chosen as examples to point out that for biaxial crystals the designations of 'positive' or 'negative' are independent of the ordering of the principal refractive indices in terms of their magnitudes. The definitions of 'positive' and 'negative' biaxial crystals will be provided later in terms of angles between the optic axes. The designation of 'positive' or 'negative' to uniaxial crystals do depend on the relative magnitudes of the principal refractive indices.

For biaxial crystals, we name the two cases of n_X the smallest and n_X the largest among the three principal refractive indices as the 'Case 1' and 'Case 2', respectively, and consider these two cases in detail next. Since these are not 'standard' definitions, we will try to define the two cases each time they occur in this book.

The propagation vector $\mathbf{k}$ is assumed to be oriented with polar and azimuthal angles θ, ϕ with respect to the principal axes X, Y, Z, as shown in Fig. 1.3. The unit vector in the direction of $\mathbf{k}$, denoted by $\hat{m}$, has components

$$m_X = \sin\theta \,\cos\phi \quad m_Y = \sin\theta \,\sin\phi \quad m_Z = \cos\theta. \quad (1.64)$$

The octants are numbered as shown in Figs. 1.4 and 1.5, i.e., the four octants with positive Z (θ between 0 and $\pi/2$) with the angle ϕ going from 0 to $\pi/2$, $\pi/2$ to π, π to $3\pi/2$ and $3\pi/2$ to 2π, are numbered 1,2,3 and 4, respectively. Similarly the four octants with negative Z (θ between $\pi/2$ and π) , with the angle ϕ going over the same quadrants in order, are numbered 5,6,7 and 8 respectively

We will assume here that $\hat{m}$ lies in the first octant, i.e., θ and ϕ both range from 0 to 90°, so that m_X, m_Y and m_Z are all positive.

To illustrate the dependence of n_s, n_f, ρ_s, ρ_f, $\hat{d}_s$, $\hat{d}_f$ and $\hat{e}_s$, $\hat{e}_f$ on the propagation direction (characterized by the angles θ and ϕ) we assume a hypothetical crystal having the values $n_X = 1.65$, $n_Y = 1.75$ and $n_Z = 1.95$ for Case 1, and $n_X = 1.95$, $n_Y = 1.75$ and $n_Z = 1.65$ for Case 2.

1.5.1 n_s and n_f

For Case 1, ($n_X < n_Y < n_Z$), substituting Eqns. 1.64 in Eqns. 1.55 and 1.56 it can be shown that n_s ranges from n_Y to n_Z and that n_f ranges from n_X to

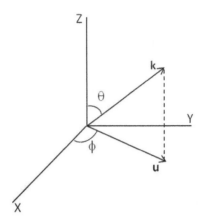

FIGURE 1.3: The orientation of **k** and **û** with respect to the principal axes X, Y and Z.

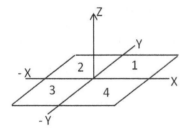

FIGURE 1.4: The four octants with positive Z, i.e., in the upper half of the XY plane. For a vector pointing in the octants 1, 2, 3 and 4, the angle θ is less than 90°, and the angle ϕ is between 0 and 90° for octant 1, between 90° and 180° for octant 2, between 180° and 270° for octant 3 and between 270 and 360° for octant 4.

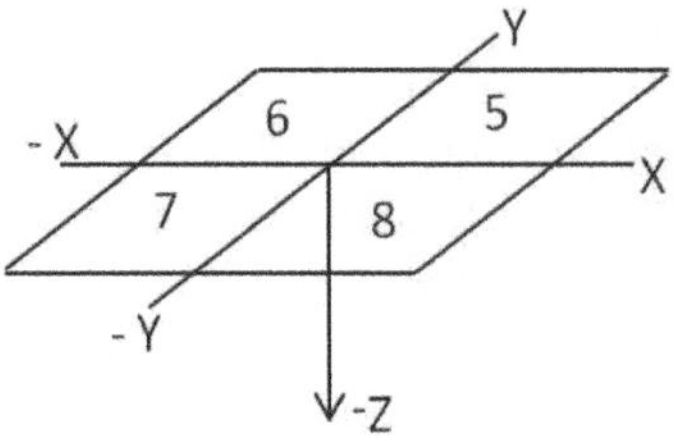

FIGURE 1.5: The four octants with negative Z, i.e., in the lower half of the XY plane. For a vector pointing in the octants 5, 6, 7 and 8, the angle θ is greater than 90°, and the angle ϕ is between 0 and 90° for octant 5, between 90° and 180° for octant 6, between 180° and 270° for octant 7 and between 270 and 360° for octant 8.

n_Y. Similarly for Case 2, $(n_X > n_Y > n_Z)$, it can be shown that n_s ranges from n_Y to n_X and that n_f ranges from n_Z to n_Y.

The dependence of the values of n_s and n_f on θ for the hypothetical crystals described above (one with $n_X < n_Y < n_Z$ and the other with $n_X > n_Y > n_Z$) are shown in Figs. 1.6 and 1.7 for a few values of ϕ.

Figure 1.6 shows that at $\phi = 0°$, as θ increases from 0 to 38°, the value of n_s is constant at n_Y, and n_f rises from n_X to n_Y. For θ increasing from 38°

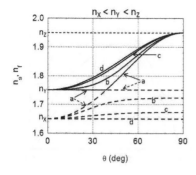

FIGURE 1.6: Dependence of n_s (shown by solid lines) and n_f (shown by dashed lines) on θ for different values of ϕ for a hypothetical crystal with $n_X < n_Y < n_Z$: (a) $\phi = 0°$, (b) $\phi = 30°$, (c) $\phi = 60°$ and (d) $\phi = 90°$.

FIGURE 1.7: Dependence of n_s (shown by solid lines) and n_f (shown by dashed lines) on θ for different values of ϕ for a hypothetical crystal with $n_X > n_Y > n_Z$: (a) $\phi = 0°$, (b) $\phi = 30°$, (c) $\phi = 60°$ and (d) $\phi = 90°$.

to 90°, n_s rises from n_Y to n_Z and n_f stays constant at n_Y. The values of θ and ϕ at which n_s and n_f are equal to each other provide the direction of the optic axis of the crystal ($\theta = 38°$, $\phi = 0°$). At values of ϕ other than 0°, n_s and n_f are not equal to each other for any value of θ.

Figure 1.7 shows that at $\phi = 0°$, as θ increases from 0 to 52°, the value of n_f is constant at n_Y, and n_s falls from n_X to n_Y. For θ increasing from 52° to 90°, n_f falls from n_Y to n_Z and n_s stays constant at n_Y. The values of θ, ϕ at which n_s and n_f are equal to each other provide the direction of the optic axis of the crystal for this case ($\theta = 52°$, $\phi = 0°$).

1.5.2 ρ_s and ρ_f

Figures 1.8 and 1.9 show the walk-off angles ρ_s and ρ_f for the two cases, $n_X < n_Y < n_Z$ and $n_X > n_Y > n_Z$, respectively, for the hypothetical crystal. For both cases, the walk-off angles are largest for $\phi = 0°$, i.e., on the YZ plane.

1.5.3 The components of $\hat{d}_s$ and $\hat{d}_f$

The values of the components of the unit vectors $\hat{d}_s$ and $\hat{d}_f$ for the two cases, $n_X < n_Y < n_Z$ and $n_X > n_Y > n_Z$ are shown in Figs. 1.10 through 1.15. Since $\tan \rho_s$ and $\tan \rho_f$ are defined to be positive, Eqns. 1.59 show that d_{sX} is positive, d_{sY} is positive and d_{sZ} is negative, while d_{fX} is positive, d_{fY} is negative and d_{fZ} is negative in Case 1, ($n_X < n_Y < n_Z$). Thus for $\hat{m}$

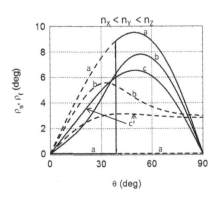

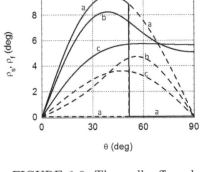

FIGURE 1.8: The walk-off angles ρ_s (solid lines) of the slow wave and ρ_f (dashed lines) of the fast wave as functions of the polar angle θ for a hypothetical crystal with $n_X < n_Y < n_Z$, for three values of ϕ: (a) $\phi = 0$, (b) $\phi = 30°$ and (c) $\phi = 60°$.

FIGURE 1.9: The walk-off angles ρ_s (solid lines) of the slow wave and ρ_f (dashed lines) of the fast wave as functions of the polar angle θ for a hypothetical crystal with $n_X > n_Y > n_{Z,}$, for three values of ϕ: (a) $\phi = 0$, (b) $\phi = 30°$ and (c) $\phi = 60°$.

in the first octant, the unit vectors $\hat{d}_s$ and $\hat{d}_f$ point in the fifth and eighth octants, respectively.

Similarly, for Case 2, $(n_X > n_Y > n_Z)$, it can be shown that n_s ranges from n_Y to n_X and that n_f ranges from n_Z to n_Y. Again, with $\tan \rho_s$ and $\tan \rho_f$ *defined* to be positive, Eqns. 1.59 show that when m is in the first octant, d_{sX} is negative, d_{sY} is positive and d_{sZ} is positive, while d_{fX} is negative, d_{fY} is negative and d_{fZ} is positive in this case, i.e., the unit vectors $\hat{d}_s$ and $\hat{d}_f$ point in the second and the third octant, respectively.

The signs of the components of the unit vectors $\hat{d}_s$ and $\hat{d}_f$ for the two cases are summarized in Table 1.1

	d_{sX}	d_{sY}	d_{sZ}	d_{fX}	d_{fY}	d_{fZ}
Case 1 $(n_X < n_Y < n_Z)$	positive	positive	negative	positive	negative	negative
Case 2 $(n_X > n_Y > n_Z)$	negative	positive	positive	negative	negative	positive

TABLE 1.1. Signs of the components of the unit vectors $\hat{d}_s$ and $\hat{d}_f$ for case 1 $(n_X < n_Y < n_Z)$ and case 2 $(n_X > n_Y > n_Z)$

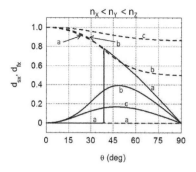

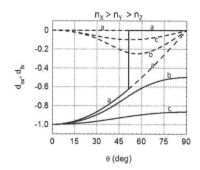

FIGURE 1.10: Dependence of d_{sX} (shown by solid lines) and d_{fX} (shown by dashed lines) on θ for different values of ϕ for a hypothetical crystal with $n_X < n_Y < n_Z$: (a) $\phi = 0$, (b) $\phi = 30°$ and (c) $\phi = 60°$.

FIGURE 1.11: Dependence of d_{sX} (shown by solid lines) and d_{fX} (shown by dashed lines) on θ for different values of ϕ for a hypothetical crystal with $n_X > n_Y > n_Z$,: (a) $\phi = 0$, (b) $\phi = 30°$ and (c) $\phi = 60°$.

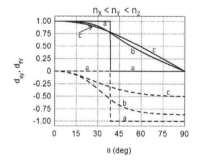

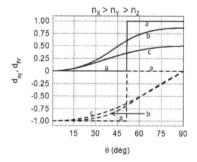

FIGURE 1.12: Dependence of d_{sY} (shown by solid lines) and d_{fY} (shown by dashed lines) on θ for different values of ϕ for a hypothetical crystal with $n_X < n_Y < n_Z$: (a) $\phi = 0°$, (b) $\phi = 30°$ and (c) $\phi = 60°$

FIGURE 1.13: Dependence of d_{sY} (shown by solid lines) and d_{fY} (shown by dashed lines) on θ for different values of ϕ for a hypothetical crystal with $n_X > n_Y > n_Z$: (a) $\phi = 0$, (b) $\phi = 30°$ and (c) $\phi = 60°$

1.5.4 The components of $\hat{e}_s$ and $\hat{e}_f$

When the propagation vector k does not lie along a principal axis or on a principal plane, the components of the unit vectors $\hat{e}_s$ and $\hat{e}_f$ are different from those of the unit vectors $\hat{d}_s$ and $\hat{d}_f$. However, when k does lie on a

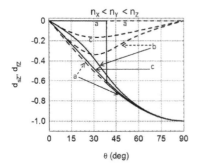

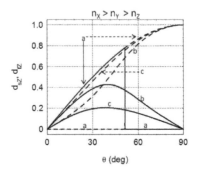

FIGURE 1.14: Dependence of d_{sZ} (shown by solid lines) and d_{fZ} (shown by dashed lines) on θ for different values of ϕ for a hypothetical crystal with $n_X < n_Y < n_Z$: a) $\phi = 0$, (b) $\phi = 30°$ and (c) $\phi = 60°$

FIGURE 1.15: Dependence of d_{sZ} (shown by solid lines) and d_{fZ} (shown by dashed lines) on θ for different values of ϕ for a hypothetical crystal with $n_X > n_Y > n_Z$: a) $\phi = 0$, (b) $\phi = 30°$ and (c) $\phi = 60°$

principal plane, the components of $\hat{e}_s$ and $\hat{e}_f$ are the same as those of $\hat{d}_s$ and $\hat{d}_f$ for certain ranges of θ or ϕ.

In Figs. 1.16 and 1.17 the components of d_s and e_s are shown as functions of θ for an azimuthal angle $\phi = 45°$. Similarly, in Figs. 1.18 and 1.19 the components of d_f and e_f are shown as functions of θ for an azimuthal angle $\phi = 45°$.

The discussion above shows how the directions of the **D** and **E** vectors can be determined for an arbitrary direction of the **k** vector in a biaxial crystal. For general values of m_X and m_Y, the expression for $\mathcal{D}$ given in Eqn. 1.56 does not reduce to a simple form so that the solutions for n_s and n_f in Eqns. 1.57 are algebraically complex, although computationally quite simple. For special cases, such as propagation along the principal axes or along principal planes, the expression for $\mathcal{D}$ and the expressions for n_s and n_f derived from it are much simpler. Also, in the special case of uniaxial crystals in which two of the principal refractive indices are equal, expressions for $\mathcal{D}$, n_s and n_f are very similar to those obtained for propagation along principal planes of biaxial crystals. These special cases will be discussed later in this chapter. Before doing that, the discussion of the biaxial crystal is continued with a description of the optic axes. The values of the components of the **D** vector have been derived in the literature in terms of the angle between the optic axes. We will re-derive those expressions and will also derive the directions of the components of the **E** field in the next three sections.

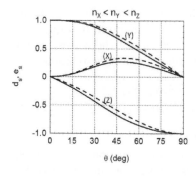

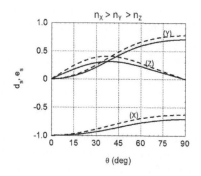

FIGURE 1.16: The components of $\hat{d}_s$ (solid lines) and $\hat{e}_s$ (dashed lines) as functions of the angle θ for $\phi = 45°$ for a hypothetical crystal with $n_X < n_Y < n_Z$: (X) indicates components $\hat{d}_{sX}$, $\hat{e}_{sX}$; (Y) indicates components $\hat{d}_{sY}$, $\hat{e}_{sY}$ and (Z) indicate the components $\hat{d}_{sZ}$, $\hat{e}_{sZ}$.

FIGURE 1.17: The components of $\hat{d}_s$ (solid lines) and $\hat{e}_s$ (dashed lines) as functions of the angle θ for $\phi = 45°$ for a hypothetical crystal with $n_X > n_Y > n_Z$: (X) indicates components d_{sX}, e_{sX}; (Y) indicates components d_{sY}, e_{sY} and (Z) indicate the components d_{sZ}, e_{sZ}.

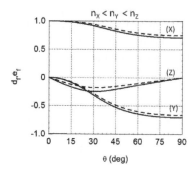

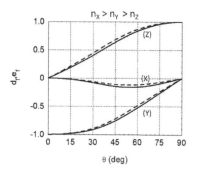

FIGURE 1.18: The components of $\hat{d}_f$ (solid lines) and $\hat{e}_f$ (dashed lines) as functions of the angle θ for $\phi = 45°$ for a hypothetical crystal with $n_X < n_Y < n_Z$: (X) indicates components $\hat{d}_{fX}$, $\hat{e}_{fX}$; (Y) indicates components $\hat{d}_{fY}$, $\hat{e}_{fY}$ and (Z) indicate the components $\hat{d}_{fZ}$, $\hat{e}_{fZ}$.

FIGURE 1.19: The components of d_f (solid lines) and e_f (dashed lines) as functions of the angle θ for $\phi = 45°$ for a hypothetical crystal with $n_X > n_Y > n_Z$: (X) indicates components d_{fX}, e_{fX}; (Y) indicates components d_{fY}, e_{fY} and (Z) indicate the components d_{fZ}, e_{fZ}.

1.6　Double Refraction And Optic Axes

A plane wave traveling in an anisotropic medium with its wave vector **k** pointing in an arbitrary direction can have two speeds c/n_s and c/n_f, with n_s and

n_f given by Eqn. 1.57. The directions of the unit vectors $\hat{e}_s$ and $\hat{e}_f$ along the electric field vectors associated with those waves are given by Eqns. 1.58 and 1.60. $\hat{s}_s$ and $\hat{s}_f$, the unit vectors in the directions of the corresponding Poynting vectors, can be obtained from $\hat{e}_s$, $\hat{e}_f$ and $\hat{m}$ using the relations

$$
\begin{aligned}
\hat{s}_s &= \hat{e}_s \times (\hat{m} \times \hat{e}_s) = \hat{m} - \hat{e}_s(\hat{e}_s \cdot \hat{m}) \\
\hat{s}_f &= \hat{e}_f \times (\hat{m} \times \hat{e}_f) = \hat{m} - \hat{e}_f(\hat{e}_f \cdot \hat{m}).
\end{aligned}
\tag{1.65}
$$

which can be derived from Eqns. 1.5 and 1.10.

Since $\hat{e}_s$ and $\hat{e}_f$ point in different directions, the directions of $\hat{s}_s$ and $\hat{s}_f$ are also different i.e., the slow and fast waves having the same wave vector **k** have different directions for energy propagation, causing the two waves to be spatially separated after traveling a certain distance through the medium. This phenomenon is given the name *double refraction* and is also called *birefringence*.

It will be shown later that when $\hat{\mathbf{k}}$ is along the principal axes X, Y or Z, the Poynting vector **S** is along $\hat{\mathbf{k}}$ for both the slow and the fast waves, and there is no double refraction. In addition there are two other directions in biaxial crystals such that light propagating with the $\hat{\mathbf{k}}$ vectors along these directions undergoes no double refraction, i.e., along these directions, $n_s = n_f$ and the speeds of the fast and the slow waves are equal. These additional directions in biaxial crystals are defined to be the directions of the *optic axes*.

The directions of the optic axes can be determined from Section 1.4.2. Equation 1.57 shows that for $n_s = n_f$, $\mathcal{D}$ must be equal to 0, i.e., from Eqn. 1.56, $\mathcal{B}^2 = 4\mathcal{A}\mathcal{C}$. For simplicity of notation, let us define here (locally) three constants a, b and c and three variables x, y and z as

$$
\begin{aligned}
a &\equiv n_X^2, \quad b \equiv n_Y^2, \quad c \equiv n_Z^2, \\
x &\equiv m_X^2, \quad y \equiv m_Y^2, \quad z \equiv m_Z^2.
\end{aligned}
\tag{1.66}
$$

Using the relation $m_X^2 + m_Y^2 + m_Z^2 = 1$, i.e., $x + y + z = 1$, we write Eqn. 1.55 as

$$
\mathcal{A} = x(a - c) + d, \quad \mathcal{B} = xbd + e, \quad \mathcal{C} = abc
\tag{1.67}
$$

where

$$
d \equiv c + y(b - c), \text{ and } \quad e \equiv ya(b - c) + c(a + b).
\tag{1.68}
$$

The condition $\mathcal{B}^2 = 4\mathcal{A}\mathcal{C}$ leads to a quadratic equation in x

$$
A_1 x^2 + B_1 x + C_1 = 0
\tag{1.69}
$$

where

$$
A_1 = b^2(a - c)^2, \quad B_1 = 2b(a - c)(e - 2ac) \text{ and } C_1 = e^2 - 4abcd.
\tag{1.70}
$$

Equation 1.69 will have real solution(s) for x if the discriminant $D_1^2 = B_1^2 - 4A_1C_1$ is nonnegative. From Eqn. 1.70 we find

$$D_1^2 = -[16b^2ac(a-c)^2(c-b)(b-a)]y^2. \tag{1.71}$$

With the values of n_X, n_Y and n_Z ordered from low to high or high to low (i.e., with n_Y in the middle) the quantity inside the square brackets in Eqn. 1.71 is always positive. The only way x can have a real solution is therefore with $y = 0$, i.e., with $m_Y = 0$.

With $m_Y = 0$, x and $z(= 1-x)$ are obtained from the solution of Eqn. 1.69:

$$x = m_X^2 = \frac{n_Z^2}{n_Y^2}\left(\frac{n_X^2 - n_Y^2}{n_X^2 - n_Z^2}\right)$$
$$z = m_Z^2 = \frac{n_X^2}{n_Y^2}\left(\frac{n_Y^2 - n_Z^2}{n_X^2 - n_Z^2}\right). \tag{1.72}$$

For these values of m_X and m_Z, the directions of the two optic axes, OA_1 and OA_2 are shown in Fig. 1.20, with the angle Ω between the optic axes and

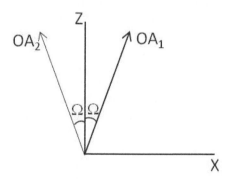

FIGURE 1.20: OA_1 and OA_2 denote the two *optic axes*. When the propagation vector **k** is along OA_1 or OA_2, the two waves associated with **k** travel with the same speed and there is no birefringence.

the Z direction can be obtained from the ratio of m_X to m_Z:

$$\tan\Omega = \frac{n_Z}{n_X}\left(\frac{n_X^2 - n_Y^2}{n_Y^2 - n_Z^2}\right)^{1/2}. \tag{1.73}$$

Since for anisotropic crystals with $n_X \neq n_Y \neq n_Z$, there are two optical axes, both lying in the XZ plane as shown above, these crystals are called *biaxial*. Biaxial crystals are designated **positive** or **negative** depending on whether the angle 2Ω between the two optic axes is less than or greater than $90°$ respectively.

Using the values of 1.65, 1.75 and 1.95 for n_X, n_Y and n_Z respectively, we find that Ω is 38°. For n_X, n_Y and n_Z respectively equal to 1.95, 1.75 and 1.65, Ω is equal to 52 °, confirming the results shown in Figs. 1.6 and 1.7.

1.6.1 Expressions for components of $\hat{d}$ in terms of the angles θ, ϕ and Ω

The directions of the electric fields of the fast and the slow waves are needed for calculating the 'effective' nonlinear optical coefficient (d_{eff}), which will be discussed in detail in the next chapter. The directions used in the literature are described by Dmitriev [2] in terms of the angles θ, ϕ and an angle δ which in turn is related to θ, ϕ and Ω. On page 27 of Ref. [2] it is pointed out that the expressions for the electric field components published between 1972 and 1975 [7], [9], [10] are incorrect and the correct forms were obtained by Lavrovskaya et al. [8]. However, there are a couple of typographical errors in Ref. [2] as well (last equation of Eqns. 2.65 in page 26 has the sign wrong and δ should be replaced by 2δ in Eqn. 2.66). The full derivation of the equation for the angle δ in Eqn. 2.66 of Ref. [2] is also hard to find (all the references lead back to a publication in Japanese [11] from 1965. Moreover, the components given by Lavrovskaya as those of the 'electric field vector' are really the components of the displacement vector **D**. The derivation of 'walk-off angles' for propagation in arbitrary directions in a biaxial crystal case is not given in any of the easily available references. Ref. [2] sets aside this task, stating (in page 31) "the inclusion of birefringence (anisotropy) in the calculation of d_{eff} for light propagation into a biaxial crystal is complicated enough and we haven't done it here." In the following these issues are addressed.

Several new unit vectors pointing at different directions will be introduced next. In an attempt to make it a little easier to follow the logic of the selection of these vectors, we summarize the procedure here: we will consider only **k** vectors lying in the first octant, i.e., with ϕ between 0 and 90°. Results for other values of ϕ can be determined using the same procedure. Since **D** is perpendicular to **k** (from Sections 1.2.1 and 1.2.2) both the slow and fast components of **D**, i.e, both $\hat{d}_s$ and $\hat{d}_f$ are perpendicular to **k** and therefore they both must lie on a plane perpendicular to **k**. The intersection of this plane (say named 'the $k_\perp$ plane') with the $k - Z$ plane containing **k** vector and the Z axis is a line, and the unit vector along this line is named $\hat{u}_1$. Thus $\hat{u}_1$ lies on the $k - Z$ plane and is perpendicular to **k**. Another unit vector, perpendicular to $\hat{u}_1$ and lying in the $k_\perp$ plane is named $\hat{u}_2$. $\hat{u}_1$ and $\hat{u}_2$ are completely determined by the angles θ and ϕ of the **k** vector. Since $\hat{d}_s$ and $\hat{d}_f$ lie on the $\hat{u}_1$-$\hat{u}_2$ plane, we *define* two mutually perpendicular unit vectors named $\hat{d}_1$ and $\hat{d}_2$, with $\hat{d}_1$ at an angle δ with $\hat{u}_1$. Cartesian components of $\hat{d}_1$ and $\hat{d}_2$ are then determined in terms of the angles θ, ϕ and δ. While the angle δ is still undetermined; it can be shown that for the two cases under consideration, (Case 1, with $n_X < n_Y < n_Z$ and Case 2, with

$n_X > n_Y > n_Z$), δ will be positive and acute if we assign $\hat{d}_s = \hat{d}_1$ and $\hat{d}_s = \hat{d}_2$ respectively. Next, the components of the unit vectors $\hat{e}_s$ and $\hat{e}_f$ of the slow and fast components of the electric fields are also determined in terms of the angles θ, ϕ, δ and the walk-off angles ρ_s and ρ_f.

The propagation vector **k** is oriented with polar and azimuthal angles θ, ϕ with respect to the principal axes X, Y, Z, as shown in Fig. 1.3 and the components of the unit vector $\hat{m}$ are given by Eqn. 1.64.

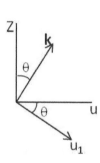
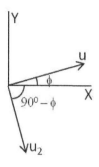

FIGURE 1.21: The orientation of **k** and **u₁** with respect to Z and **u**

FIGURE 1.22: The orientation of the vectors $\hat{u}$ and $\hat{u}_2$ with respect to X and Y

As shown in Fig. 1.3, the unit vector along the projection of **k** on the XY plane is denoted by $\hat{u}$, so that

$$\hat{u} = \hat{X} \cos\phi + \hat{Y} \sin\phi. \tag{1.74}$$

Figure 1.21 shows that the unit vector $\hat{u}_1$, lying on the $k - Z$ plane and perpendicular to **k**, can be written as

$$
\begin{aligned}
\hat{u}_1 &= \hat{u} \cos\theta - \hat{Z} \sin\theta \\
&= \hat{X} \cos\theta\cos\phi + \hat{Y} \cos\theta\sin\phi - \hat{Z} \sin\theta
\end{aligned} \tag{1.75}
$$

using Eqn. 1.74.

$\hat{u}_2$ denotes the unit vector perpendicular to both $\hat{u}_1$ and **k**. $\hat{u}_2$ must then also be perpendicular to $\hat{u}$ and Z and can be obtained by taking the cross product of $\hat{u}_1$ and **k** or any two of the vectors $\hat{u}_1$, $\hat{u}$, **k** and $\hat{Z}$ in the Zu plane. Taking the cross product of the unit vectors $\hat{u}$ and $\hat{Z}$ we find

$$
\begin{aligned}
\hat{u}_2 &= \hat{u} \times \hat{Z} \\
&= \begin{vmatrix} \hat{X} & \hat{Y} & \hat{Z} \\ \cos\phi & \sin\phi & 0 \\ 0 & 0 & 1 \end{vmatrix} \\
&= \hat{X} \sin\phi - \hat{Y} \cos\phi
\end{aligned} \tag{1.76}
$$

showing that $\hat{u}_2$ lies on the XY plane, along with $\hat{u}$ (Fig. 1.22).

In the plane u_1-u_2 formed by the vectors $\hat{u}_1$ and $\hat{u}_2$, another pair of mutually perpendicular unit vectors $\hat{d}_1$ and $\hat{d}_2$ are drawn, with δ denoting the angle between $\hat{d}_1$ and $\hat{u}_1$ as shown in Fig. 1.23.

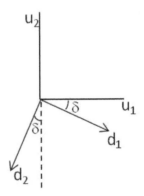

FIGURE 1.23: The orientation of $\mathbf{d_1}$ and $\mathbf{d_2}$ with respect to $\mathbf{u_1}$ and $\mathbf{u_2}$

Figure 1.23 shows that the unit vectors $\hat{d}_1$ and $\hat{d}_2$ can be expressed in the X, Y, Z coordinate system as

$$
\begin{aligned}
\hat{d}_1 &= \hat{u}_1 \, \cos\delta - \hat{u}_2 \, \sin\delta \\
&= \hat{X} \, (\cos\theta \, \cos\phi \, \cos\delta - \sin\phi \, \sin\delta) + \hat{Y} \, (\cos\theta \, \sin\phi \, \cos\delta + \cos\phi \, \sin\delta) \\
&\quad - \hat{Z} \, \sin\theta \cos\delta
\end{aligned}
\tag{1.77}
$$

and

$$
\begin{aligned}
\hat{d}_2 &= -\hat{u}_1 \, \sin\delta - \hat{u}_2 \, \cos\delta \\
&= -\hat{X} \, (\cos\theta \, \cos\phi \, \sin\delta + \sin\phi \, \cos\delta) - \hat{Y} \, (\cos\theta \, \sin\phi \, \sin\delta - \cos\phi \, \cos\delta) \\
&\quad + \hat{Z} \, \sin\theta \sin\delta
\end{aligned}
\tag{1.78}
$$

using Eqns. 1.75 and 1.76. The components of $\hat{d}_1$ and $\hat{d}_2$ are then given by

$$
\begin{aligned}
d_{1X} &= \cos\theta \, \cos\phi \, \cos\delta - \sin\phi \, \sin\delta \\
d_{1Y} &= \cos\theta \, \sin\phi \, \cos\delta + \cos\phi \, \sin\delta \\
d_{1Z} &= -\sin\theta \, \cos\delta
\end{aligned}
\tag{1.79}
$$

and

$$
\begin{aligned}
d_{2X} &= -\cos\theta \, \cos\phi \, \sin\delta - \sin\phi \, \cos\delta \\
d_{2Y} &= -\cos\theta \, \sin\phi \, \sin\delta + \cos\phi \, \cos\delta \\
d_{2Z} &= \sin\theta \sin\delta.
\end{aligned}
\tag{1.80}
$$

1.6.2 Relating the angle δ to Ω, θ and ϕ

To relate the angle δ to Ω, θ and ϕ we start from the Eqns. 1.59 and express the refractive indices n_X, n_Y and n_Z in terms of the Cartesian components of the unit vectors $\hat{d}_s$ as

$$n_X^2 = \frac{n_s^2 d_{sX}}{d_{sX} + m_X \tan \rho_s} \quad , \quad n_Y^2 = \frac{n_s^2 d_{sY}}{d_{sY} + m_Y \tan \rho_s} \quad \text{and}$$

$$n_Z^2 = \frac{n_s^2 d_{sZ}}{d_{sZ} + m_Z \tan \rho_s}. \tag{1.81}$$

Inserting these values of n_X, n_Y and n_Z in Eqn. 1.73 we obtain (after a little algebra)

$$\cot^2 \Omega = \frac{d_{sX}}{d_{sZ}} \frac{m_Z d_{sY} - m_Y d_{sZ}}{m_Y d_{sX} - m_X d_{sY}}. \tag{1.82}$$

The angle δ has not yet been defined except for the specification that it is the angle between some vector $\hat{d}_1$ and the vector $\hat{u}_1$. δ gets defined when we identify the vector $\hat{d}_1$ with the unit vector $\hat{d}_s$, i.e., when we require

$$\hat{d}_s = \hat{d}_1. \tag{1.83}$$

Then using Eqns. 1.79 and 1.64 we obtain

$$\begin{aligned} m_Z d_{sY} - m_Y d_{sZ} &= m_Z d_{1Y} - m_Y d_{1Z} \\ &= \cos\theta(\cos\theta\sin\phi\cos\delta + \cos\phi\sin\delta) \\ &\quad + \sin\theta\sin\phi(\sin\theta\cos\delta) \\ &= \sin\phi\cos\delta + \cos\theta\cos\phi\sin\delta, \end{aligned} \tag{1.84}$$

and

$$\begin{aligned} m_Y d_{sX} - m_X d_{sy} &= m_Y d_{1X} - m_X d_{1X} \\ &= \sin\theta\sin\phi(\cos\theta\cos\phi\cos\delta - \sin\phi\sin\delta) \tag{1.85} \\ &\quad - \sin\theta\cos\phi(\cos\theta\sin\phi\cos\delta + \cos\phi\sin\delta) \\ &= -\sin\theta\sin\delta \end{aligned} \tag{1.86}$$

and

$$\begin{aligned} \frac{d_{sX}}{d_{sZ}} &= \frac{d_{1X}}{d_{1Z}} \\ &= -\frac{\cos\theta\cos\phi\cos\delta - \sin\phi\sin\delta}{\sin\theta\cos\delta} \end{aligned} \tag{1.87}$$

inserting which in Eqn. 1.82 we get

$$\cot^2 \Omega = \frac{(\cos\theta\,\cos\phi\,\cos\delta - \sin\phi\,\sin\delta)(\cos\theta\,\cos\phi\,\sin\delta + \sin\phi\,\cos\delta)}{\sin^2\theta\,\sin\delta\,\cos\delta} \tag{1.88}$$

which can be solved for the angle δ

$$\cot 2\delta = \frac{\cot^2 \Omega \ \sin^2 \theta + \sin^2 \phi - \cos^2 \theta \cos^2 \phi}{\cos \theta \sin 2\phi}. \tag{1.89}$$

If instead of the choice made in Eqn. 1.83, we make the alternate choice

$$\hat{d}_s = \hat{d}_2 \tag{1.90}$$

and go through the same algebra as above, we get the same expression for δ as given in Eqn. 1.89.

The dependence of δ on θ is shown for the two cases of $n_X < n_Y < n_Z$ and $n_X > n_Y > n_Z$ in Figs. 1.24 and 1.25 for three values of the angle ϕ.

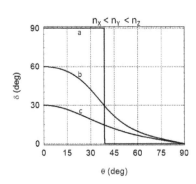

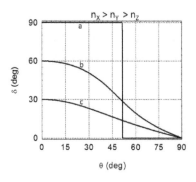

FIGURE 1.24: The angle δ is plotted against θ for three values of the angle ϕ for case 1, with $n_X = 1.65$, $n_Y = 1.75$ and $n_Z = 1.95$. (a) $\phi = 0°$, (b) $\phi = 30°$ and (c) $\phi = 60°$. At $\phi = 90°$, $\delta = 0$ (not shown here).

FIGURE 1.25: The angle δ is plotted against θ for three values of the angle ϕ for case 2, with $n_X = 1.95$, $n_Y = 1.75$ and $n_Z = 1.65$. (a) $\phi = 0°$, (b) $\phi = 30°$ and (c) $\phi = 60°$. At $\phi = 90°$, $\delta = 0$ (not shown here).

For a given crystal with Ω known from Eqn. 1.73, δ can be calculated from Eqn. 1.89 for any values of θ and ϕ. A series of values of δ, both positive and negative, separated by 90° can satisfy Eqn. 1.89. We adopt the convention here of choosing for δ only the lowest positive value among the series of values. For example, say, $\delta = 40°$ is a solution of Eqn. 1.89. Then the angles of 130°, 220°, 310°, etc. as well as -50°, - 140°, - 230° etc. are also solutions. From all these solutions, *only* $\delta = 40°$ is chosen for the calculation of the components of $\hat{d}$. This is different from the convention adopted in Ref. [8] and cited in Ref. [2] where negative values of δ that satisfy Eqn. 1.89 (and lie between 0 and - 90°) are chosen to be associated with case 2 ($n_X > n_Y > n_Z$).

With δ chosen to be positive, we see from Eqns. 1.79 and 1.80 that at $\phi = 0$, the components (d_{1X}, d_{1Y}, d_{1Z}) and (d_{2X}, d_{2Y}, d_{2Z}) are (positive, positive, negative) and (negative, positive, positive) respectively. Thus, comparing with the signs given in Table 1.1 we see that with $\hat{d}_s$ identified with $\hat{d}_1$, $\hat{d}_f$ needs to be along $-\hat{d}_2$ for Case 1, $(n_X < n_Y < n_Z)$. Similarly, if $\hat{d}_s$ identified with $\hat{d}_2$, $\hat{d}_f$ needs to be along $-\hat{d}_1$ for case 2, $(n_X > n_Y > n_Z)$.

Writing out the expressions for the components of $\hat{d}_s$ and $\hat{d}_f$ explicitly, we have the following sets of equations for the two cases:

Case 1: $n_X < n_Y < n_Z$

$$\hat{d}_s = \hat{d}_1, \qquad \hat{d}_f = -\hat{d}_2$$

$$
\begin{aligned}
d_{sX} &= \cos\theta \ \cos\phi \ \cos\delta - \sin\phi \ \sin\delta \\
d_{sY} &= \cos\theta \ \sin\phi \ \cos\delta + \cos\phi \ \sin\delta \\
d_{sZ} &= -\sin\theta \ \cos\delta
\end{aligned}
\tag{1.91}
$$

and

$$
\begin{aligned}
d_{fX} &= \cos\theta \ \cos\phi \ \sin\delta + \sin\phi \ \cos\delta \\
d_{fY} &= \cos\theta \ \sin\phi \ \sin\delta - \cos\phi \ \cos\delta \\
d_{fZ} &= -\sin\theta \sin\delta.
\end{aligned}
\tag{1.92}
$$

Case 2: $n_X > n_Y > n_Z$

$$\hat{d}_s = \hat{d}_2, \qquad \hat{d}_f = -\hat{d}_1$$

$$
\begin{aligned}
d_{sX} &= -\cos\theta \ \cos\phi \ \sin\delta - \sin\phi \ \cos\delta \\
d_{sY} &= -\cos\theta \ \sin\phi \ \sin\delta + \cos\phi \ \cos\delta \\
d_{sZ} &= \sin\theta \sin\delta
\end{aligned}
\tag{1.93}
$$

and

$$
\begin{aligned}
d_{fX} &= -\cos\theta \ \cos\phi \ \cos\delta + \sin\phi \ \sin\delta \\
d_{fY} &= -\cos\theta \ \sin\phi \ \cos\delta - \cos\phi \ \sin\delta \\
d_{fZ} &= \sin\theta \ \cos\delta.
\end{aligned}
\tag{1.94}
$$

1.6.3 Directions of E and S

As discussed earlier, for light propagating in a biaxial crystal with propagation vector $\mathbf{k}$ in a general direction there are two oscillation directions of the electric fields, with unit vectors $\hat{e}_s$ and $\hat{e}_f$ which are different from the unit vectors $\hat{d}_s$ and $\hat{d}_f$ along the electric displacements. Since the Poynting vector ($\mathbf{S}$) is perpendicular to the electric field, for a given $\mathbf{k}$ there are also two directions of $\mathbf{S}$ along which the light energy travels. Suppose the unit vectors along the two possible directions of $\mathbf{S}$ are denoted by $\hat{s}_s$ and $\hat{s}_f$. It was shown in Secs. 1.2.1 and 1.2.2 that the vectors $\hat{e}_s$, $\hat{d}_s$, $\hat{s}_s$ and $\mathbf{k}$ lie in one plane (named say, P_s) and so do the vectors $\hat{e}_f$, $\hat{d}_f$, $\hat{s}_f$ and $\mathbf{k}$, (named say, P_f). Since $\mathbf{k}$, $\hat{d}_s$ and $\hat{d}_f$ are mutually perpendicular, the planes P_s and P_f are also perpendicular to each other.

The directions of the vectors $\hat{e}_s$, and $\hat{s}_s$ in the plane P_s and the vectors $\hat{e}_f$, and $\hat{s}_f$ in the plane P_f are shown in Figs. 1.26 and 1.27.

As shown earlier in Sec. 1.4.3, $\hat{d}_s$ and $\hat{d}_f$ are always perpendicular to each other, but in general $\hat{e}_s$ is not perpendicular $\hat{e}_f$. The directions of $\hat{e}_s$ and $\hat{e}_f$ can be completely determined from the directions of $\hat{d}_s$ and $\hat{d}_f$, as shown in Figs. 1.26 and 1.27.

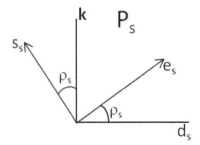

FIGURE 1.26: For the slow wave, the unit vectors $\hat{s}_s$, $\hat{e}_s$ and $\hat{d}_s$ along the directions of the Poynting vector, the electric field and the displacement vector, respectively.

FIGURE 1.27: For the fast wave, the unit vectors $\hat{s}_f$, $\hat{e}_f$ and $\hat{d}_f$ along the directions of the Poynting vector, the electric field and the displacement vector, respectively.

If ρ_s and ρ_f denote the angles between $\hat{e}_s$ and $\hat{d}_s$ and between $\hat{e}_f$ and $\hat{d}_f$ respectively, Figs. 1.26 and 1.27 show that

$$\begin{aligned}
\hat{e}_s &= \hat{d}_s \cos\rho_s + \hat{m} \sin\rho_s \\
\hat{s}_s &= -\hat{d}_s \sin\rho_s + \hat{m} \cos\rho_s \\
\hat{e}_f &= \hat{d}_f \cos\rho_f + \hat{m} \sin\rho_f \\
\hat{s}_f &= -\hat{d}_f \sin\rho_f + \hat{m} \cos\rho_f
\end{aligned} \tag{1.95}$$

where as before, $\hat{m}$ is the unit vector along $\mathbf{k}$. Since the Cartesian components of $\hat{d}_s$, $\hat{d}_f$ and $\hat{m}$ are known (from Eqns. 1.91 to 1.94 and 1.64), the components of $\hat{e}_s$, $\hat{s}_s$, $\hat{e}_f$ and $\hat{s}_f$ can be obtained from Eqns. 1.95. Thus, for Case 1, $(n_X < n_Y < n_Z)$

$$\begin{aligned}
e_{sX} &= \cos\theta \,\cos\phi \,\cos\delta \,\cos\rho_s - \sin\phi \,\sin\delta \,\cos\rho_s + \sin\theta \,\cos\phi \,\sin\rho_s \\
e_{sY} &= \cos\theta \,\sin\phi \,\cos\delta \,\cos\rho_s + \cos\phi \,\sin\delta \,\cos\rho_s + \sin\theta \,\sin\phi \,\sin\rho_s \\
e_{sZ} &= -\sin\theta \,\cos\delta \,\cos\rho_s + \cos\theta \,\sin\rho_s
\end{aligned} \tag{1.96}$$

$$\begin{aligned}
s_{sX} &= -\cos\theta \,\cos\phi \,\cos\delta \,\sin\rho_s + \sin\phi \,\sin\delta \,\sin\rho_s + \sin\theta \,\cos\phi \,\cos\rho_s \\
s_{sY} &= -\cos\theta \,\sin\phi \,\cos\delta \,\sin\rho_s - \cos\phi \,\sin\delta \,\sin\rho_s + \sin\theta \,\sin\phi \,\cos\rho_s \\
s_{sZ} &= \sin\theta \,\cos\delta \,\sin\rho_s + \cos\theta \,\cos\rho_s
\end{aligned} \tag{1.97}$$

$$e_{fX} = \cos\theta \ \cos\phi \ \sin\delta \ \cos\rho_f + \sin\phi \ \cos\delta \ \cos\rho_f + \sin\theta \ \cos\phi \ \sin\rho_f$$
$$e_{fY} = \cos\theta \ \sin\phi \ \sin\delta \ \cos\rho_f - \cos\phi \ \cos\delta \ \cos\rho_f + \sin\theta \ \sin\phi \ \sin\rho_f$$
$$e_{fZ} = -\sin\theta \ \sin\delta \ \cos\rho_f + \cos\theta \ \sin\rho_f \tag{1.98}$$

$$s_{fX} = -\cos\theta \ \cos\phi \ \sin\delta \ \sin\rho_f - \sin\phi \ \cos\delta \ \sin\rho_f + \sin\theta \ \cos\phi \ \cos\rho_f$$
$$s_{fY} = -\cos\theta \ \sin\phi \ \sin\delta \ \sin\rho_f + \cos\phi \ \cos\delta \ \sin\rho_f + \sin\theta \ \sin\phi \ \cos\rho_f$$
$$s_{fZ} = \sin\theta \ \sin\delta \ \sin\rho_f + \cos\theta \ \cos\rho_f \tag{1.99}$$

and for Case 2, $(n_X > n_Y > n_Z)$

$$e_{sX} = -\cos\theta \ \cos\phi \ \sin\delta \ \cos\rho_s - \sin\phi \ \cos\delta \ \cos\rho_s + \sin\theta \ \cos\phi \ \sin\rho_s$$
$$e_{sY} = -\cos\theta \ \sin\phi \ \sin\delta \ \cos\rho_s + \cos\phi \ \cos\delta \ \cos\rho_s + \sin\theta \ \sin\phi \ \sin\rho_s$$
$$e_{sZ} = \sin\theta \ \sin\delta \ \cos\rho_s + \cos\theta \ \sin\rho_s \tag{1.100}$$

$$s_{sX} = \cos\theta \ \cos\phi \ \sin\delta \ \sin\rho_s + \sin\phi \ \cos\delta \ \sin\rho_s + \sin\theta \ \cos\phi \ \cos\rho_s$$
$$s_{sY} = \cos\theta \ \sin\phi \ \sin\delta \ \sin\rho_s - \cos\phi \ \cos\delta \ \sin\rho_s + \sin\theta \ \sin\phi \ \cos\rho_s$$
$$s_{sZ} = -\sin\theta \ \sin\delta \ \sin\rho_s + \cos\theta \ \cos\rho_s \tag{1.101}$$

$$e_{fX} = -\cos\theta \ \cos\phi \ \cos\delta \ \cos\rho_f + \sin\phi \ \sin\delta \ \cos\rho_f + \sin\theta \ \cos\phi \ \sin\rho_f$$
$$e_{fY} = -\cos\theta \ \sin\phi \ \cos\delta \ \cos\rho_f - \cos\phi \ \sin\delta \ \cos\rho_f + \sin\theta \ \sin\phi \ \sin\rho_f$$
$$e_{fZ} = \sin\theta \ \cos\delta \ \cos\rho_f + \cos\theta \ \sin\rho_f \tag{1.102}$$

$$s_{fX} = \cos\theta \ \cos\phi \ \cos\delta \ \sin\rho_f - \sin\phi \ \sin\delta \ \sin\rho_f + \sin\theta \ \cos\phi \ \cos\rho_f$$
$$s_{fY} = \cos\theta \ \sin\phi \ \cos\delta \ \sin\rho_f + \cos\phi \ \sin\delta \ \sin\rho_f + \sin\theta \ \sin\phi \ \cos\rho_f$$
$$s_{fZ} = -\sin\theta \ \cos\delta \ \sin\rho_f + \cos\theta \ \cos\rho_f. \tag{1.103}$$

1.6.4   The walk-off angles ρ_s and ρ_f

The walk-off angles ρ_s and ρ_f expressed in Eqn. 1.60 in terms of n_s and n_f can be used in the last section to find the components of $\hat{e}$ and $\hat{s}$. Here we provide alternate expressions for ρ_s and ρ_f in terms of the components of $\hat{d}$.

From Eqn. 1.32 we find

$$\frac{e_{sZ}}{e_{sY}} = \frac{n_Y^2}{n_Z^2} \frac{d_{sZ}}{d_{sY}}$$
$$\frac{e_{fZ}}{e_{fY}} = \frac{n_Y^2}{n_Z^2} \frac{d_{fZ}}{d_{fY}}. \tag{1.104}$$

Using Eqns. 1.95 we obtain

$$\frac{d_{sZ}\,\cos\rho_s + m_Z\,\sin\rho_s}{d_{sY}\,\cos\rho_s + m_Y\,\sin\rho_s} = \frac{n_Y^2\,d_{sZ}}{n_Z^2\,d_{sY}}$$

$$\frac{d_{fZ}\,\cos\rho_f + m_Z\,\sin\rho_f}{d_{fY}\,\cos\rho_f + m_Y\,\sin\rho_f} = \frac{n_Y^2\,d_{fZ}}{n_Z^2\,d_{fY}}. \tag{1.105}$$

To derive Eqns. 1.105 we started from the ratio of the Z and Y components of the $\hat{e}$ and $\hat{d}$ fields in Eqn. 1.104. However any other ratios (such as that of the Y to the X components or the X to the Z components could equally well be taken to obtain the walk-off angles ρ_s and ρ_f. Using Eqns. 1.105 and similar equations for the other ratios, we obtain

$$
\begin{aligned}
\tan\rho_s &= \frac{\left(n_Z^2 - n_Y^2\right)\,d_{sZ}d_{sY}}{m_Y n_Y^2 d_{sZ} - m_Z n_Z^2 d_{sY}} = \frac{\left(n_Z^2 - n_X^2\right)\,d_{sZ}d_{sX}}{m_X n_X^2 d_{sZ} - m_Z n_Z^2 d_{sX}} \\[2mm]
&= \frac{\left(n_X^2 - n_Y^2\right)\,d_{sX}d_{sY}}{m_Y n_Y^2 d_{sX} - m_X n_X^2 d_{sY}} \\[2mm]
\tan\rho_f &= \frac{\left(n_Z^2 - n_Y^2\right)\,d_{fZ}d_{fY}}{m_Y n_Y^2 d_{fZ} - m_Z n_Z^2 d_{fY}} = \frac{\left(n_Z^2 - n_X^2\right)\,d_{fZ}d_{fX}}{m_X n_X^2 d_{fZ} - m_Z n_Z^2 d_{fX}} \\[2mm]
&= \frac{\left(n_X^2 - n_Y^2\right)\,d_{fX}d_{fY}}{m_Y n_Y^2 d_{fX} - m_X n_X^2 d_{fY}}
\end{aligned}
$$

$$\tag{1.106}$$

which can be expressed in terms of n_X, n_Y, n_Z and the angles θ, ϕ and δ using Eqns. 1.91 - 1.94.

1.6.5   An interim summary

Before moving to the next section, we summarize here what has been described thus far and present a preview of what will come next in this chapter:

1. For light traveling in an anisotropic crystal, the directions of the displacement vector **D** and the electric field vector **E** are in general different. The angle between the **D** and **E** vectors is the 'walk-off' angle ρ, given in Eqn. 1.48, which is also the angle between the directions of the Poynting vector **S** and the propagation vector **k**.

2. Light traveling in a crystal with known values of the principal refractive indices n_X, n_Y, n_Z will have two waves (a *slow* wave and a *fast* wave) propagating with two speeds (c/n_s and c/n_f, respectively) associated with each direction of the propagation vector **k**. The values of n_s and n_f are given in Eqn. 1.57.

3. The displacement vectors of the slow wave and the fast wave are perpendicular to each other. The unit vectors $\hat{d}_s$ and $\hat{d}_f$ of the two waves along these perpendicular directions are given in Eqns. 1.59.

4. In general, the electric field directions of the *slow* and the *fast* waves are not perpendicular to each other. The unit vectors $\hat{e}_s$ and $\hat{e}_f$ of the two waves along the directions of the two electric field vectors are given in Eqns. 1.58.

5. The values of the walk-off angles of the *slow* and the *fast* waves are different in general, and are obtained from Eqns. 1.60

6. When the propagation vector **k** lies along a particular direction on the $Z - X$ plane, the n_s and n_f values are equal, and light propagates as in an isotropic medium with no beam walk-off. This direction is called the direction of the 'optic axis'. The value of the angle between the optic axis and the Z axis is denoted by Ω and given in Eqn. 1.73 in terms of the n_X, n_Y and n_Z values.

7. The unit vectors $\hat{d}_s$ and $\hat{d}_f$ are given in terms of angles in Eqns. 1.91, 1.92, 1.93, 1.94.

8. The unit vectors $\hat{e}_s$ and $\hat{e}_f$ as well as $\hat{s}_s$ and $\hat{s}_f$, which are the unit vectors along the directions of the Poynting vectors of the *slow* and the *fast* waves respectively, are given in terms of angles in Eqns. 1.96 through 1.103.

Next, the special cases of propagation with the k vector along the principal axes and along the principal planes of a biaxial crystal are described. The case of uniaxial crystals is treated after that and finally the propagation equation with beam walk-off is derived.

1.7 Propagation Along The Principal Axes And Along The Principal Planes

1.7.1 Introduction

From the expressions for the components of the $\hat{d}$, $\hat{e}$, and $\hat{s}$ vectors for a general direction of the propagation vector **k**, the results for the special cases of propagation along the principal axes and along the principal planes can be obtained. However, the unit vectors $\hat{e}_s$, $\hat{e}_f$, $\hat{d}_s$ and $\hat{d}_f$ are determined here directly from the expressions for n_s and n_f so as to have a check on the results obtained from the general case, and also to have a direct method of determining the walk-off angles and the relative orientations of the $\hat{d}$, $\hat{e}$, and $\hat{s}$ vectors. Thus results will be obtained in this section for the cases of $n_X < n_Z$ and $n_X > n_Z$, for propagation directions along the X, Y, Z axes and along the YZ, ZX and XY planes. For propagation along the ZX plane ($\phi = 0$), the two cases of $\theta < \Omega$ and $\theta > \Omega$ are distinct and need to be considered separately. The results for these special cases of propagation are tabulated at the end of this section.

1.7.2   Propagation along the principal axes X, Y and Z

Say light is propagating along the Z axis, i.e., $\mathbf{k}$ is parallel to Z. Then $\theta = 0°$ and ϕ is undefined. The equations of the last section can still be used, assuming *small* but non zero value of θ and finding the limits of the component values as θ goes to 0. However, instead of undertaking that algebraic complication, we find the directions of the $\mathbf{D}$ and $\mathbf{E}$ vectors in a different way.

When $\mathbf{k}$ is parallel to Z, $m_X = m_Y = 0$ and $m_Z = 1$. Since $\hat{m} \cdot \hat{d} = 0$ (from Sec. 1.2.1 and 1.2.2), $\hat{\mathbf{d}}$ lies in the $X - Y$ plane, i.e., $d_Z = 0$, implying $e_Z = 0$ (from Eqn. 1.30).

For Case 1, $(n_X < n_Y < n_Z)$, Eqns. 1.55 and 1.56 give

$$\begin{aligned}
\mathcal{A} &= n_Z^2 \\
\mathcal{B} &= n_Z^2(n_X^2 + n_Y^2) \\
\mathcal{C} &= n_X^2\, n_Y^2\, n_Z^2 \\
\mathcal{D} &= n_Z^2(n_Y^2 - n_X^2).
\end{aligned}$$
$$(1.107)$$

Using Eqn. 1.57 the possible values of n are obtained as

$$n_s = n_Y \qquad n_f = n_X.$$
$$(1.108)$$

With $n_s = n_Y$, and $m_X = 0$, we obtain $d_{sX} = 0$ and $e_{sX} = 0$ from Eqns. 1.59 and 1.58. Since d_{sZ} and e_{sZ} are also equal to zero, d_{sY} and e_{sY} are the only nonzero components of $\hat{d}_s$ and $\hat{e}_s$. Similarly, since $n_f = n_X$, and $m_Y = 0$, we obtain $d_{fY} = 0$ and $e_{fY} = 0$ from Eqns. 1.59 and 1.58. With d_{fZ} and e_{fZ} also equal to zero, d_{fX} and e_{fX} are the only nonzero components of $\hat{d}_f$ and $\hat{e}_f$. Thus, the electric field direction of the slow wave is along Y and that of the fast wave is along X, as shown in Fig. 1.28.

Following the same arguments for case 2, i.e., for $n_X > n_Y > n_Z$, the electric field direction of the slow wave is along X and that of the fast wave is along Y, as shown in Fig. 1.29.

When $\mathbf{k}$ is along the X or Y axes, the same arguments as those presented here can be used with appropriate substitution of variables to obtain the directions of the $\hat{d}_s$, $\hat{e}_s$, $\hat{d}_f$ and $\hat{e}_f$ as summarized in Table 1.2:

Direction of $\mathbf{k}$	Case 1, $n_X < n_Y < n_Z$		Case 2, $n_X > n_Y > n_Z$	
	$\hat{d}_s, \hat{e}_s$	$\hat{d}_f, \hat{e}_f$	$\hat{d}_s, \hat{e}_s$	$\hat{d}_f, \hat{e}_f$
X	Z	Y	Y	Z
Y	Z	X	X	Z
Z	Y	X	X	Y

TABLE 1.2. The directions of the $\hat{d}_s$, $\hat{e}_s$, $\hat{d}_f$ and $\hat{e}_f$ for $\mathbf{k}$ along the X, Y or Z axes

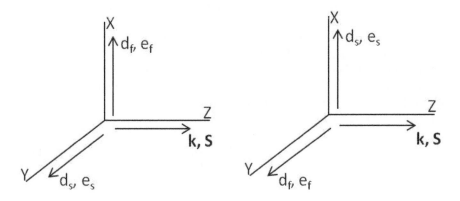

FIGURE 1.28: Propagation with **k** along the principal axis Z for Case 1, $n_X < n_Y < n_Z$

FIGURE 1.29: Propagation with **k** along the principal axis Z for Case 2, $n_X > n_Y > n_Z$

1.7.3 Propagation along the principal plane YZ

Another special case of light propagation in an anisotropic crystal is that of propagation along a principal plane, such as the XY, YZ or ZX planes. We start with the propagation vector **k** lying in the YZ plane, so that $m_X = 0$ and $m_Y^2 + m_Z^2 = 1$. With $m_X = 0$, Eqns. 1.55 give

$$\mathcal{A} = n_Y^2 m_Y^2 + n_Z^2 m_Z^2, \quad \mathcal{B} = n_Y^2 n_Z^2 + n_X^2 \mathcal{A}, \quad \text{and} \quad \mathcal{C} = n_X^2 n_Y^2 n_Z^2. \quad (1.109)$$

The two cases of $n_X < n_Y < n_Z$ and $n_X > n_Y > n_Z$ are considered separately in the next two subsections.

1.7.4 k along YZ plane, Case 1: $n_X < n_Y < n_Z$

In this case, from 1.56 $\mathcal{D} = n_Y^2 n_Z^2 - n_X^2 \mathcal{A}$, since $\mathcal{D}$ is defined to be positive and $\mathcal{A}$ takes the values from n_Y^2 to n_Z^2 as m_Y goes from 1 to 0. Thus, the possible values of n (from Eqn. 1.57) are

$$n_s = \frac{n_Y \, n_Z}{(n_Y^2 \, m_Y^2 + n_Z^2 \, m_Z^2)^{1/2}} \quad \text{and} \quad n_f = n_X. \quad (1.110)$$

As m_Y goes from 0 to 1, n_s increases in value from n_Y to n_Z, and is always greater than n_X. Since $m_X = 0$ and $n_s \neq n_X$, from the first lines of Eqns. 1.58 and 1.59 we obtain $e_{sX} = d_{sX} = 0$, i.e., the unit vectors $\hat{e}_s$ and $\hat{d}_s$ lie on the YZ plane, along with the propagation vector **k**. The unit vector $\hat{d}_f$, perpendicular to both **k** and to $\hat{d}_s$, must therefore lie along the X axis. Thus $d_{fY} = d_{fZ} = 0$, requiring $\rho_f = 0$ (from Eqn. 1.59), which in turn requires $e_{fY} = e_{fZ} = 0$. Thus, for **k** along the YZ plane and $n_X < n_Y < n_Z$ the **D** and **E** vectors of the fast wave are along the X axis, and those of the slow wave lie on the YZ plane.

Since n_s goes from n_Y to n_Z, i.e., for this case n_s is less than n_Z and greater than n_Y, Eqns. 1.58 and 1.59 show that e_{sY} and d_{sY} are positive and e_{sZ} and d_{sZ} are negative, i.e., the vectors $\mathbf{D}$ and $\mathbf{E}$ lie in the fourth quadrant of the YZ plane as shown in Fig. 1.30.

With the angle between $\mathbf{k}$ and the Z axis denoted by θ, $\hat{d}_s$ makes an angle θ with the Y axis, so that $\tan\theta = |d_{sZ}|/|d_{sY}|$. If θ' is the angle between $\hat{e}_s$ and the Y direction, then $\tan\theta' = |e_{sZ}|/|e_{sY}|$.

From Eqn. 1.32 we have

$$\left|\frac{E_Z}{E_Y}\right| = \frac{n_Y^2}{n_Z^2}\left|\frac{D_Z}{D_Y}\right| \tag{1.111}$$

so that

$$\tan\theta' = \frac{n_Y^2}{n_Z^2}\tan\theta. \tag{1.112}$$

For n_Y smaller than n_Z, θ' is smaller than θ, and $\hat{e}_s$ points between the directions of $\hat{d}_s$ and Y, as shown in Fig. 1.30.

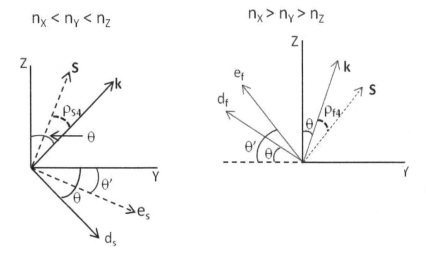

FIGURE 1.30: Directions of $\hat{d}_s$ and $\hat{e}_s$ for Case 1: $n_X < n_Y < n_Z$

FIGURE 1.31: Directions of $\hat{d}_s$ and $\hat{e}_s$ for Case 2: $n_X > n_Y > n_Z$

Denoting by ρ_{s4} the angle between $\hat{e}_s$ and $\hat{d}_s$, we have $\theta' = \theta - \rho_{s4}$. Using Eqn. 1.112

$$\tan(\theta - \rho_{s4}) = \left(\frac{n_Y}{n_Z}\right)^2\tan\theta \tag{1.113}$$

which can be solved for ρ_{s4}:

$$\tan\rho_{s4} = \frac{(1 - p_4)\tan\theta}{1 + p_4\tan^2\theta} \tag{1.114}$$

where $p_4 \equiv (n_Y/n_Z)^2$ is less than 1.

The unit vectors $\hat{d}_s$ and $\hat{e}_s$ are given (from Fig. 1.30) by

$$
\begin{aligned}
\hat{d}_s &= \cos\theta\hat{Y} - \sin\theta\hat{Z} \\
\hat{e}_s &= \cos(\theta - \rho_{s4})\hat{Y} - \sin(\theta - \rho_{s4})\hat{Z}.
\end{aligned} \tag{1.115}
$$

Here and in the next few sub-sections below we have anticipated the notation which will be introduced in the next chapter with the numbers 4, 5 and 6 assigned to the combinations of coordinates YZ, ZX and XY respectively.

1.7.5 k along YZ plane, Case 2: $n_X > n_Y > n_Z$

In this case, 1.56 gives $\mathcal{D} = n_X^2 \mathcal{A} - n_Y^2 n_Z^2$, since $\mathcal{D}$ is defined to be positive and $\mathcal{A}$ takes the values from n_Y^2 to n_Z^2 as m_Y goes from 0 to 1. Thus, the possible values of n from Eqn. 1.57 are

$$
n_s = n_X \qquad \text{and} \qquad n_f = \frac{n_Y\, n_Z}{(n_Y^2\, m_Y^2 + n_Z^2\, m_Z^2)^{1/2}}. \tag{1.116}
$$

As m_Y goes from 0 to 1, n_f decreases from n_Y to n_Z, and is always smaller than n_X. Since $m_X = 0$ and $n_f \neq n_X$, from the first lines of Eqns. 1.58 and 1.59 we obtain $e_{fX} = d_{fX} = 0$, i.e., the unit vectors $\hat{d}_f$ and $\hat{e}_f$ lie on the YZ plane, along with the propagation vector **k**. The unit vector $\hat{d}_s$, perpendicular to both **k** and to $\hat{d}_f$ must therefore lie along the X axis. Thus $d_{sY} = d_{sZ} = 0$, which requires $\rho_s = 0$ (from Eqn. 1.59), which in turn requires $e_{sY} = e_{sZ} = 0$. Thus, for **k** along the YZ plane and $n_X > n_Y > n_Z$ the **D** and **E** vectors of the slow wave are along the X axis, and those of the fast wave lie on the YZ plane.

Since n_f goes from n_Z to n_Y, i.e., for this case n_f is less than n_Y and greater than n_Z, Eqns. 1.58 and 1.59 show that e_{fY} and d_{fY} are negative and e_{fZ} and d_{fZ} are positive, i.e., the vectors $\hat{d}_f$ and $\hat{e}_f$ lie in the second quadrant of the YZ plane as shown in Fig. 1.31.

Since the angle between **k** and the Z axis is denoted by θ, $\hat{d}_f$ makes an angle θ with the -Y direction, so that $\tan\theta = |d_{fZ}|/|d_{fY}|$. If θ' is the angle between $\hat{e}_f$ and the -Y direction, then $\tan\theta' = |e_{fZ}|/|e_{fY}|$.

Equation 1.112 shows that for n_Y bigger than n_Z, θ' is larger than θ, so that $\hat{e}_f$ points between the directions of $\hat{d}_f$ and Z, as shown in Fig. 1.31.

Denoting by ρ_{f4} the angle between $\hat{e}_f$ and $\hat{d}_f$, we have $\theta' = \theta + \rho_{f4}$. Using Eqn. 1.112

$$
\tan(\rho_{f4} + \theta) = \left(\frac{n_Y}{n_Z}\right)^2 \tan\theta \tag{1.117}
$$

which can be solved for ρ_{f4}:

$$
\tan\rho_{f4} = \frac{(p_4 - 1)\tan\theta}{1 + p_4 \tan^2\theta} \tag{1.118}
$$

where $p_4 = (n_Y/n_Z)^2$ is greater than 1.

The unit vectors $\hat{d}_f$ and $\hat{e}_f$ are given (from Fig. 1.31) by

$$\begin{aligned} \hat{d}_f &= -\cos\theta\hat{Y} + \sin\theta\hat{Z} \\ \hat{e}_f &= -\cos(\theta + \rho_{f4})\hat{Y} + \sin(\theta + \rho_{f4})\hat{Z}. \end{aligned} \qquad (1.119)$$

Equations 1.114 and 1.118 show that the angles ρ_{s4} and ρ_{f4} can be denoted by the same symbol ρ_4, if ρ_4 is defined as

$$\rho_4 = \tan^{-1}\frac{|p_4 - 1|\tan\theta}{1 + p_4\tan^2\theta} = \tan^{-1}\frac{|n_Y^2 - n_Z^2|\sin\theta\cos\theta}{n_Y^2\sin^2\theta + n_Z^2\cos^2\theta}. \qquad (1.120)$$

1.7.6 Propagation along the principal plane ZX

For propagation along the ZX plane an additional complication arises because of the choice of n_Y as having a value between n_X and n_Z. With $m_Y = 0$, Eqns. 1.55 and 1.56 give

$$\mathcal{A} = n_X^2 m_X^2 + n_Z^2 m_Z^2, \quad \mathcal{B} = n_X^2 n_Z^2 + n_Y^2\mathcal{A} \quad \text{and} \quad \mathcal{C} = n_X^2 n_Y^2 n_Z^2 \quad (1.121)$$

and $\mathcal{D} = \pm(n_Y^2\mathcal{A} - n_X^2 n_Z^2)$, where the sign chosen depends on the value of m_Z that makes $\mathcal{D}$ positive. For m_Z varying from 0 to 1, the two possible values of n are denoted by n_1 and n_2

$$n_1 = \frac{n_X\, n_Z}{(n_X^2\, m_X^2 + n_Z^2\, m_Z^2)^{1/2}} \qquad \text{and} \qquad n_2 = n_Y. \qquad (1.122)$$

To determine the assignment of n_1 and n_2 to n_s and n_f we consider the four combinations of cases $n_X < n_Y < n_Z$ and $n_X > n_Y > n_Z$ with the conditions $\theta < \Omega$ and $\theta > \Omega$.

1.7.7 k along ZX plane, Case 1a: $n_X < n_Y < n_Z, \theta < \Omega$

As the angle θ goes from 0 to Ω, the value of n_1 increases from n_X to n_Y so n_2 is greater than n_1. Thus, for this range of θ, $n_s = n_2 = n_Y$ and $n_f = n_1 = \dfrac{n_X\, n_Z}{(n_X^2\, m_X^2 + n_Z^2\, m_Z^2)^{1/2}}$. At $\theta = \Omega$, n_1 and n_2 are equal to each other and light propagates as in an isotropic medium with no birefringence and with refractive index n_Y.

Since n_f goes from n_X to n_Y, i.e., for this case n_f is greater than n_X, Eqns. 1.58 and 1.59 show that e_{fX} and d_{fX} are positive. Since n_Z is larger than both n_X and n_Y, n_f is smaller than n_Z, so both e_{fZ} and d_{fZ} are negative. The vectors $\hat{d}_f$ and $\hat{e}_f$ therefore lie in the fourth quadrant of the XZ plane, as shown in Fig. 1.32.

Since the angle between **k** and the Z axis is denoted by θ, $\hat{d}_f$ makes an angle θ with the X direction, so that $\tan\theta = |d_{fZ}|/|d_{fX}|$. If θ' is the angle between $\hat{e}_f$ and the X direction, then $\tan\theta' = |\hat{e}_{fZ}|/|\hat{e}_{fX}|$.

$$n_X < n_Y < n_Z$$
$$\theta < \Omega$$

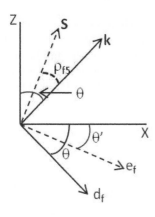

$$n_X < n_Y < n_Z$$
$$\theta > \Omega$$

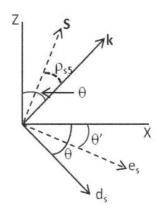

FIGURE 1.32: Directions of $\hat{d}$ and $\hat{e}$ for Case 1: $n_X < n_Y < n_Z$

FIGURE 1.33: Directions of $\hat{d}$ and $\hat{e}$ for Case 2: $n_X > n_Y > n_Z$

From Eqn. 1.32 we have

$$\left|\frac{E_Z}{E_X}\right| = \left(\frac{n_X}{n_Z}\right)^2 \left|\frac{D_Z}{D_X}\right| \tag{1.123}$$

so that

$$\tan\theta' = \left(\frac{n_X}{n_Z}\right)^2 \tan\theta. \tag{1.124}$$

Equation 1.124 shows that for n_X smaller than n_Z, θ' is smaller than θ, so that $\hat{e}_f$ points between the directions of $\hat{d}_f$ and X, as shown in Fig. 1.32.

Denoting by ρ_{f5} is the angle between $\hat{e}_f$ and $\hat{d}_f$, we have $\theta' = \theta - \rho_{f5}$. Using Eqn. 1.124

$$\tan(\theta - \rho_{f5}) = \left(\frac{n_X}{n_Z}\right)^2 \tan\theta \tag{1.125}$$

which can be solved for ρ_{f5}:

$$\tan\rho_{f5} = \frac{(1 - p_5)\tan\theta}{1 + p_5 \tan^2\theta} \tag{1.126}$$

where $p_5 \equiv (n_X/n_Z)^2$ is less than 1.

The unit vectors $\hat{d}_f$ and $\hat{e}_f$ are given (from Fig. 1.32) by

$$\begin{aligned}
\hat{d}_f &= \cos\theta\hat{X} - \sin\theta\hat{Z} \\
\hat{e}_f &= \cos(\theta - \rho_{f5})\hat{X} - \sin(\theta - \rho_{f5})\hat{Z}.
\end{aligned} \tag{1.127}$$

1.7.8   k along ZX plane, Case 1b: $n_X < n_Y < n_Z$, $\theta > \Omega$

For θ increasing from Ω to $90°$, n_1 increases from n_Y to n_Z, whereas $n_2 = n_Y$ so $n_1 > n_2$. Thus, for this range of θ, $n_s = n_1 = \dfrac{n_X \, n_Z}{(n_X^2 \, m_X^2 + n_Z^2 \, m_Z^2)^{1/2}}$ and $n_f = n_2 = n_Y$.

Since n_s goes from n_Y to n_Z, and n_s is larger than n_X and smaller than n_Z. Equations 1.58 and 1.59 show that in this case e_{sX} and d_{sX} are positive and e_{sZ} and d_{sZ} are negative, i.e., the vectors $\hat{d}_s$ and $\hat{e}_s$ lie in the fourth quadrant of the XZ plane, as shown in Fig. 1.33. The angle θ' between $\hat{e}_s$ and the X direction is equal to $\theta - \rho_{s5}$, where ρ_{s5} is given by Eqn. 1.130. The unit vectors $\hat{d}_s$ and $\hat{e}_s$ are then given (from Fig. 1.33) by

$$
\begin{aligned}
\hat{d}_s &= \cos\theta \hat{X} - \sin\theta \hat{Z} \\
\hat{e}_s &= \cos(\theta - \rho_{s5})\hat{X} - \sin(\theta - \rho_{s5})\hat{Z}.
\end{aligned}
\tag{1.128}
$$

1.7.9   k along ZX plane, Case 2a: $n_X > n_Y > n_Z$, $\theta < \Omega$

As the angle θ goes from 0 to Ω, the value of n_1 decreases from n_X to n_Y, whereas $n_2 = n_Y$, so n_1 is bigger than n_2. Thus, for this range of θ, $n_s = n_1 = \dfrac{n_X \, n_Z}{(n_X^2 \, m_X^2 + n_Z^2 \, m_Z^2)^{1/2}}$ and $n_f = n_2 = n_Y$. Again, at $\theta = \Omega$, n_1 and n_2 are equal to each other and light propagates as in an isotropic medium with no birefringence and with refractive index equal to n_Y.

Since n_s, which goes from n_X to n_Y, is smaller than n_X, we find from Eqns. 1.58 and 1.59 that e_{sX} and d_{sX} are negative. Since n_s is larger than n_Z, both e_{sZ} and d_{sZ} are positive, i.e., the vectors $\hat{d}_s$ and $\hat{e}_s$ lie in the second quadrant of the XZ plane, as shown in Fig. 1.34.

Since the angle between $\mathbf{k}$ and the Z axis is denoted by θ, $\hat{d}_s$ makes an angle θ with the $-X$ direction, so that $\tan\theta = |d_{sZ}|/|d_{sX}|$. If θ' is the angle between $\hat{e}_s$ and the $-X$ direction, $\tan\theta' = |\hat{e}_{sZ}|/|\hat{e}_{sX}|$.

Equation 1.124 shows that for n_X bigger than n_Z, θ' is bigger than θ, so that $\hat{e}_s$ points between the directions of $\hat{d}_s$ and Z, as shown in Fig. 1.34.

Denoting by ρ_{s5} the angle between $\hat{e}_s$ and $\hat{d}_s$, we have $\theta' = \theta + \rho_{s5}$. Using Eqn. 1.124

$$
\tan(\rho_{s5} + \theta) = \left(\frac{n_X}{n_Z}\right)^2 \tan\theta
\tag{1.129}
$$

which can be solved for ρ_{s5}:

$$
\tan\rho_{s5} = \frac{(p_5 - 1)\tan\theta}{1 + p_5 \tan^2\theta}
\tag{1.130}
$$

where $p_5 = (n_X/n_Z)^2$ is greater than 1.

The unit vectors $\hat{d}_s$ and $\hat{e}_s$ are given (from Fig. 1.34) by

$$
\begin{aligned}
\hat{d}_s &= -\cos\theta \hat{X} + \sin\theta \hat{Z} \\
\hat{e}_s &= -\cos(\theta + \rho_{s5})\hat{X} + \sin(\theta + \rho_{s5})\hat{Z}.
\end{aligned}
\tag{1.131}
$$

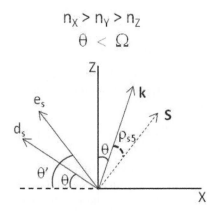

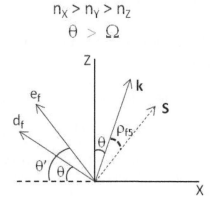

FIGURE 1.34: Directions of $\hat{d}$ and $\hat{e}$ for Case 1: $n_X < n_Y < n_Z$

FIGURE 1.35: Directions of $\hat{d}$ and $\hat{e}$ for Case 2: $n_X > n_Y > n_Z$

1.7.10 k along ZX plane, Case 2b: $n_X > n_Y > n_Z, \theta > \Omega$

For θ increasing from Ω to $90°$, n_1 decreases from n_Y to n_Z, whereas n_2 stays equal to n_Y, so that $n_2 > n_1$. Thus, for this range of θ, $n_s = n_2 = n_Y$ and

$$n_f = n_1 = \frac{n_X \, n_Z}{(n_X^2 \, m_X^2 + n_Z^2 \, m_Z^2)^{1/2}}.$$

With n_f going from n_Y to n_Z, we have $n_f < n_X$, so that from Eqn. 1.58 and 1.59 we find that e_{fX} and d_{fX} are negative. Since n_f is bigger than n_Z, both e_{fZ} and d_{fZ} are positive, i.e., the vectors $\hat{d}_f$ and $\hat{e}_f$ lie in the second quadrant of the XZ plane, as shown in Fig. 1.35.

The angle θ' between $\hat{e}_f$ and the X direction is equal to $\theta + \rho_{f5}$, where ρ_{f5} is given by Eqn. 1.126. The unit vectors $\hat{d}_f$ and $\hat{e}_f$ are then given (from Fig. 1.35) by

$$\begin{aligned}
\hat{d}_f &= -\cos\theta \hat{X} + \sin\theta \hat{Z} \\
\hat{e}_f &= -\cos(\theta + \rho_{f5})\hat{X} + \sin(\theta + \rho_{f5})\hat{Z}.
\end{aligned} \qquad (1.132)$$

Equation 1.126 and 1.130 show that the angles ρ_{f5} and ρ_{s5} can be denoted by the same symbol ρ_5, defined as

$$\rho_5 = \tan^{-1}\frac{|p_5 - 1|\tan\theta}{1 + p_5\tan^2\theta} = \tan^{-1}\frac{|n_Z^2 - n_X^2|\sin\theta\cos\theta}{n_X^2\sin^2\theta + n_Z^2\cos^2\theta}. \qquad (1.133)$$

1.7.11 Propagation along the principal plane XY

When the propagation vector **k** lies in the XY plane, $m_Z = 0$ and $m_X^2 + m_Y^2 = 1$, so that Eqns. 1.55 give

$$\mathcal{A} = n_X^2 m_X^2 + n_Y^2 m_Y^2, \quad \mathcal{B} = n_X^2 n_Y^2 + n_Z^2 \mathcal{A} \quad \text{and} \quad \mathcal{C} = n_X^2 n_Y^2 n_Z^2. \qquad (1.134)$$

We consider the following two cases:

1.7.12   k along XY plane, Case 1: $n_X < n_Y < n_Z$

In this case, from 1.56, we get $\mathcal{D} = n_Z^2 \mathcal{A} - n_X^2 n_Y^2$, since $\mathcal{D}$ is defined to be positive and $\mathcal{A}$ takes the values from n_X^2 to n_Y^2 as m_Y goes from 0 to 1. Thus, the possible values of n (from Eqn. 1.57) are

$$n_s = n_Z \qquad \text{and} \qquad n_f = \frac{n_X\, n_Y}{(n_X^2\, m_X^2 + n_Y^2\, m_Y^2)^{1/2}}. \tag{1.135}$$

As m_X goes from 0 to 1, n_f takes the values from n_X to n_Y, so that $n_f \geq n_X$. Since $m_Z = 0$ and $n_f \neq n_Z$, from the third lines of Eqns. 1.58 and 1.59 we obtain $e_{fZ} = d_{fZ} = 0$, i.e., the unit vectors $\hat{d}_f$ and $\hat{e}_f$ lie on the XY plane, along with the propagation vector **k**. The unit vector $\hat{d}_s$, which is perpendicular to both **k** and to $\hat{d}_f$ must therefore lie along the Z axis. Thus $d_{sX} = d_{sY} = 0$, which requires $\rho_s = 0$ (from Eqn. 1.59), which in turn requires $e_{sX} = e_{sY} = 0$. So, for **k** along the XY plane and $n_X < n_Y < n_Z$ the **D** and **E** vectors of the slow wave are along the Z axis, and those of the fast wave lie on the XY plane.

Since n_f goes from n_X to n_Y, n_f is greater than n_X and smaller than n_Y, so Eqns. 1.58 and 1.59 show that e_{fX} and d_{fX} are positive and e_{fY} and d_{fY} are negative, i.e., the vectors **D** and **E** lie in the fourth quadrant of the XY plane as shown in Fig. 1.36.

$n_X < n_Y < n_Z$

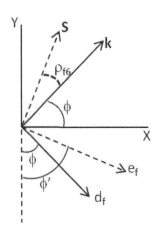

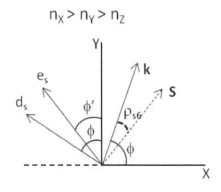

FIGURE 1.36: Directions of $\hat{d}_f$ and $\hat{e}_f$ for Case 1: $n_X < n_Y < n_Z$

FIGURE 1.37: Directions of $\hat{d}_s$ and $\hat{e}_s$ for Case 2: $n_X > n_Y > n_Z$

Since the angle between **k** and the X axis is denoted by ϕ, $\hat{d}_f$ makes an angle ϕ with the $-Y$ direction, so that $\tan\phi = |d_{fX}|/|d_{fY}|$. If ϕ' is the angle between $\hat{e}_f$ and the $-Y$ direction, then $\tan\phi' = |e_{fX}|/|e_{fY}|$.

From Eqn. 1.32 we have

$$\left|\frac{E_X}{E_Y}\right| = \left(\frac{n_Y}{n_X}\right)^2 \left|\frac{D_X}{D_Y}\right| \tag{1.136}$$

so that

$$\tan\phi' = \left(\frac{n_Y}{n_X}\right)^2 \tan\phi. \tag{1.137}$$

For n_X smaller than n_Y, ϕ' is larger than ϕ, so that $\hat{d}_f$ points between the directions of $\hat{e}_f$ and $-Y$, as shown in Fig. 1.36.

Denoting the angle between $\hat{e}_f$ and $\hat{d}_f$ by ρ_{f6}, we have $\phi' = \phi + \rho_{f6}$. Using Eqn. 1.137

$$\tan(\phi + \rho_{f6}) = \left(\frac{n_Y}{n_X}\right)^2 \tan\phi \tag{1.138}$$

which can be solved for ρ_{f6}:

$$\tan\rho_{f6} = \frac{(p_6 - 1)\tan\phi}{1 + p_6 \tan^2\phi} \tag{1.139}$$

where $p_6 \equiv (n_Y/n_X)^2$ is greater than 1.

From Fig. 1.36 the unit vectors $\hat{d}_f$ and $\hat{e}_f$ are given by

$$\begin{aligned}
\hat{d}_f &= \sin\phi\hat{X} - \cos\phi\hat{Y} \\
\hat{e}_f &= \sin(\phi + \rho_{f6})\hat{X} - \cos(\phi + \rho_{f6})\hat{Y}.
\end{aligned} \tag{1.140}$$

1.7.13 k along XY plane, Case 2: $n_X > n_Y > n_Z$

In this case, $\mathcal{D} = n_X^2 n_Y^2 - n_Z^2 \mathcal{A}$ from 1.56, since $\mathcal{D}$ is defined to be positive and $\mathcal{A}$ takes the values from n_Y^2 to n_X^2 as m_X goes from 0 to 1. Thus, the possible values of n from Eqn. 1.57 are

$$n_s = \frac{n_X\, n_Y}{(n_X^2\, m_X^2 + n_Y^2\, m_Y^2)^{1/2}} \quad \text{and} \quad n_f = n_Z. \tag{1.141}$$

As m_X goes from 0 to 1, n_s takes the values from n_X to n_Y, and is always larger than n_Z. Since $m_Z = 0$ and $n_s \neq n_z$, from the third lines of Eqns. 1.58 and 1.59 we obtain $e_{sZ} = d_{sZ} = 0$, i.e., the unit vectors $\hat{d}_s$ and $\hat{e}_s$ lie on the XY plane, along with the propagation vector $\mathbf{k}$. The unit vector $\hat{d}_f$, which is perpendicular to both $\mathbf{k}$ and to $\hat{d}_s$ must therefore lie along the Z axis. Thus $d_{fX} = d_{fY} = 0$, which requires $\rho_f = 0$ (from Eqn. 1.59), which in turn requires $e_{fX} = e_{fY} = 0$. So for $\mathbf{k}$ along the XY plane and $n_X > n_Y > n_Z$, the $\mathbf{D}$ and $\mathbf{E}$ vectors of the fast wave are along the Z axis, and those of the slow wave lie on the XY plane.

n_s goes from n_Y to n_X, which means $n_s \leq n_X$. Eqns. 1.58 and 1.59 then show that e_{sX} and d_{sX} are negative. Since $n_s \geq n_Y$, e_{sY} and d_{sY} are positive, i.e., the vectors $\hat{d}_s$ and $\hat{e}_s$ lie in the second quadrant of the XY plane as shown in Fig. 1.37.

Since the angle between $\mathbf{k}$ and the X axis is denoted by ϕ, $\hat{d}_s$ makes an angle ϕ with the Y direction, so that $\tan \phi = |d_{sX}|/|d_{sY}|$. If ϕ' is the angle between $\hat{e}_s$ and the Y direction, then $\tan \phi' = |\hat{e}_{sX}|/|\hat{e}_{sY}|$.

Equation 1.137 shows that for n_Y smaller than n_X, ϕ' is smaller than ϕ, so that $\hat{e}_s$ points between the directions of $\hat{d}_s$ and Y, as shown in Fig. 1.37.

Denoting by ρ_{s6} the angle between $\hat{e}_s$ and $\hat{d}_s$, we have $\phi' = \phi - \rho_{s6}$. Using Eqn. 1.137

$$\tan(\phi - \rho_{s6}) = \left(\frac{n_Y}{n_X}\right)^2 \tan \phi \qquad (1.142)$$

which can be solved for ρ_{s6}:

$$\tan \rho_{s6} = \frac{(1 - p_6) \tan \phi}{1 + p_6 \tan^2 \phi} \qquad (1.143)$$

where $p_6 = (n_Y/n_X)^2$ is smaller than 1.

From Fig. 1.37 the unit vectors $\hat{d}_s$ and $\hat{e}_s$ are given by

$$\begin{aligned} \hat{d}_s &= -\sin\phi \hat{X} + \cos\phi \hat{Y} \\ \hat{e}_s &= -\sin(\phi - \rho_{s6})\hat{X} + \cos(\phi - \rho_{s6})\hat{Y}. \end{aligned} \qquad (1.144)$$

Equations 1.139 and 1.143 show that the angles ρ_{f6} and ρ_{s6} can be denoted by the same symbol ρ_6, if ρ_6 is defined as

$$\rho_6 = \tan^{-1} \frac{|p_6 - 1| \tan \phi}{1 + p_5 \tan^2 \phi} = \tan^{-1} \frac{|n_X^2 - n_Y^2| \sin \phi \cos \phi}{n_Y^2 \sin^2 \phi + n_X^2 \cos^2 \phi}. \qquad (1.145)$$

1.7.14   Summary of the cases of propagation along principal planes

Tables 1.3 and 1.4 summarize the results of this section. The values of the angle δ for the different cases are obtained from Eqn. 1.89. For each case, the n_s and n_f values are first calculated using Eqns. 1.57. For the θ, ϕ and δ values for each case, Eqns. 1.91, 1.92, 1.93 and 1.94 are used to find the components $d_{sX}, d_{sY}, d_{sZ}, d_{fX}, d_{fY}$ and d_{fZ}. From these values of the components of the unit vectors $\hat{d}_s$ and $\hat{d}_f$, angles ρ_s and ρ_f are found using Eqns. 1.106. Using the values of ρ_s and ρ_f along with those of the components of the unit vectors $\hat{d}_s$ and $\hat{d}_f$ and the angles δ in Eqns. 1.96, 1.98, 1.100 and 1.102 components of the unit vectors $\hat{e}_s$ and $\hat{e}_f$ are obtained.

	k along YZ plane $(\phi = 90°, \delta = 0)$		k along XY plane $(\theta = 90°, \delta = 0)$	
	Case 1	Case 2	Case 1	Case 2
	$n_X < n_Y < n_Z$	$n_X > n_Y > n_Z$	$n_X < n_Y < n_Z$	$n_X > n_Y > n_Z$
n_s	$(n_Y n_Z)/\sqrt{\mathcal{A}_{YZ}}$	n_X	n_Z	$(n_X n_Y)/\sqrt{\mathcal{A}_{XY}}$
n_f	n_X	$(n_Y n_Z)/\sqrt{\mathcal{A}_{YZ}}$	$(n_X n_Y)/\sqrt{\mathcal{A}_{XY}}$	n_Z
ρ_s	ρ_4	0	0	ρ_6
ρ_f	0	ρ_4	ρ_6	0
e_{sX}	0	-1	0	$-\sin(\phi - \rho_6)$
e_{sY}	$\cos(\theta - \rho_4)$	0	0	$\cos(\phi - \rho_6)$
e_{sZ}	$-\sin(\theta - \rho_4)$	0	-1	0
e_{fX}	1	0	$\sin(\phi + \rho_6)$	0
e_{fY}	0	$-\cos(\theta + \rho_4)$	$-\cos(\phi + \rho_6)$	0
e_{fZ}	0	$\sin(\theta + \rho_4)$	0	1

$$\mathcal{A}_{YZ} = n_Y^2 \sin^2 \theta + n_Z^2 \cos^2 \theta; \quad \mathcal{A}_{XY} = n_Y^2 \sin^2 \phi + n_X^2 \cos^2 \phi.$$

$$\rho_4 = \tan^{-1}\left(\frac{|n_Y^2 - n_Z^2| \sin\theta \cos\theta}{\mathcal{A}_{YZ}} \right),$$

$$\rho_6 = \tan^{-1}\left(\frac{|n_X^2 - n_Y^2| \sin\phi \cos\phi}{\mathcal{A}_{XY}} \right)$$

TABLE 1.3. n_s, n_f, ρ_s, ρ_f, $\hat{e}_s$ and $\hat{e}_f$ for k along the YZ and the XY planes. The components of the unit vector $\hat{d}$ are obtained from the corresponding components of the unit vector $\hat{e}$, with the walk-off angles ρ_4 or ρ_6 set equal to zero.

1.8 Uniaxial Crystals

When two of the principal refractive indices, (by definition n_X and n_Y) are equal, Eqn. 1.73 and Fig. 1.20 show that the two optic axes coalesce into one axis parallel to the Z direction. Such crystals with only *one* optic axis are of course called *uniaxial*. The two equal indices n_X and n_Y in uniaxial crystals are given the name n_o, so that

$$D_X = n_o^2 E_X \qquad D_Y = n_o^2 E_Y \qquad D_Z = n_Z^2 E_Z. \qquad (1.146)$$

A crystal with $n_Z > n_o$ is called *positive* uniaxial and if $n_Z < n_o$ it is called *negative* uniaxial. We saw in the last section that for a biaxial crystal, n_s and n_f take simple forms when the propagation vector $\hat{k}$ is along special directions, such as along the principal axes or along the principal planes. For uniaxial crystals, n_s and n_f have such simple expressions for *arbitrary* directions of propagation, because of the additional symmetry condition. These expressions are similar to those for propagation along principal planes in biaxial crystals,

	k along ZX plane $(\phi = 0)$			
	Case 1, $n_X < n_Y < n_Z$		Case 2, $n_X > n_Y > n_Z$	
	$\theta < \Omega,\ \delta = 90°$	$\theta > \Omega,\ \delta = 0°$	$\theta < \Omega,\ \delta = 90°$	$\theta > \Omega,\ \delta = 0°$
n_s	n_Y	$(n_X n_Z)/\sqrt{\mathcal{A}_{ZX}}$	$(n_X n_Z)/\sqrt{\mathcal{A}_{ZX}}$	n_Y
n_f	$(n_X n_Z)/\sqrt{\mathcal{A}_{ZX}}$	n_Y	n_Y	$(n_X n_Z)/\sqrt{\mathcal{A}_{ZX}}$
ρ_s	0	ρ_5	ρ_5	0
ρ_f	ρ_5	0	0	ρ_5
e_{sX}	0	$\cos(\theta - \rho_5)$	$-\cos(\theta + \rho_5)$	0
e_{sY}	1	0	0	1
e_{sZ}	0	$-\sin(\theta - \rho_5)$	$\sin(\theta + \rho_5)$	0
e_{fX}	$\cos(\theta - \rho_5)$	0	0	$-\cos(\theta + \rho_5)$
e_{fY}	0	-1	-1	0
e_{fZ}	$-\sin(\theta - \rho_5)$	0	0	$\sin(\theta + \rho_f)$

$$\mathcal{A}_{ZX} = n_X^2 \sin^2\theta + n_Z^2 \cos^2\theta$$
$$\rho_5 = \tan^{-1}\left(\frac{|n_Z^2 - n_X^2|\sin\theta\cos\theta}{\mathcal{A}_{ZX}}\right)$$

TABLE 1.4. n_s, n_f, ρ_s, ρ_f, $\hat{e}_s$ and $\hat{e}_f$ for **k** along the ZX plane. The components of the unit vector $\hat{d}$ are obtained from the corresponding components of the unit vector $\hat{e}$, with the walk-off angle ρ_5 set equal to zero.

but are sufficiently different in notation and usage, making it worthwhile to re-derive them in detail here.

With $n_X = n_Y = n_o$ inserted in Eqns. 1.55 we get

$$\mathcal{A} = n_o^2(m_X^2 + m_Y^2) + n_Z^2 m_Z^2, \quad \mathcal{B} = n_o^2(n_Z^2 + \mathcal{A}) \quad \text{and} \quad \mathcal{C} = n_o^4 n_Z^2 \quad (1.147)$$

from which we obtain

$$\mathcal{D}^2 = n_o^4(n_Z^2 - \mathcal{A})^2. \tag{1.148}$$

Since from Eqn. 1.147 $\mathcal{A} = (n_o^2 \sin^2\theta + n_Z^2 \cos^2\theta)$, $\mathcal{A}$ takes the values between n_Z^2 and n_o^2. With $\mathcal{D}$ defined to be positive, we have the following two cases shown in Table 1.5:

Defining

$$n_e(\theta) \equiv \frac{n_o n_Z}{\sqrt{\mathcal{A}}} = \frac{n_o n_Z}{\sqrt{n_o^2 \sin^2\theta + n_Z^2 \cos^2\theta}} \tag{1.149}$$

the two waves traveling with speeds c/n_o and $c/n_e(\theta)$ in the uniaxial medium are called the *ordinary wave* and the *extraordinary wave* respectively. In a positive uniaxial crystal the extraordinary wave is the slow wave, and the ordinary wave is the fast wave. For negative uniaxial crystals the situation is reversed, with the extraordinary wave being the fast wave and the ordinary wave the slow wave. Field directions for these two waves are determined next.

The subscripts o and e are used on the field variables to designate the

	Positive uniaxial $n_Z > n_o$	Negative uniaxial $n_Z < n_o$
$\mathcal{D}$	$n_o^2(n_Z^2 - \mathcal{A})$	$n_o^2(\mathcal{A} - n_Z^2)$
n_s	$(n_o n_Z)/\sqrt{\mathcal{A}}$	n_o
n_f	n_o	$(n_o n_Z)/\sqrt{\mathcal{A}}$

$$\mathcal{A} = n_o^2 \sin^2 \theta + n_Z^2 \cos^2 \theta$$

TABLE 1.5. The values of $\mathcal{D}$, n_s and n_f for positive and negative uniaxial crystals

cases of the *ordinary* and the *extraordinary* waves, respectively. Thus $\hat{d}_o$, $\hat{e}_o$, $\hat{d}_e$ and $\hat{e}_e$ denote the unit vectors along the directions of the displacement and the electric field vectors, respectively. In a positive uniaxial crystal, $\hat{d}_s = \hat{d}_e$, $\hat{e}_s = \hat{e}_e$, $\hat{d}_f = \hat{d}_o$, $\hat{e}_f = \hat{e}_o$ and in negative uniaxial crystals, the situation is reversed, with $\hat{d}_s = \hat{d}_o$, $\hat{e}_s = \hat{e}_o$, $\hat{d}_f = \hat{d}_e$, $\hat{e}_f = \hat{e}_e$.

1.8.1 Field directions of the D and E vectors for extraordinary and ordinary waves

To determine the values of the components of the unit vectors $\hat{d}_o$, $\hat{e}_o$, $\hat{d}_e$ and $\hat{e}_e$ along the principal dielectric axes, we start with the facts that **k** and $\hat{d}$ are perpendicular and that **k**, $\hat{d}$ and $\hat{e}$ lie in one plane, as shown in Sec. 1.2.1 and 1.2.2. Denoting by ρ the angle between $\hat{d}$ and $\hat{e}$ we show below that there are two values of ρ possible, one equal to 0 and the other non-zero. The non-zero value of ρ is found as a function of n_Z, n_o and θ and it is shown that when $\rho \neq 0$, the refractive index n is equal to $n_e(\theta)$, i.e., for this case, $\hat{d} = \hat{d}_e$. The values of the components of d_e are found after that. Next, it is shown that when $\rho = 0$, n is equal to n_o, i.e., for this case, $\hat{d} = \hat{d}_o$. Lastly, the values of the components of d_o are found and the results are summarized in Table 1.6.

Redrawing Figs. 1.26 and 1.27 as Fig. 1.38 without the subscripts we rewrite Eqns. 1.95 as

$$\hat{e} = \hat{m} \sin \rho + \hat{d} \cos \rho \tag{1.150}$$

from which we obtain

$$e_X = m_X \sin \rho + d_X \cos \rho$$
$$e_Y = m_Y \sin \rho + d_Y \cos \rho$$
$$e_Z = m_Z \sin \rho + d_Z \cos \rho. \tag{1.151}$$

For a uniaxial crystal, we obtain from Eqn. 1.32

$$\frac{e_X}{e_Y} = \frac{d_X}{d_Y}$$
$$\frac{e_Z}{e_Y} = \frac{n_o^2 d_Z}{n_Z^2 d_Y} \tag{1.152}$$

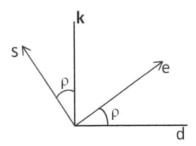

FIGURE 1.38: The unit vectors $\hat{e}$, $\hat{d}$ and $\hat{s}$ for the propagation vector **k**. ρ is the walk-off angle.

and using Eqn. 1.151 we find

$$\sin \rho (m_X\, d_Y - m_Y\, d_X) = 0 \tag{1.153}$$

and

$$\tan \rho = \frac{(n_Z^2 - n_o^2)\, d_Z d_Y}{n_o^2 d_Z m_Y - n_Z^2 d_Y m_Z}. \tag{1.154}$$

There are two possible solutions of Eqn. 1.153: $\rho = 0$ and $\rho \neq 0$. We consider the case of nonzero ρ first.

1.8.2 $\rho \neq 0$ Case (extraordinary wave)

When $\rho \neq 0$, according to Eqn. 1.153,

$$m_X\, d_Y = m_Y\, d_X \tag{1.155}$$

which shows that the Z component of the cross product $\hat{m} \times \hat{d}$, given by $m_X\, d_Y - m_Y\, d_X$ is equal to 0. The vector $\hat{m} \times \hat{d}$ is therefore perpendicular to Z, as well as to $\hat{m}$ and $\hat{d}$. This means $\hat{m}$, $\hat{d}$ and Z are three vectors all perpendicular to the same vector - therefore $\hat{m}$, $\hat{d}$ and Z must be coplanar. By definition, vector u shown in Fig. 1.3 also lies on the plane containing $\hat{m}$ and Z.

The plane containing the propagation vector **k** and the axis Z is defined as the *principal plane* in Ref. [2], although the planes XY, YZ and ZX are also called the *principal planes* in the same book. To avoid possible confusion arising from the use of the same name, we will define the **k** $- Z$ plane here the '**k** $- Z$ principal plane'. The directions of the unit vectors d and e on the **k** $- Z$ principal plane are shown in Figs. 1.39 and 1.40 for the cases of positive and negative axial crystals respectively.

In Figs. 1.39 and 1.40, the quadrants into which d and e point are determined by the following method. Since $\hat{m}$ and $\hat{d}$ are perpendicular to each other, we have

$$m_X\, d_X + m_Y\, d_Y + m_Z\, d_Z = 0. \tag{1.156}$$

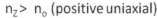

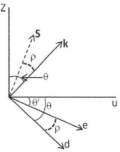

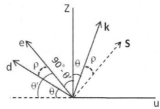

FIGURE 1.39: The orientation of the unit vectors $\hat{d}$ and $\hat{e}$ for the extraordinary wave traveling in a positive uniaxial crystal.

FIGURE 1.40: The orientation of the fields $\hat{d}$ and $\hat{e}$ for the extraordinary wave traveling in a negative uniaxial crystal.

Eliminating d_X in Eqns. 1.155 and 1.156, we obtain the relation between d_Y and d_Z as

$$d_Y = -\frac{m_Y m_Z}{m_X^2 + m_Y^2} d_Z. \tag{1.157}$$

Following the same argument, we obtain the relation between d_X and d_Z as

$$d_X = -\frac{m_X m_Z}{m_X^2 + m_Y^2} d_Z. \tag{1.158}$$

For the **k** vector in the first octant of the XYZ coordinate axes, m_X, m_Y and m_Z are all positive, so Eqn. 1.157 shows that d_Y and d_Z must have opposite signs. Eqn. 1.154 then shows that in a positive uniaxial crystal $(n_Z > n_o)$, for $\tan \rho$ to be positive, d_Z must be negative and therefore d_Y must be positive. Similarly, in a negative uniaxial crystal $(n_Z < n_o)$, for $\tan \rho$ to be positive, d_Z must be positive and therefore d_Y must be negative. These directions are indicated in Figs. 1.39 and 1.40, showing that for the cases of positive and negative uniaxial crystals, $\hat{d}$ lies in the fourth and second quadrant of the $u - Z$ coordinate system, respectively.

The corresponding directions of the unit vector $\hat{e}$ are found as follows: From Eqn. 1.30 it can be shown (through a bit of algebra) that

$$\frac{e_Z}{e_u} = \frac{n_o^2}{n_Z^2} \frac{d_Z}{d_u} \tag{1.159}$$

where d_u and e_u are the components of $\hat{d}$ and $\hat{e}$ in the direction of the vector $\hat{u}$. Since the angle between $\mathbf{k}$ and the Z axis is denoted by θ, $\hat{d}$ makes an angle θ with the $\hat{u}$ direction. If θ' is the angle between the vector $\hat{e}$ and the u direction, then Eqn. 1.159 shows that for a positive uniaxial crystal, i.e., with

n_Z larger than n_o, θ is larger than θ', so that $\hat{e}$ points between the directions of $\hat{d}$ and $\hat{u}$, as shown in Fig. 1.39.

The Poynting vector $\mathbf{S} = \mathbf{E} \times \mathbf{H}$ lies in the same plane as $\mathbf{E}$, $\mathbf{D}$ and $\mathbf{k}$. The beam walk-off angle ρ between the vectors $\mathbf{S}$ and $\mathbf{k}$, i.e., between the energy propagation direction and the direction of the propagation vector, is also the angle between $\hat{d}$ and $\hat{e}$, and we have $\rho = \theta - \theta'$. Using Eqn. 1.159

$$\tan \theta' = \tan(\theta - \rho) = \left(\frac{n_o}{n_Z}\right)^2 \tan \theta \tag{1.160}$$

which can be solved for ρ:

$$\tan \rho = \frac{(1 - p_1) \tan \theta}{1 + p_1 \tan^2 \theta} \tag{1.161}$$

where $p_1 \equiv (n_o/n_Z)^2$.

For the negative uniaxial crystal case, Eqn. 1.159 shows that for n_o larger than n_Z, θ is smaller than θ', so that $\hat{e}$ points between the directions of $\hat{d}$ and Z, as shown in Fig. 1.40. In this case the walk-off angle $\rho = \theta' - \theta$, and using Eqn. 1.159

$$\tan \theta' = \tan(\theta + \rho) = p_1 \tan \theta \tag{1.162}$$

which can be solved for ρ:

$$\tan \rho = \frac{(p_1 - 1) \tan \theta}{1 + p_1 \tan^2 \theta}. \tag{1.163}$$

1.8.3   Another expression relating ρ and θ

With the angles θ and θ' defined as above, the walk-off angle ρ is equal to $\pm(\theta - \theta')$, with the upper and lower signs applicable for positive and negative uniaxial crystals respectively. Defining a sign parameter s, such that $s = +1$ for a positive and $s = -1$ for a negative uniaxial crystal, we have

$$\rho s = \theta - \theta' \quad \text{i.e.,} \quad \theta' = \theta - \rho s \tag{1.164}$$

so that

$$\tan \theta' = \tan(\theta - \rho s) = p_1 \tan \theta \tag{1.165}$$

which can be solved for ρ:

$$\tan \rho = s\frac{(1 - p_1) \tan \theta}{1 + p_1 \tan^2 \theta} \tag{1.166}$$

ρ can also be expressed as

$$\rho = \tan^{-1} \frac{|n_Z^2 - n_o^2| \tan \theta}{1 + p_1 \tan^2 \theta} = \tan^{-1} \frac{|n_Z^2 - n_o^2| \sin \theta \cos \theta}{n_o^2 \sin^2 \theta + n_Z^2 \cos^2 \theta}. \tag{1.167}$$

Using the relation $\cos^2 \rho = 1/(1 + \tan^2 \rho)$ we obtain from Eqn. 1.166

$$\cos \rho = \frac{n_Z^2 \cos^2 \theta + n_o^2 \sin^2 \theta}{(n_Z^4 \cos^2 \theta + n_o^4 \sin^2 \theta)^{1/2}}. \tag{1.168}$$

The refractive index n for the case of $\rho \neq 0$

With D and E denoting the magnitudes of the **D** and **E** vectors, Eqn. 1.32 can be re-written for uniaxial crystals as

$$
\begin{aligned}
D\, d_X &= \varepsilon_0 n_o^2 E e_X \\
D\, d_Y &= \varepsilon_0 n_o^2 E e_Y \\
D\, d_Z &= \varepsilon_0 n_Z^2 E e_Z
\end{aligned}
\tag{1.169}
$$

which can be rewritten as

$$
\begin{aligned}
e_X &= \frac{D}{\varepsilon_0 E}\frac{d_X}{n_o^2} \\
e_Y &= \frac{D}{\varepsilon_0 E}\frac{d_Y}{n_o^2} \\
e_Z &= \frac{D}{\varepsilon_0 E}\frac{d_Z}{n_Z^2}.
\end{aligned}
\tag{1.170}
$$

Squaring and summing the components of the unit vector $\hat{e}$ in Eqn. 1.170 we obtain

$$
\begin{aligned}
1 &= \frac{D^2}{\varepsilon_0^2 E^2}\left\{\frac{d_X^2 + d_Y^2}{n_o^4} + \frac{d_Z^2}{n_Z^4}\right\} \\
&= \frac{D^2}{\varepsilon_0^2 E^2}\left\{\frac{\cos^2\theta}{n_o^4} + \frac{\sin^2\theta}{n_Z^4}\right\}
\end{aligned}
\tag{1.171}
$$

where we have used Eqns. 1.157 and 1.158, with $d_Z = \pm\sin\theta$ as shown above to obtain

$$
d_X^2 + d_Y^2 = \cos^2\theta.
\tag{1.172}
$$

From Eqn. 1.171 we obtain

$$
\frac{D}{\varepsilon_0 E} = \frac{n_o^2 n_Z^2}{(n_Z^4 \cos^2\theta + n_o^4 \sin^2\theta)^{1/2}}.
\tag{1.173}
$$

Using Eqn. 1.50, 1.168 and 1.173 we obtain the refractive index n

$$
\begin{aligned}
n^2 &= \frac{D}{\varepsilon_0 E}\frac{1}{\cos\rho} \\
&= \frac{n_Z^2 n_o^2}{n_Z^2 \cos^2\theta + n_o^2 \sin^2\theta}
\end{aligned}
\tag{1.174}
$$

so that from the definition of $n_e(\theta)$ in Eqn. 1.149 we get $n = n_e(\theta)$ when the walk-off angle ρ is not equal to zero.

The D and E components of the extraordinary wave

Figure 1.39 shows that for positive uniaxial crystals,

$$
\begin{aligned}
\hat{e} &= \hat{u}\cos\theta' - \hat{Z}\sin\theta' \\
\hat{d} &= \hat{u}\cos\theta - \hat{Z}\sin\theta
\end{aligned}
\tag{1.175}
$$

and using Eqn. 1.74 we obtain for positive uniaxial crystals

$$
\begin{aligned}
\hat{e} &= \hat{X}\cos\theta'\cos\phi + \hat{Y}\cos\theta'\sin\phi - \hat{Z}\sin\theta' \\
\hat{d} &= \hat{X}\cos\theta\cos\phi + \hat{Y}\cos\theta\sin\phi - \hat{Z}\sin\theta.
\end{aligned}
\tag{1.176}
$$

Similarly, Fig. 1.40 shows that for negative uniaxial crystals,

$$
\begin{aligned}
\hat{e} &= -\hat{u}\cos\theta' + \hat{Z}\sin\theta' \\
\hat{d} &= -\hat{u}\cos\theta + \hat{Z}\sin\theta
\end{aligned}
\tag{1.177}
$$

and using Eqn. 1.74 again, we obtain for negative uniaxial crystals

$$
\begin{aligned}
\hat{e} &= -\hat{X}\cos\theta'\cos\phi - \hat{Y}\cos\theta'\sin\phi + \hat{Z}\sin\theta' \\
\hat{d} &= -\hat{X}\cos\theta\cos\phi - \hat{Y}\cos\theta\sin\phi + \hat{Z}\sin\theta.
\end{aligned}
\tag{1.178}
$$

Thus the unit vectors in the directions of the fields $\mathbf{D}_e$ and $\mathbf{E}_e$ can be written in a compact form as

$$
\begin{aligned}
\hat{e}_e &= (s\cos\theta'\cos\phi,\, s\cos\theta'\sin\phi,\, -s\sin\theta') \\
\hat{d}_e &= (s\cos\theta\cos\phi,\, s\cos\theta\sin\phi,\, -s\sin\theta)
\end{aligned}
\tag{1.179}
$$

where s is equal to $+1$ or -1 for the positive and negative uniaxial crystal cases, respectively, and $\theta' = \theta - s\rho$.

1.8.4 $\rho = 0$ Case (ordinary wave)

Equation 1.154 shows that d_Z is equal to 0 for ρ equal to 0. From Eqn. 1.30, e_Z must then also be equal to 0, and the $\hat{d}$ and $\hat{e}$ unit vectors lie on the XY plane. Since from Eqn. 1.30

$$
\begin{aligned}
\hat{e} &= e_X\hat{X} + e_Y\hat{Y}, \\
\hat{d} &= \varepsilon_0 n_o^2(e_X\hat{X} + e_Y\hat{Y})
\end{aligned}
\tag{1.180}
$$

$\hat{d}$ and $\hat{e}$ are parallel.

From Eqn. 1.170 with $e_Z = 0$, we obtain $D = \varepsilon_0 n_o^2 E$, whereas from Eqn. 1.50 with $\rho = 0$, $D = \varepsilon_0 n^2 E$, thus for this case $n = n_o$ and light propagates as the *ordinary wave*. To summarize, for the ordinary wave, the unit vectors $\hat{d}$ and $\hat{e}$, designated $\hat{d}_o$ and $\hat{e}_o$, are parallel (walk-off angle is zero) and they lie on the XY plane as shown in Fig. 1.41.

Since $\hat{d}_o$ is perpendicular to $\mathbf{k}$, when $\hat{k}_u$ lies in the first quadrant of the XY

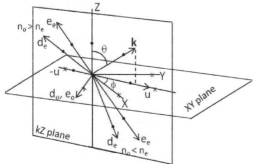

FIGURE 1.41: Polarization directions of unit vectors $\hat{d}$ and $\hat{e}$ of the ordinary and extraordinary waves. The *ordinary* fields lie on the XY plane and are shown by the cross-marks on the lines. The *extraordinary* fields lie on the $k - Z$ plane and are shown by dots on the lines.

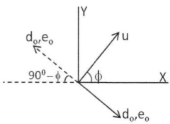

FIGURE 1.42: Possible directions of unit vectors $\hat{d}$ and $\hat{e}$ of the ordinary wave on the XY plane.

coordinate plane, $\hat{d}_o$ can be in the second or the fourth quadrant, as shown in Fig. 1.42 by the dashed and solid arrows, respectively. Equations 1.58 do not provide any guidance for the selection of the quadrant here, as it did in the case of propagation along the principal planes of the biaxial crystal. Following the choice established in earlier work (Sec. 3.5 of Ref. [12]), the direction of $\hat{d}$ is chosen here to be in the fourth quadrant so that for the wave with propagation vector **k** traveling with speed c/n_o the unit vectors along $\hat{e}$ and $\hat{d}$ are given by

$$\hat{e}_o = (\sin \phi, -\cos \phi, 0)$$
$$\hat{d}_o = (\sin \phi, -\cos \phi, 0). \tag{1.181}$$

1.8.5 Two special cases: $\theta = 0$ and $\theta = 90°$

From Eqns. 1.161 and 1.163, when **k** is along the Z axis, i.e., for $\theta = 0$, the walk-off angle ρ is also equal to 0, and the $\hat{d}$ and $\hat{e}$ vectors are parallel. With $\theta = 0$, $n_e(\theta) = n_o$ and the angle ϕ becomes indeterminate. The unit vector $\hat{d}$ can be along any direction on the XY plane and the crystal behaves as an isotropic material with refractive index n_o. This is in contrast with the case of a biaxial crystal, in which $\hat{d}$ is along the X or Y axes when **k** is along Z.

When **k** lies on the XY plane, i.e., $\theta = 90°$, the walk-off angle ρ for the extraordinary wave is equal to 0, from Eqn. 1.168. From Eqns. 1.179, the unit vector $\hat{d}_e$ for the extraordinary wave is along the positive or negative Z axis, depending on whether it is a negative or positive uniaxial crystal, respectively.

The value of $n_e(\theta)$ is n_Z, irrespective of the angle ϕ. In a biaxial crystal, for **k** lying on the XY plane, the walk-off angle is not zero except for $\phi = 0$ or $90°$.

Results for the slow and fast components of the electric fields for an uniaxial medium with propagation vector **k** at an angle θ with the Z axis are summarized in Table 1.6.

	Positive uniaxial $n_Z > n_o$	negative uniaxial $n_Z < n_o$
n_s	$(n_Z\,n_o)/\sqrt{\mathcal{A}}$	n_o
n_f	n_o	$(n_Z\,n_o)/\sqrt{\mathcal{A}}$
ρ	$\tan^{-1}\dfrac{\lvert n_Z^2 - n_o^2\rvert \sin\theta\cos\theta}{\mathcal{A}}$	
e_{sX}	$\cos(\theta - \rho)\cos\phi$	$\sin\phi$
e_{sY}	$\cos(\theta - \rho)\sin\phi$	$-\cos\phi$
e_{sZ}	$-\sin(\theta - \rho)$	0
e_{fX}	$\sin\phi$	$-\cos(\theta + \rho)\cos\phi$
e_{fY}	$-\cos\phi$	$-\cos(\theta + \rho)\sin\phi$
e_{fZ}	0	$\sin(\theta + \rho)$

$$\mathcal{A} = n_o^2 \sin^2\theta + n_Z^2 \cos^2\theta$$

TABLE 1.6. n_s, n_f, ρ, $\hat{e}_s$ and $\hat{e}_f$ for **k** along an arbitrary direction in an uniaxial crystal. The components of the unit vector $\hat{d}$ are obtained from the corresponding components of the unit vector $\hat{e}$, with the walk-off angle ρ set equal to zero.

1.9 Propagation Equation In Presence Of Walk-off

The propagation equation for the electric field component of light is expressed conveniently in a 'laboratory coordinate system' (xyz) with the z direction along the propagation vector **k** and the transverse variation of the field being on the xy plane. Polarization components of the electric fields have thus far been determined with respect to the principal dielectric axes (XYZ) of the crystal. To derive the equation describing light propagation in presence of beam walk-off, it is necessary to perform a transformation between the xyz and XYZ coordinate systems.

1.9.1 Transformation between laboratory and crystal coordinate systems

For light propagating with propagation vector **k** having polar coordinates (θ, ϕ) with respect to the principal dielectric axes in an anisotropic crystal with known values of n_X, n_Y and n_Z, components of the vectors $\hat{m}$, $\hat{d}$ and $\hat{e}$ along the X, Y and Z directions for the two allowed polarization directions (denoted by indices s and f) are known from Eqns. 1.64, 1.91, 1.92, 1.93, 1.94 and 1.95. For a given beam of light, the laboratory coordinate system (xyz) can be defined with $\hat{z}$ equal to $\hat{m}$, and x equal to the unit vector $\hat{d}$ for one of the polarizations. The unit vector $\hat{y}$ can then be determined from the cross product $\hat{z} \times \hat{x}$, i.e., we have

$$
\begin{aligned}
\hat{x} &= \hat{d} = d_X \hat{X} + d_Y \hat{Y} + d_Z \hat{Z} \\
\hat{z} &= \hat{m} = m_X \hat{X} + m_Y \hat{Y} + m_Z \hat{Z} \\
\hat{y} &= (m_Y \, d_z - m_Z \, d_Y) \, \hat{X} + (m_Z \, d_X - m_X \, d_Z) \, \hat{Y} + (m_X \, d_Y - m_Y \, d_X) \, \hat{Z}
\end{aligned}
\tag{1.182}
$$

and from Eqn. 1.95

$$
\hat{e} = \cos\rho \, \hat{d} + \sin\rho \, \hat{m} = \cos\rho \, \hat{x} + \sin\rho \, \hat{z}.
\tag{1.183}
$$

with the subscripts s or f added to the unit vectors $\hat{d}$ or $\hat{e}$ as appropriate. For example, if $n_X < n_Y < n_Z$ and the slow component is being considered, the values of d_X, d_Y and d_Z will be given by d_{sX}, d_{sY} and d_{sZ} in Eqn. 1.91.

1.9.2 The propagation equation in presence of walk-off

Since **k** has been chosen to be in the z direction, $\mathbf{k} \cdot \mathbf{r} = kz$ so Eqns. 1.12 and 1.13 can be rewritten as

$$
\begin{aligned}
\widetilde{\mathbf{E}} &= \mathbf{E}e^{-i\omega t} = \mathbf{A}\psi e^{-i\omega t} \\
\widetilde{\mathbf{D}} &= \mathbf{D}e^{-i\omega t} = \mathfrak{D}\psi e^{-i\omega t}
\end{aligned}
\tag{1.184}
$$

where

$$
\psi = e^{ikz}, \quad \text{i.e.,} \quad \frac{\partial \psi}{\partial z} = ik\psi \quad \text{and} \quad \nabla\psi = ik\psi\hat{z}.
\tag{1.185}
$$

Inserting the identity $\nabla \times (\nabla \times \mathbf{E}) = \nabla(\nabla \cdot \mathbf{E}) - \nabla^2 \mathbf{E}$ in Eqn. 1.39 we obtain

$$
\nabla(\nabla \cdot \mathbf{E}) - \nabla^2 \mathbf{E} = \mu_0 \omega^2 \mathbf{D}.
\tag{1.186}
$$

Using $\mathbf{E} = \psi\mathbf{A}$ from Eqn. 1.184,

$$
\begin{aligned}
\nabla \cdot \mathbf{E} = \nabla \cdot (\psi\mathbf{A}) &= \psi(\nabla \cdot \mathbf{A} + ikA_z) \\
&= \psi(f + ikA_z)
\end{aligned}
\tag{1.187}
$$

where we have defined

$$f \equiv \nabla \cdot \mathbf{A}$$
$$= \frac{\partial A_x}{\partial x} + \frac{\partial A_z}{\partial z}. \tag{1.188}$$

since $\mathbf{A}$ lies on the xz plane (Eqn. 1.183). Since $\mathbf{D}$ is along the x direction, and from Eqn. 1.184, $\mathbf{D}$ is parallel to $\mathfrak{D}$, the z component of $\mathfrak{D}$, i.e., $\mathfrak{D}_z$ is equal to 0.

Thus

$$\nabla \cdot \mathbf{D} = \nabla \cdot (\psi \mathfrak{D})$$
$$= \psi (\nabla \cdot \mathfrak{D}) + \mathfrak{D} \cdot \nabla \psi$$
$$= \psi (\nabla \cdot \mathfrak{D}) + ik \mathfrak{D}_z$$
$$= \psi (\nabla \cdot \mathfrak{D}). \tag{1.189}$$

Thus the Maxwell's equation $\nabla \cdot \mathbf{D} = 0$ requires $\nabla \cdot \mathfrak{D}$ to be zero as well.

In an isotropic medium, $\mathbf{A}$ is parallel to $\mathfrak{D}$, thus $\nabla \cdot \mathfrak{D} = 0$ requires $\nabla \cdot \mathbf{A}$ to be equal to zero, so that $f = \nabla \cdot \mathbf{A} = 0$. In anisotropic media, in which $\mathfrak{D}$ and $\mathbf{A}$ are in general not parallel to each other, f is non-zero. However, for small anisotropy, the value of f is small, and the gradient of f can usually be ignored, i.e., $\nabla f \approx 0$. This is the key assumption in the derivation of the propagation equation below.

Using $\nabla f \approx 0$ and Eqns. 1.187 and 1.185

$$\nabla (\nabla \cdot \mathbf{E}) = \nabla (\nabla \cdot (\psi \mathbf{A})) = \nabla (\psi f) + ik \nabla (\psi A_z)$$
$$\approx (f + ik A_z) \nabla \psi + ik \psi \nabla A_z$$
$$= ik \psi \{ (f + ik A_z) \hat{z} + \nabla A_z \}. \tag{1.190}$$

Under the paraxial approximation, i.e., ignoring the $\partial^2 / \partial z^2$ term we get

$$\nabla^2 \mathbf{E} = \nabla^2 (\psi \mathbf{A}) = \psi \left(\nabla_T^2 \mathbf{A} + 2ik \frac{\partial \mathbf{A}}{\partial z} - k^2 \mathbf{A} \right). \tag{1.191}$$

If A and $\mathfrak{D}$ denote the magnitudes of the vectors $\mathbf{A}$ and $\mathfrak{D}$ respectively, we have (using Eqn. 1.183)

$$\mathbf{A} = \hat{x} A_x + \hat{z} A_z$$
$$= (\hat{x} \cos \rho + \hat{z} \sin \rho) A, \tag{1.192}$$

and

$$f = \cos \rho \frac{\partial A}{\partial x} + \sin \rho \frac{\partial A}{\partial z} \tag{1.193}$$

Inserting Eqns. 1.190 and 1.191 in Eqn. 1.186 (and canceling the common ψ) we get

$$ik \{ (f + ik A_z) \hat{z} + \nabla A_z \}$$
$$- (\nabla_T^2 \mathbf{A} + 2ik \frac{\partial \mathbf{A}}{\partial z} - k^2 \mathbf{A}) = \mu_0 \omega^2 \mathfrak{D}. \tag{1.194}$$

Taking the dot products of the vectors $\mathbf{A}$, $\mathfrak{D}$ and ∇A_z with unit vector $\hat{e}$ given in Eqn. 1.183 we get

$$\hat{e} \cdot \mathbf{A} = A, \qquad \hat{e} \cdot \mathfrak{D} = \mathfrak{D} \cos \rho \qquad (1.195)$$

and

$$
\begin{aligned}
\hat{e} \cdot \nabla A_z &= \cos \rho \frac{\partial A_z}{\partial x} + \sin \rho \frac{\partial A_z}{\partial z} \\
&= \cos \rho \sin \rho \frac{\partial A}{\partial x} + \sin^2 \rho \frac{\partial A}{\partial z}.
\end{aligned}
\qquad (1.196)
$$

Using Eqns. 1.196 and 1.195 in Eqn. 1.194 we get

$$
ik \sin \rho (f + ikA_z) \;+\; ik \left(\cos \rho \sin \rho \frac{\partial A}{\partial x} + \sin^2 \rho \frac{\partial A}{\partial z} \right)
$$
$$
- \left(\nabla_T^2 A + 2ik \frac{\partial A}{\partial z} - k^2 A \right) = \mu_0 \omega^2 \hat{e} \cdot \mathfrak{D} \quad (1.197)
$$

where the right hand side of Eqn. 1.197 is intentionally left in this form.

With $A_x = A \cos \rho$ and $A_z = A \sin \rho$ (from Eqn. 1.192), and inserting Eqn. 1.193 in Eqn. 1.197 we obtain

$$
ik \sin \rho (\cos \rho \frac{\partial A}{\partial x} \;+\; \sin \rho \frac{\partial A}{\partial z} + ik \sin \rho A)
$$
$$
+ \quad ik \sin \rho (\cos \rho \frac{\partial A}{\partial x} + \sin \rho \frac{\partial A}{\partial z})
$$
$$
- \quad (\nabla_T^2 A + 2ik \frac{\partial A}{\partial z} - k^2 A) = \mu_0 \omega^2 \hat{e} \cdot \mathfrak{D}. \quad (1.198)
$$

Rearranging and collecting the terms in Eqn. 1.197

$$
2ik \sin \rho \cos \rho \frac{\partial A}{\partial x} \quad - \quad 2ik \cos^2 \rho \frac{\partial A}{\partial z} - \nabla_T^2 A
$$
$$
= \quad \mu_0 \omega^2 \hat{e} \cdot \mathfrak{D} - k^2 A \cos^2 \rho
$$
$$
= \quad 0 \qquad (1.199)
$$

where the last equality follows from Eqn. 1.50 with $k = n\omega/c$ and the second relation in Eqn. 1.195. Rewriting Eqn. 1.199 with the terms rearranged, we get

$$
\frac{\partial A}{\partial z} = \frac{i}{2k \cos^2 \rho} \nabla_T^2 A + \tan \rho \frac{\partial A}{\partial x} \qquad (1.200)
$$

which is the linear propagation equation in presence of walk-off.

Bibliography

[1] M. Born and E. Wolf, *Principles of optics, 7th Edition*, Cambridge University Press, Cambridge, 1999.

[2] V. G. Dmitriev, G. G. Gurzadyan, D. N. Nikogosyan, *Handbook of Nonlinear Optical Crystals*, Springer, Berlin, 1999.

[3] J. D. Jackson, *Classical Electrodynamics*, John Wiley and Sons, Inc., New York, 1975.

[4] S. A. Akhmanov and S. Yu. Nikitin, *Physical Optics*, Clarendon Press, Oxford, 1997.

[5] A. Yariv, *Quantum Electronics*, John Wiley and Sons, Inc., 1978.

[6] J. F. Nye, *Physical properties of crystals : their representation by tensors and matrices*, Imprint Oxford [Oxfordshire] Clarendon Press, New York, Oxford University Press, 1985.

[7] H. Ito and H. Inaba, Optical Properties and UV N_2 Laser-Pumped Parametric Fluorescence in $LiCOOH \cdot H_2O$, *IEEE J. Quantum Electron.* **8**, 612, 1972.

[8] O. I. Lavrovskaya, N. I. Pavlova and A. V. Tarasov, Second harmonic generation of light from an YAG:Nd^{3+} laser in an optically biaxial crystal $KTiOPO_4$, *Sov. Phys. Crystallogr.* **31**, 678, 1986.

[9] H. Ito, H. Naito and H. Inaba, New Phase-Matchable Nonlinear Optical Crystals of the Formate Family, *IEEE J. Quantum Electron.* **10**, 247–252, 1974.

[10] H. Ito, H. Naito and H. Inaba, Generalized study on angular dependence of induced second-order nonlinear optical polarizations and phase matching in biaxial crystals, *J. Appl. Phys.* **46**, 3992–3998, 1975.

[11] S. Tsuboi, *Henko Kenbikyo (Polarizing Microscope)*, Iwanami, Tokyo (page 99) 1965.

[12] F. Zernike and J. E. Midwinter, *Applied Nonlinear Optics*, John Wiley and Sons, New York, 1973.

2

Nonlinear Optical Processes

This chapter

1. Describes different second order nonlinear optical processes

2. Introduces the second order susceptibility tensor d, describes the symmetry conditions it satisfies and introduces the 3×6 d matrix

3. Expresses the components of the nonlinear polarization vector in terms of the d matrix and the components of the electric fields

4. Introduces the concept of phase matching

5. Calculates the phase matching angles and walk-off angles for a biaxial crystal for propagation along principal planes and in a general direction

2.1 Introduction

Nonlinear optical phenomena are inherent in the quantum mechanical description of the interaction of electromagnetic radiation and matter, as has been shown using higher order perturbation theory by Dirac [1] in 1927 and Göppert-Mayer [2] in 1931. The field of nonlinear optics blossomed after the invention of the laser in 1960 made it easy to obtain monochromatic light with irradiance exceeding hundreds of kilowatts per cm^2. Demonstration of second harmonic generation of light by Franken and coworkers [3] and two-photon absorption by Kaiser and Garrett [4], both in 1961, ushered in this new field with its many potentials. Theoretical understanding of the observed processes was provided by Kleinman [5] as well as Armstrong and others [6] in 1962. Using a semiclassical treatment (with light field treated classically), they described how second and third order nonlinear polarizations arise from the light induced dipole moments in dielectric media, and following them we introduce here the nonlinear polarizations that are needed for the description of laser beam propagation through nonlinear optical media.

The relation between the components of the electric field vector $\widetilde{\mathbf{E}}$ of a light beam and the 'polarization' (the average dipole moment density) vector

$\widetilde{\mathbf{P}}$ in a linear medium is given by

$$\widetilde{P}_i = \varepsilon_0 \chi_{ij} \widetilde{E}_j \tag{2.1}$$

as was shown in the Chapter 1, Eqns. 1.24 and 1.27.

When the irradiance of light is of the order of megawatts per cm^2 or higher, its electric field amplitude (of the order of 10^6 V/m) is not completely negligible compared to the atomic field strengths (say of order 10^9 V/m), making the relationship between $\widetilde{\mathbf{P}}$ and $\widetilde{\mathbf{E}}$ no longer strictly linear. Assuming that the relationship can be expressed as a power series, which is justified by the quantum mechanical calculation of strong field light-matter interaction using perturbation theory, $\widetilde{\mathbf{P}}$ is expressed as

$$\begin{aligned} \widetilde{\mathbf{P}} &= \epsilon_0(\chi.\widetilde{\mathbf{E}} + \chi^{(2)} : \widetilde{\mathbf{E}}\,\widetilde{\mathbf{E}} + \chi^{(3)} : \widetilde{\mathbf{E}}\,\widetilde{\mathbf{E}}\,\widetilde{\mathbf{E}} + .. \\ &\equiv \widetilde{\mathbf{P}}_L + \widetilde{\mathbf{P}}_{NL}^2 + \widetilde{\mathbf{P}}_{NL}^3 + ... \end{aligned} \tag{2.2}$$

where χ, $\chi^{(2)}$ and $\chi^{(3)}$ denote the linear, second order and third order susceptibilities, respectively. The products of the χs with combinations of the field $\widetilde{\mathbf{E}}$ shown in Eqn. 2.2 are tensor products, χ, $\chi^{(2)}$ and $\chi^{(3)}$ being tensors of ranks 1, 2 and 3 respectively. The field $\widetilde{\mathbf{E}}$ is not necessarily of a single frequency and in general can be a combination of fields oscillating at many different frequencies. If $\widetilde{\mathbf{E}}$ consists of fields oscillating at frequencies ω_1, ω_2, ω_3, ω_4 etc., then in the linear case, i.e., when only the $\chi.\widetilde{E}$ term in Eqn. 2.2 is appreciable, the polarization $\widetilde{\mathbf{P}}$ would also be oscillating at these frequencies. If in addition the $\chi^{(2)}$ term (the second term in Eqn. 2.2) is large enough, $\widetilde{\mathbf{P}}$ will also oscillate at combinations of the frequencies, such as $\omega_1 + \omega_2$, $2\omega_1$, $2\omega_1 - \omega_2$ etc., taken two frequencies at a time, from the quadratic dependence of this term on the electric field. If, in addition, the third term in Eqn. 2.2 is also appreciable, the polarization would also be oscillating at the combinations of the frequencies taken three at a time, such as $\omega_1 + \omega_2 + \omega_3$, $3\omega_1$, $3\omega_2$, $\omega_1 + \omega_2$ - ω_3 etc. The oscillating polarization, i.e., the macroscopically averaged oscillating induced dipoles, would radiate at their frequencies of oscillations, thereby generating light at the new frequencies. In this chapter we discuss the processes arising from the presence of the second order susceptibility in nonlinear optical media.

2.2 Second Order Susceptibility

Suppose two monochromatic and linearly polarized beams of light having frequencies ω_1 and ω_2 spatially overlap in a material having non-negligible second order nonlinearity characterized by the tensor $\chi^{(2)}$. The total electric field can be expressed as

$$\widetilde{\mathbf{E}} = \mathbf{A}_1 e^{i(k_1 z - \omega_1 t)} + \mathbf{A}_2^{i(k_2 z - \omega_2 t)} + \text{cc} \tag{2.3}$$

where $\mathbf{A}_1$ and $\mathbf{A}_2$ are the slowly varying complex amplitudes of the incident beams. Components of the second order nonlinear polarization $\tilde{P}^{(2)}$ (with the subscript NL dropped for convenience) are then given by

$$\tilde{P}_i^{(2)} = \varepsilon_0 \chi_{ijk}^{(2)} \tilde{E}_j \tilde{E}_k \qquad (2.4)$$

where the indices i, j and k run over the Cartesian coordinates X, Y and Z. Frequency dependence of the components of $\chi^{(2)}$ are left implicit. Also, whenever the indices i, j, k are used repeatedly in one side of an equation, as in Eqn. 2.4, the Einstein summation convention (i.e., summation over the repeated indices) is assumed to be valid.

As the indices j and k each take the values of X, Y and Z, there are nine combinations of j,k: XX, YY, ZZ, YZ, ZX, XY, ZY, XZ and YX. The right hand side of Eqn. 2.3 consists of 4 terms (with amplitudes $\mathbf{A}_1$, $\mathbf{A}_2$, $\mathbf{A}_1^*$, $\mathbf{A}_2^*$), for each component i in Eqn. 2.4, there are 36 terms on the right hand side. Instead of writing out all 36 combinations for each i, and then doing this for the three values of i, we write out only two combinations XX and XY explicitly, from which all the others can be easily inferred.

With $\mathbf{E} \equiv \mathbf{A}e^{ikz}$ for frequencies ω_1 and ω_2 (with appropriate subscripts 1 and 2) Eqn. 2.3 is re-written as

$$\tilde{\mathbf{E}} = \mathbf{E}_1 e^{-i\omega_1 t} + \mathbf{E}_2^{-i\omega_2 t} + \text{cc}. \qquad (2.5)$$

Inserting Eqn. 2.5 in Eqn. 2.4 we obtain

$$\tilde{P}_i^{(2)} = \varepsilon_0 \chi_{ijk}^{(2)} (E_{1j}e^{-i\omega_1 t} + E_{2j}e^{-i\omega_2 t} + \text{cc})(E_{1k}e^{-i\omega_1 t} + E_{2k}e^{-i\omega_2 t} + \text{cc}). \qquad (2.6)$$

When $j = k = X$, say, one of the terms on the right hand side of Eqn. 2.6 is

$$\chi_{iXX}^{(2)}(E_{1X}e^{-i\omega_1 t} + E_{2X}e^{-i\omega_2 t} + \text{cc})(E_{1X}e^{-i\omega_1 t} + E_{2X}e^{-i\omega_2 t} + \text{cc})$$
$$= \chi_{iXX}^{(2)}(E_{1X}^2 e^{-2i\omega_1 t} + E_{2X}^2 e^{-2i\omega_2 t} + 2E_{1X}E_{2X}e^{-i(\omega_1+\omega_2)t}$$
$$+ 2E_{1X}E_{2X}^* e^{-i(\omega_1-\omega_2)t} + |E_{1X}|^2 + |E_{2X}|^2 + \text{cc}). \qquad (2.7)$$

Similarly, when $j \neq k$, say, $j = X$, and $k = Y$, on the right hand side of Eqn. 2.6 is

$$\chi_{iXY}^{(2)}(E_{1X}e^{-i\omega_1 t} + E_{2X}e^{-i\omega_2 t} + \text{cc})(E_{1Y}e^{-i\omega_1 t} + E_{2Y}e^{-i\omega_2 t} + \text{cc})$$
$$= \chi_{iXY}^{(2)}\{E_{1X}E_{1Y}e^{-2i\omega_1 t} + E_{2X}E_{2Y}e^{-2i\omega_2 t}$$
$$+ (E_{1X}E_{2Y} + E_{2X}E_{1Y})e^{-i(\omega_1+\omega_2)t}$$
$$+ (E_{1X}E_{2Y}^* + E_{1Y}E_{2X}^*)e^{-i(\omega_1-\omega_2)t} + E_{1X}E_{1Y}^* + E_{2X}E_{2Y}^* + \text{cc}\}. \qquad (2.8)$$

Equations 2.7 and 2.8 show that the second order nonlinear interaction of two fields at frequencies ω_1 and ω_2 can give rise to polarization components oscillating at frequencies of $2\omega_1$, $2\omega_2$, $\omega_1 + \omega_2$, $\omega_1 - \omega_2$ as well as a constant, non-oscillating polarization. The polarization components oscillating at a given frequency can give rise to electric fields oscillating at the same frequency, so oscillating fields are generated at $2\,\omega_1$, $2\,\omega_2$, $\omega_1+\omega_2$, $\omega_1-\omega_2$. Various nonlinear optical phenomena arise from the oscillation of the generated polarization at different frequencies:

SHG

The phenomenon of generating light oscillating at twice the frequency of an incident beam is known as *Second Harmonic Generation* (SHG). In Eqns. 2.7 and 2.8, the terms with amplitudes E_{1X}^2 and $E_{1X}E_{1Y}$ give rise to the second harmonic of frequency ω_1. Terms with amplitudes E_{2X}^2 and $E_{2X}E_{2Y}$ give rise to the second harmonic of frequency ω_2. The term with amplitude $E_{1X}E_{2Y} + E_{2X}E_{1Y}$ also gives rise to SHG when $\omega_1 = \omega_2$.

OR

Generation of a zero frequency component of electric field from an oscillating field is known as *Optical Rectification* (OR). In Eqn. 2.7 and 2.8, the terms with amplitudes $|E_{1X}|^2$, $|E_{2X}|^2$, $E_{1X}E_{1Y}^* + E_{2X}E_{2Y}^*$ correspond to this case. As in the SHG case, the phenomena of OR can take place even when only one frequency of light is initially present in the medium.

EO

The phenomenon of modulation of a light beam by an applied zero (or low) frequency electric field is known as the *linear eletro-optic effect*(EO) or *Pockel's effect*. In Eqns. 2.7 and 2.8, the terms with amplitudes $E_{1X}E_{2X}$, $E_{1X}E_{2X}^*$, $E_{1X}E_{2Y} + E_{2X}E_{1Y}$ and $E_{1X}E_{2Y}^* + E_{1Y}E_{2X}^*$, with $\omega_2 = 0$ all correspond to the case of EO.

SFG and DFG

The phenomena of generation of light at frequencies $\omega_1 + \omega_2$ and $\omega_1 - \omega_2$ are termed *Sum Frequency Generation* (SFG) and *Difference frequency Generation* (DFG), respectively. In Eqns. 2.7 and 2.8, the terms with amplitudes $E_{1X}E_{2X}$ and $E_{1X}E_{2Y} + E_{1Y}E_{2X}$ give rise to SFG and the terms $E_{1X}E_{2X}^*$ and $E_{1X}E_{2Y}^* + E_{1Y}E_{2X}^*$ give rise to DFG. In the degenerate case, i.e., when the two frequencies ω_1 and ω_2 of the initial light beams are the same, SFG and DFG reduce to SHG and OR, respectively.

OPA

Various nonlinear optical phenomena have been described above assuming that two beams of *arbitrary* frequencies ω_1 and ω_2 are interacting in a nonlinear medium. Suppose the frequencies of two light beams initially present are designated ω_3 and ω_2, with $\omega_3 > \omega_2$. Then a nonlinear polarization oscillating at frequency ω_1 can arise through difference frequency generation, where $\omega_1 = \omega_3 - \omega_2$. This polarization can give rise to a light beam at the frequency ω_1, which in turn can nonlinearly interact with the beam at frequency ω_3 to give rise to a nonlinear polarization at frequency ω_2. This polarization oscillating at frequency ω_2 can cause an amplification of the light beam at frequency ω_2 that was initially present. This process of amplification of a field at a lower frequency (ω_2) made possible by the nonlinear frequency mixing with a wave at a higher frequency (ω_3) is called optical parametric amplification (OPA).

OPG

The phenomenon of optical parametric generation (OPG) can be thought of as the optical parametric amplification of noise. According to a postulate by Nernst [7], any medium, including vacuum, is filled with zero-point electromagnetic radiation of all frequencies which can be thought of as noise. Suppose light at a single frequency ω_3 is initially present. A component of the noise radiation which is oscillating at frequency ω_2 can produce amplification at frequency ω_2 through the parametric amplification process described above, thereby eventually generating detectable amount of light at frequency ω_2. The generated radiation oscillating at frequency ω_2 in turn can provide parametric amplification at the frequency ω_1, and generate detectable amount of light at that frequency. Thus, 'one' incident beam of light (at frequency ω_3) generates 'two' new frequencies (at frequencies ω_1 and ω_2), showing that the 'OPG' process is the reverse of the 'SFG' process, while for both, the relationship

$$\omega_1 + \omega_2 = \omega_3 \tag{2.9}$$

is maintained.

In the OPG process, for a given incident frequency ω_3, there are of course an infinite number of possible combinations for the two lower frequencies, ω_1 and ω_2. The particular values of ω_1 and ω_2 that are actually generated depend on the 'phase matching condition' which will be described later.

OPO

If the phenomenon of optical parametric generation (OPG) takes place inside an optical resonator, in which the gain experienced by the beam of light at one or both of the generated frequencies equals or exceeds the losses in a round-trip through the resonator, oscillation occurs and coherent emission can be obtained. This phenomenon is called optical parametric oscillation (OPO).

The three frequencies involved in an OPA, OPO or an OPG, in the order of increasing frequencies (i.e., decreasing wavelengths), are usually designated the 'idler', the 'signal' and the 'pump' frequencies.

In the degenerate parametric generation case, i.e., with $\omega_1 = \omega_2$, the generated signal and idler frequencies are both equal to $\omega_3/2$. This phenomenon of generating light at half the frequency of an incident beam is called subharmonic generation or degenerate parametric generation.

2.3 Properties of $\chi^{(2)}$

In a material with a center of symmetry, if the directions of the two initial fields are reversed, the polarization component generated by these fields will also be reversed, by definition of having the center of symmetry. That is, the signs of all three vectors in Eqn. 2.4 need to be changed together

$$-\widetilde{P}_i^{(2)} = \chi_{ijk}^{(2)}(-\widetilde{E}_j)(-\widetilde{E}_k) \quad \text{(in materials with center of symmetry)}. \tag{2.10}$$

Equations 2.4 and 2.10 can both be simultaneously valid only if $\chi_{ijk}^{(2)}$ is equal to zero. Liquids and gases, which are isotropic materials having centers of symmetry, do not display second order nonlinearities except in regions of discontinuities such as surfaces. Some solids, like many metals, also have crystalline structure with a center of symmetry and therefore the second order susceptibility is zero for them. In the rest of the discussion of second order nonlinearities, only noncentrosymmetric crystalline solids will be considered.

$\chi_{ijk}^{(2)}$ is a third rank tensor having 27 components, with the indices i, j and k each taking three values X, Y and Z. However, for each second order nonlinear optical phenomenon (SHG, SFG, DFG, EO and OR) a reduction in the number of relevant $\chi^{(2)}$ components can be brought about by invoking an intrinsic permutation symmetry.

The intrinsic permutation symmetry can be understood by expanding the summation over the indices j and k implicit in Eqn. 2.4. Writing out, say only the sum frequency components, Eqns. 2.7 and 2.8 provide the $j = X, k = X$ and $j = X, k = Y$ terms

$$\widetilde{P}_i^{\omega_1+\omega_2} = e^{-i(\omega_1+\omega_2)t}\{2\chi_{iXX}^{(2)}E_{1X}E_{2X} + \chi_{iXY}^{(2)}(E_{1X}E_{2Y} + E_{2X}E_{1Y}) + ...\} \tag{2.11}$$

Extending to other possible combinations of j and k, the ellipsis in Eqn. 2.11 can be easily completed :

$$\begin{aligned}
\widetilde{P}_i^{\omega_1+\omega_2} =\ & e^{-i(\omega_1+\omega_2)t}\{2\chi_{iXX}^{(2)}E_{1X}E_{2X} + 2\chi_{iYY}^{(2)}E_{1Y}E_{2Y} + 2\chi_{iZZ}^{(2)}E_{1Z}E_{2Z} \\
+\ & \chi_{iXY}^{(2)}(E_{1X}E_{2Y} + E_{2X}E_{1Y}) + \chi_{iYX}^{(2)}(E_{1Y}E_{2X} + E_{2Y}E_{1X}) \\
+\ & \chi_{iYZ}^{(2)}(E_{1Y}E_{2Z} + E_{2Y}E_{1Z}) + \chi_{iZY}^{(2)}(E_{1Z}E_{2Y} + E_{2Z}E_{1Y}) \\
+\ & \chi_{iZX}^{(2)}(E_{1Z}E_{2X} + E_{2Z}E_{1X}) + \chi_{iXZ}^{(2)}(E_{1X}E_{2Z} + E_{2X}E_{1Z})\}.
\end{aligned} \tag{2.12}$$

Collecting the terms, we get

$$\begin{aligned}
P_i^{\omega_1+\omega_2} =\ & e^{-i(\omega_1+\omega_2)t}\{2\chi_{iXX}^{(2)}E_{1X}E_{2X} + 2\chi_{iYY}^{(2)}E_{1Y}E_{2Y} + 2\chi_{iZZ}^{(2)}E_{1Z}E_{2Z} \\
+\ & (\chi_{iXY}^{(2)} + \chi_{iYX}^{(2)})(E_{1X}E_{2Y} + E_{2X}E_{1Y}) \\
+\ & (\chi_{iYZ}^{(2)} + \chi_{iZY}^{(2)})(E_{1Y}E_{2Z} + E_{2Y}E_{1Z}) \\
+\ & (\chi_{iZX}^{(2)} + \chi_{iXZ}^{(2)})(E_{1Z}E_{2X} + E_{2X}E_{1Z})\}.
\end{aligned} \tag{2.13}$$

Since the terms $\chi_{ijk}^{(2)}$ (for $j \neq k$) always occur in the combination $\chi_{ijk}^{(2)} + \chi_{ikj}^{(2)}$, which is symmetric under the interchange of the indices k and j, the convention adopted in nonlinear optics has been to assume that

$$\chi_{ijk}^{(2)} = \chi_{ikj}^{(2)}. \tag{2.14}$$

As pointed out in Ref. [8], the condition shown in Eqn. 2.14 is chosen just as a matter of convenience. Another possible choice could have been a new set of $\chi^{(2)}$, designated by, say, $\chi_{ijk}^{(2)new}$ with

$$\chi_{ijk}^{(2)new} = 2\chi_{ikj}^{(2)} \quad \text{and} \quad \chi_{ikj}^{(2)new} = 0. \tag{2.15}$$

In this book, we will stick with the well established convention shown in Eqn. 2.14.

2.3.1 Properties of $\chi^{(2)}$ away from resonance

In general, the second order susceptibility $\chi^{(2)}$ depends on the oscillation frequencies of the three interacting electric fields. Away from any material resonances, that is, at frequencies where the absorption of light is low, the values of $\chi^{(2)}$ are largely frequency independent. In the discussion here, the frequency independence of $\chi^{(2)}$ is implied even when not explicitly stated. Also, at frequencies of light much smaller than material resonant frequencies, $\chi^{(2)}$ is a real quantity.

2.3.2 Kleinman's symmetry

Kleinman [5] pointed out that when $\chi^{(2)}$ is independent of the frequencies of the interacting electric fields and polarizations, the indices of $\chi_{ijk}^{(2)}$ can be freely permuted, leading to

$$\begin{aligned} \chi_{ijk}^{(2)} &= \chi_{jki}^{(2)} = \chi_{kij}^{(2)} \\ &= \chi_{ikj}^{(2)} = \chi_{jki}^{(2)} = \chi_{kij}^{(2)}. \end{aligned} \tag{2.16}$$

As pointed out by Zernike and Midwinter [2], Kleinman's symmetry condition has been experimentally verified within measurement accuracy. The symmetry condition however is not usually valid for optical rectification, electro-optic effect or DFG of far infrared wavelengths because in these cases absorption bands usually lie in between the frequencies involved in the nonlinear interaction.

2.4 d coefficients and the contracted notation

Along with $\chi^{(2)}$, the letter d is also used to designate the second order nonlinear optical susceptibility tensor, with the relationship between d and $\chi^{(2)}$ given by

$$d_{ijk} = \frac{1}{2}\chi_{ijk}^{(2)}. \tag{2.17}$$

For a given i, there are nine possible combinations of j and k: $XX, YY, ZZ,$ YZ, ZY, ZX, XZ, XY and YX. Due to intrinsic permutation symmetry only the six combinations, XX, YY, ZZ, YZ, ZX and XY are independent.

Adopting the convention of using the numbers 1,2 and 3 in place of the coordinates X, Y and Z, respectively, the permutation symmetry can be used to replace the two indices j and k in the coefficient ijk by an index l, with l taking six values from 1 to 6 corresponding to the six combinations of X, Y and Z for j and k, as shown in the Table 2.1. Thus there are 18 independent

j and k combinations	j	k	l
XX	1	1	1
YY	2	2	2
ZZ	3	3	3
YZ	2	3	4
ZX	3	1	5
XY	1	2	6

TABLE 2.1. The index l is defined in terms of the indices j and k. For j and k, the numbers 1,2 and 3 represent coordinates X, Y and Z respectively.

components of the d_{ijk} tensor which can be expressed as a 3 x 6 matrix d_{il} with i ranging from 1 to 3 and l from 1 to 6:

$$d_{il} = \begin{pmatrix} d_{11} & d_{12} & d_{13} & d_{14} & d_{15} & d_{16} \\ d_{21} & d_{22} & d_{23} & d_{24} & d_{25} & d_{26} \\ d_{31} & d_{32} & d_{33} & d_{34} & d_{35} & d_{36} \end{pmatrix}. \tag{2.18}$$

Thus, for example,

$$d_{XYZ} = d_{XZY} = d_{14}. \tag{2.19}$$

2.4.1   d coefficients under Kleinman symmetry

When Kleinman's symmetry is valid, the eighteen components of the d_{il} matrix are reduced to ten independent components. In the uncontracted notation, Kleinman's symmetry condition states

$$\begin{aligned} d_{ijk} &= d_{jki} = d_{kij} \\ &= d_{ikj} = d_{kji} = d_{jik}. \end{aligned}$$

When i, j and k are all different, we have

$$d_{XYZ} = d_{YZX} = d_{ZXY}, \quad \text{i.e.,} \quad d_{14} = d_{25} = d_{36}. \tag{2.20}$$

When two of the components of the i, j and k indices are the same and Kleinman's symmetry is valid, we get the relations

$$
\begin{aligned}
d_{YXX} &= d_{XXY} & \text{i.e.,} \quad d_{21} &= d_{16} \\
d_{YYZ} &= d_{ZYY} & \text{i.e.,} \quad d_{24} &= d_{32} \\
d_{YXY} &= d_{XXY} & \text{i.e.,} \quad d_{26} &= d_{16} \\
d_{ZXX} &= d_{XZX} & \text{i.e.,} \quad d_{31} &= d_{15} \\
d_{ZYZ} &= d_{YZZ} & \text{i.e.,} \quad d_{34} &= d_{23} \\
d_{ZZX} &= d_{XZZ} & \text{i.e.,} \quad d_{35} &= d_{13}.
\end{aligned}
\tag{2.21}
$$

With the eight relationships given in Eqns. 2.20 and 2.21 that hold under Kleinman's symmetry (KS) condition, the d matrix reduces to

$$
d_{(KS)} = \begin{pmatrix}
d_{11} & d_{12} & d_{13} & d_{14} & d_{15} & d_{16} \\
\underline{d}_{16} & d_{22} & d_{23} & d_{24} & \underline{d}_{14} & \underline{d}_{16} \\
\underline{d}_{15} & \underline{d}_{24} & d_{33} & \underline{d}_{23} & \underline{d}_{13} & \underline{d}_{14}.
\end{pmatrix}
\tag{2.22}
$$

where the eight underlined coefficients are repeats of previously shown ones.

Furthermore, material point group symmetry reduces the number of independent components of the d_{il} matrix for a given crystal to a much smaller value, as shown below.

2.5   The Non-Zero d Coefficients Of Biaxial Crystals

Crystals belonging to the triclinic, monoclinic and orthorhombic systems are biaxial in their optical properties. For a triclinic crystal (point group 1) the d matrix has in general 18 nonzero elements. Table 2.2 lists the nonzero elements of the d matrix for the 5 biaxial crystal classes, 1,2, m, 222 and mm2 [10]: When Kleinman's symmetry condition is *not* assumed, the elements of the d matrix for a given class of biaxial crystal are all independent of each other. When 8 Kleinman's symmetry conditions given in Eqns. 2.20 and 2.21 hold, the number of *independent* non-zero elements of the d matrix is reduced for each of the crystal classes. For example, in triclinic crystals, (class 1), the Kleinman symmetry conditions reduce the number of independent components from 18 to 10. Similarly, for crystals belonging to class 2, the 8 nonzero elements listed in Table 2.2 reduce to 4 independent components, d_{14}, d_{16}, d_{22} and d_{23}. For crystals belonging to the class m, the 8 nonzero elements listed in Table 2.2 reduce to 7 independent components, d_{26} being equal to d_{12}. For crystals belonging to the class 222, the 3 nonzero elements listed in Table 2.2 reduce to 1 independent component, d_{14}; and for crystals belonging to the class mm2, the 5 nonzero elements listed reduce to 3 independent components, d_{15}, d_{24} and d_{33}.

Crystal System	Point Group	Nonzero elements of d
Triclinic	1	All 18
Monoclinic	2	d_{13}, d_{14}, d_{16} $d_{21}, d_{22}, d_{23}, d_{25}$ d_{34}, d_{36}
	m	$d_{11}, d_{12}, d_{13}, d_{15}$ d_{24}, d_{26} $d_{31}, d_{32}, d_{33}, d_{35}$
Orthorhombic	222	d_{14} d_{25} d_{36}
	mm2	d_{15} d_{24} d_{31}, d_{32}, d_{33}

TABLE 2.2. Biaxial crystal classes and the non-zero elements of the d_{il} matrix

2.6   The Non-Zero d Coefficients Of Uniaxial Crystals

Crystals belonging to the tetragonal, trigonal (rhombohedral) and hexagonal systems are uniaxial in their optical properties. Table 2.3 lists the nonzero elements of the d matrix and the relations between the elements for the 13 uniaxial crystal classes.

Kleinman's symmetry condition, when valid, further reduces the number of independent elements in each crystal class. For example, there are 7 nonzero elements in the crystal class 4, and three symmetry relations relating the elements - so there are 4 independent elements d_{14}, d_{15}, d_{31} and d_{33}. Under Kleinman's symmetry condition $d_{31} = d_{15}$, reducing the number of independent elements to 3: d_{14}, d_{15} and d_{33}. Similarly, for trigonal crystals belonging to class 3, there are 13 nonzero elements related by 7 symmetry conditions, resulting in 6 independent components: d_{11}, d_{14}, d_{15}, d_{16}, d_{31} and d_{33}. Again, under Kleinman's symmetry condition $d_{31} = d_{15}$, the number of independent elements is reduced to 5.

The polarization vectors for the second-order nonlinear optical processes are expressed below in terms of the d matrices and the incident field components.

Crystal System	Point Group	Nonzero elements of d	Relations between the nonzero elements
Tetragonal	4	d_{14}, d_{15} d_{24}, d_{25} d_{31}, d_{32}, d_{33}	$d_{24} = d_{15}$ $d_{25} = -d_{14}$ $d_{32} = -d_{31}$
	$\bar{4}$	d_{14}, d_{15} d_{24}, d_{25} d_{31}, d_{32}, d_{36}	$d_{24} = -d_{15}$ $d_{25} = d_{14}$ $d_{32} = -d_{31}$
	422	d_{14} d_{25}	$d_{25} = -d_{14}$
	4mm	d_{15} d_{24} d_{31}, d_{32}, d_{33}	$d_{24} = d_{15}$ $d_{32} = d_{31}$
	$\bar{4}$2m	d_{14} d_{25} d_{36}	$d_{25} = d_{14}$
Trigonal	3	$d_{11}, d_{12}, d_{14}, d_{15}, d_{16}$ $d_{21}, d_{22}, d_{24}, d_{25}, d_{26}$ d_{31}, d_{32}, d_{33}	$d_{12} = -d_{11}$ $d_{21} = d_{16}$ $d_{22} = -d_{16}$ $d_{24} = d_{15}$ $d_{25} = -d_{14}$ $d_{26} = d_{12}$ $d_{32} = d_{31}$
	32	d_{11}, d_{12}, d_{14} d_{25}, d_{26}	$d_{12} = -d_{11}$ $d_{25} = -d_{14}$ $d_{26} = d_{12}$
	3m	d_{15}, d_{16} d_{21}, d_{22}, d_{24} d_{31}, d_{32}, d_{33}	$d_{21} = d_{16}$ $d_{22} = -d_{16}$ $d_{24} = d_{15}$ $d_{32} = d_{31}$

TABLE 2.3. The nonzero elements of the d matrix and the relations between the elements for the uniaxial crystal classes belonging to the tetragonal and trigonal crystal systems.

Crystal System	Point Group	Nonzero elements of d	Relations between the nonzero elements
Hexagonal	6	d_{14}, d_{15} d_{24}, d_{25} d_{31}, d_{32}, d_{33}	$d_{24} = d_{15}$ $d_{25} = -d_{14}$ $d_{32} = -d_{31}$
	6mm	d_{15} d_{24} d_{31}, d_{32}, d_{33}	$d_{24} = d_{15}$ $d_{32} = d_{31}$
	622	d_{14} d_{25}	$d_{25} = -d_{14}$
	$\bar{6}$m2	d_{16} d_{21}, d_{22}	$d_{22} = -d_{21} = -d_{16}$
	$\bar{6}$	d_{11}, d_{12}, d_{16} d_{21}, d_{22}, d_{26}	$d_{12} = -d_{11}$ $d_{22} = -d_{21} = -d_{16}$ $d_{26} = d_{12}$

TABLE 2.4. The nonzero elements of the d matrix and the relations between the elements for the uniaxial crystal classes belonging to the hexagonal crystal system.

2.7 Nonlinear polarizations

2.7.1 Nondegenerate sum frequency generation

When $\omega_1 \neq \omega_2$, collecting terms oscillating at frequency $\omega_1 + \omega_2$ from Eqns. 2.7 and 2.8 and adding the terms for other possible combinations of indices j and k in χ_{ijk}, we obtain the polarization component oscillating at frequency $\omega_1 + \omega_2$

to be

$$
\begin{aligned}
P_i^{\omega_1+\omega_2} &= \varepsilon_0 \{ 2\chi_{iXX}^{(2)} E_{1X} E_{2X} + 2\chi_{iYY}^{(2)} E_{1Y} E_{2Y} + 2\chi_{iZZ}^{(2)} E_{1Z} E_{2Z} \\
&\quad + \chi_{iXY}^{(2)} (E_{1X} E_{2Y} + E_{2X} E_{1Y}) + \chi_{iYX}^{(2)} (E_{1Y} E_{2X} + E_{2Y} E_{1X}) \\
&\quad + \chi_{iYZ}^{(2)} (E_{1Y} E_{2Z} + E_{2Y} E_{1Z}) + \chi_{iZY}^{(2)} (E_{1Z} E_{2Y} + E_{2Z} E_{1Y}) \\
&\quad + \chi_{iZX}^{(2)} (E_{1Z} E_{2X} + E_{2Z} E_{1X}) + \chi_{iXZ}^{(2)} (E_{1X} E_{2Z} + E_{2X} E_{1Z}) \} \\
&= 4\varepsilon_0 \{ d_{i1} E_{1X} E_{2X} + d_{i2} E_{1Y} E_{2Y} + d_{i3} E_{1Z} E_{2Z} \\
&\quad + d_{i4} (E_{1Y} E_{2Z} + E_{2Y} E_{1Z}) + d_{i5} (E_{1Z} E_{2X} + E_{2Z} E_{1X}) \\
&\quad + d_{i6} (E_{1X} E_{2Y} + E_{2X} E_{1Y}) \}.
\end{aligned}
\tag{2.23}
$$

using the contracted notation for d and the intrinsic permutation symmetry conditions. Eqn. 2.23 can be written in the matrix form as

$$
\begin{pmatrix} P_X^{\omega_1+\omega_2} \\ P_Y^{\omega_1+\omega_2} \\ P_Z^{\omega_1+\omega_2} \end{pmatrix} = 4\varepsilon_0 \begin{pmatrix} d_{11} & d_{12} & d_{13} & d_{14} & d_{15} & d_{16} \\ d_{21} & d_{22} & d_{23} & d_{24} & d_{25} & d_{26} \\ d_{31} & d_{32} & d_{33} & d_{34} & d_{35} & d_{36} \end{pmatrix} \begin{pmatrix} E_{1X} E_{2X} \\ E_{1Y} E_{2Y} \\ E_{1Z} E_{2Z} \\ E_{1Y} E_{2Z} + E_{2Y} E_{1Z} \\ E_{1Z} E_{2X} + E_{2Z} E_{1X} \\ E_{1X} E_{2Y} + E_{2X} E_{1Y} \end{pmatrix}.
\tag{2.24}
$$

The expression for the sum frequency polarization given in Eqn. 2.24 is valid only for the nondegenerate case, i.e., for $\omega_1 \neq \omega_2$. In the degenerate case, when ω_1 and ω_2 are equal, in addition to the term with amplitude $2E_{1X} E_{2X}$ in Eqn. 2.7, the terms E_{1X}^2 and E_{2X}^2 also oscillate with frequency $2\omega_1$, and similarly there are two additional terms in Eqn. 2.8. These terms will be added below in the expression of the SHG polarization.

2.7.2 Difference frequency generation

Collecting terms oscillating at frequency $\omega_1 - \omega_2$ from Eqns. 2.7 and 2.8 and adding the terms for other possible combinations j and k indices of χ_{ijk}, we obtain the polarization component oscillating at frequency $\omega_1 - \omega_2$ to be

$$
\begin{aligned}
P_i^{\omega_1-\omega_2} &= \varepsilon_0 \{ 2\chi_{iXX}^{(2)} E_{1X} E_{2X}^* + 2\chi_{iYY}^{(2)} E_{1Y} E_{2Y}^* + 2\chi_{iZZ}^{(2)} E_{1Z} E_{2Z}^* \\
&\quad + \chi_{iXY}^{(2)} (E_{1X} E_{2Y}^* + E_{2X}^* E_{1Y}) + \chi_{iYX}^{(2)} (E_{1Y} E_{2X}^* + E_{2Y}^* E_{1X}) \\
&\quad + \chi_{iYZ}^{(2)} (E_{1Y} E_{2Z}^* + E_{2Y}^* E_{1Z}) + \chi_{iZY}^{(2)} (E_{1Z} E_{2Y}^* + E_{2Z}^* E_{1Y}) \\
&\quad + \chi_{iZX}^{(2)} (E_{1Z} E_{2X}^* + E_{2Z}^* E_{1X}) + \chi_{iXZ}^{(2)} (E_{1X} E_{2Z}^* + E_{2X}^* E_{1Z}) \} \\
&= 4\varepsilon_0 \{ d_{i1} E_{1X} E_{2X}^* + d_{i2} E_{1Y} E_{2Y}^* + d_{i3} E_{1Z} E_{2Z}^* \\
&\quad + d_{i4} (E_{1Y} E_{2Z}^* + E_{2Y}^* E_{1Z}) + d_{i5} (E_{1Z} E_{2X}^* + E_{2Z}^* E_{1X}) \\
&\quad + + d_{i6} (E_{1X} E_{2Y}^* + E_{2X}^* E_{1Y}) \}
\end{aligned}
\tag{2.25}
$$

using the contracted notation for d and the intrinsic permutation symmetry conditions. Eqn. 2.26 can be written in the matrix form as

$$
\begin{pmatrix} P_x^{\omega_1-\omega_2} \\ P_y^{\omega_1-\omega_2} \\ P_z^{\omega_1-\omega_2} \end{pmatrix} = 4\varepsilon_0 \begin{pmatrix} d_{11} & d_{12} & d_{13} & d_{14} & d_{15} & d_{16} \\ d_{21} & d_{22} & d_{23} & d_{24} & d_{25} & d_{26} \\ d_{31} & d_{32} & d_{33} & d_{34} & d_{35} & d_{36} \end{pmatrix} \begin{pmatrix} E_{1X}E_{2X}^* \\ E_{1Y}E_{2Y}^* \\ E_{1Z}E_{2Z}^* \\ E_{1Y}E_{2Z}^* + E_{2Y}^*E_{1Z} \\ E_{1Z}E_{2X}^* + E_{2Z}^*E_{1X} \\ E_{1X}E_{2Y}^* + E_{2X}^*E_{1Y} \end{pmatrix}.
$$
$$(2.26)$$

2.7.3   Second harmonic generation (SHG)

To obtain the expressions for the nonlinear polarization giving rise to second harmonic generation, the frequencies ω_1 and ω_2 in Eqn. 2.5 can be set equal to each other (and renamed ω). In harmonic conversion processes, the light beam oscillating at the lowest frequency is called the *fundamental* beam and the integer multiples of the lowest frequency are the *harmonics*. Collecting the terms oscillating at frequency $2\omega_1$ (which is equal to 2ω) from Eqns. 2.7 and 2.8 and adding terms for other possible combinations of indices j and k in χ_{ijk}, we obtain the polarization component oscillating at frequency 2ω to be

$$
\begin{aligned}
P_i^{2\omega} &= \varepsilon_0\{\chi_{iXX}^{(2)}(E_{1X}+E_{2X})^2 + \chi_{iYY}^{(2)}(E_{1Y}+E_{2Y})^2 \\
&\quad + \chi_{iZZ}^{(2)}(E_{1Z}+E_{2Z})^2 \\
&\quad + 2\chi_{iXY}^{(2)}(E_{1X}+E_{2X})(E_{1Y}+E_{2Y}) \\
&\quad + 2\chi_{iXY}^{(2)}(E_{1Y}+E_{2Y})(E_{1Z}+E_{2Z}) \\
&\quad + 2\chi_{iXY}^{(2)}(E_{1Z}+E_{2Z})(E_{1X}+E_{2X})\} \\
\\
&= 2\varepsilon_0\{d_{i1}(E_{1X}+E_{2X})^2 + d_{i2}(E_{1Y}+E_{2Y})^2 \\
&\quad + d_{i3}(E_{1Z}+E_{2Z})^2 \\
&\quad + 2d_{i4}(E_{1Y}+E_{2Y})(E_{1Z}+E_{2Z}) \\
&\quad + 2d_{i5}(E_{1Z}+E_{2Z})(E_{1X}+E_{2X}) \\
&\quad + 2d_{i6}(E_{1X}+E_{2X})(E_{1Y}+E_{2Y})\}
\end{aligned}
$$
$$(2.27)$$

where the contracted notation for d is used and the intrinsic permutation symmetry conditions are invoked. To derive Eqn. 2.27 from Eqns. 2.7 and 2.8, we need to remember that for $\omega_1 = \omega_2$, $2\,\omega_1 = 2\omega_2 = \omega_1 + \omega_2$, so the terms oscillating at $2\,\omega_1$, $2\,\omega_2$ and $\omega_1 + \omega_2$ all need to be included.

Equation 2.27 can be written in matrix form as

$$
\begin{pmatrix} P_X^{2\omega} \\ P_Y^{2\omega} \\ P_Z^{2\omega} \end{pmatrix} = 2\varepsilon_0 \begin{pmatrix} d_{11} & d_{12} & d_{13} & d_{14} & d_{15} & d_{16} \\ d_{21} & d_{22} & d_{23} & d_{24} & d_{25} & d_{26} \\ d_{31} & d_{32} & d_{33} & d_{34} & d_{35} & d_{36} \end{pmatrix} \begin{pmatrix} (E_{1X} + E_{2X})^2 \\ (E_{1Y} + E_{2Y})^2 \\ (E_{1Z} + E_{2Z})^2 \\ 2(E_{1Y} + E_{2Y})(E_{1Z} + E_{2Z}) \\ 2(E_{1Z} + E_{2Z})(E_{1X} + E_{2X}) \\ 2(E_{1X} + E_{2X})(E_{1Y} + E_{2Y}) \end{pmatrix}.
$$
$$(2.28)$$

2.7.4 Optical rectification

The non-oscillating, i.e., zero frequency terms in Eqns. 2.7 and 2.8 give rise to optical rectification. Collecting those terms from the expressions for χ_{iXX} and χ_{iXY}, and adding the terms for other possible combinations j and k indices of χ_{ijk}, we obtain the non-oscillating polarization component generated by one incident field oscillating at frequency ω_1 to be

$$
\begin{aligned}
P_i^0 &= \varepsilon_0 \{ 2\chi_{iXX}^{(2)} |E_{1X}|^2 + 2\chi_{iYY}^{(2)} |E_{1Y}|^2 + 2\chi_{iZZ}^{(2)} |E_{1Z}|^2 \\
&+ \chi_{iXY}^{(2)}(E_{1X}E_{1Y}^* + E_{1X}^*E_{1Y}) + \chi_{iYX}^{(2)}(E_{1Y}E_{1X}^* + E_{1Y}^*E_{1X}) \\
&+ \chi_{iYZ}^{(2)}(E_{1Y}E_{1Z}^* + E_{1Y}^*E_{1Z}) + \chi_{iZY}^{(2)}(E_{1Z}E_{1Y}^* + E_{1Z}^*E_{1Y}) \\
&+ \chi_{iZX}^{(2)}(E_{1Z}E_{1X}^* + E_{1Z}^*E_{1X}) + \chi_{iXZ}^{(2)}(E_{1X}E_{1Z}^* + E_{1X}^*E_{1Z}) \} \\
&= 4\varepsilon_0 \{ d_{i1}|E_{1X}|^2 + d_{i2}|E_{1Y}|^2 + d_{i3}|E_{1Z}|^2 \\
&+ d_{i4}(E_{1Y}E_{1Z}^* + E_{1Y}^*E_{1Z}) + d_{i5}(E_{1Z}E_{1X}^* + E_{1Z}^*E_{1X}) \\
&+ + d_{i6}(E_{1X}E_{1Y}^* + E_{1X}^*E_{1Y}) \}
\end{aligned}
$$
$$(2.29)$$

using the contracted notation for d and the intrinsic permutation symmetry conditions. Eqn. 2.29 can be written in matrix form as

$$
\begin{pmatrix} P_X^0 \\ P_Y^0 \\ P_Z^0 \end{pmatrix} = 4\varepsilon_0 \begin{pmatrix} d_{11} & d_{12} & d_{13} & d_{14} & d_{15} & d_{16} \\ d_{21} & d_{22} & d_{23} & d_{24} & d_{25} & d_{26} \\ d_{31} & d_{32} & d_{33} & d_{34} & d_{35} & d_{36} \end{pmatrix} \begin{pmatrix} |E_{1X}|^2 \\ |E_{1Y}|^2 \\ |E_{1Z}|^2 \\ E_{1Y}E_{1Z}^* + E_{1Y}^*E_{1Z} \\ E_{1Z}E_{1X}^* + E_{1Z}^*E_{1X} \\ E_{1X}E_{1Y}^* + E_{1X}^*E_{1Y} \end{pmatrix}.
$$
$$(2.30)$$

2.7.5 Convention used for numbering the three interacting beams of light

In the frequency mixing processes of non-degenerate SFG, DFG and OPG/OPA/OPO there are three interacting beams with three different frequencies. A convention is adopted here to use the subscripts 1,2 and 3 to refer to the three waves in increasing order of frequency, i.e., 1 will denote the lowest and 3 the highest frequency wave. With this notation, the relationship shown in Eqn. 2.9 is valid not just for sum frequency generation but for all the three wave mixing processes.

SHG can also be considered a *three*-wave mixing process, as it is the degenerate case of SFG, with the frequencies of the two incident beams the same. For the sake of simplicity, we will consider SHG as a *two*-wave mixing process with the subscripts p and s used for the lower and higher frequencies, respectively. Thus, for SHG, the Eqn. 2.9 is to be replaced by

$$\omega_s = \omega_p + \omega_p = 2\omega_p. \tag{2.31}$$

The notation used in this book for three wave mixing processes is summarized in Table 2.5:

Nonlinear process	Incident wave(s)	Generated wave(s)
SFG	ω_1 , ω_2	ω_3
OPG, OPO	ω_3	ω_1 , ω_2
DFG, OPA	ω_3 , ω_2 (or ω_1)	ω_1 (or ω_2)
SHG (Second Harmonic Generation)	ω_p ($= \omega$)	ω_s ($=2\omega$)
ShG (Sub-harmonic Generation)	ω_p ($= \omega$)	ω_s ($=\omega/2$)

TABLE 2.5. The notation used in this book for the various nonlinear frequency mixing processes.

2.7.6 Summary of polarization components for non-degenerate three wave mixing

The processes of SFG, DFG, as well as OPA, OPG and OPO all involve the mixing of three frequencies ω_1, ω_2 and ω_3. With the convention $\omega_1 < \omega_2 < \omega_3$, and $\omega_3 = \omega_1 + \omega_2$, the polarization generated at the three frequencies in each of these processes can be summarized in a compact form.

Say $\mathbf{E}_1$, $\mathbf{E}_2$ and $\mathbf{E}_3$ denote the field amplitudes at these frequencies with unit vectors $\hat{e}(\omega_1)$, $\hat{e}(\omega_2)$ and $\hat{e}(\omega_3)$ along their respective directions, i.e.,

$$\mathbf{E}_1 = \hat{e}(\omega_1)E_1$$
$$\mathbf{E}_2 = \hat{e}(\omega_2)E_2$$
$$\mathbf{E}_3 = \hat{e}(\omega_3)E_3. \tag{2.32}$$

The Cartesian components of the polarization vectors at the three frequencies can then be determined from the SFG and DFG expressions given in Eqns. 2.24

and 2.26 above to be

$$
\begin{aligned}
P_i^{\omega_3} &= 4\varepsilon_0 d_{il} u_l(\omega_1, \omega_2) E_1 E_2 \\
P_i^{\omega_2} &= 4\varepsilon_0 d_{il} u_l(\omega_1, \omega_3) E_1^* E_3 \\
P_i^{\omega_1} &= 4\varepsilon_0 d_{il} u_l(\omega_2, \omega_3) E_2^* E_3
\end{aligned}
\tag{2.33}
$$

where $u(\omega_A, \omega_B)$ is a 6×1 column matrix defined as

$$
u(\omega_A, \omega_B) = \begin{pmatrix}
e_X(\omega_A) e_X(\omega_B) \\
e_Y(\omega_A) e_Y(\omega_B) \\
e_Z(\omega_A) e_Z(\omega_B) \\
e_Y(\omega_A) e_Z(\omega_B) + e_Z(\omega_A) e_Y(\omega_B) \\
e_Z(\omega_A) e_X(\omega_B) + e_X(\omega_A) e_Z(\omega_B) \\
e_X(\omega_A) e_Y(\omega_B) + e_Y(\omega_A) e_X(\omega_B)
\end{pmatrix}.
\tag{2.34}
$$

As before, Einstein summation convention over repeated indices is assumed in Eqns. 2.33).

2.7.7 Summary of polarization components for degenerate three wave mixing (SHG and degenerate parametric mixing)

When frequencies ω_1 and ω_2 are equal we have the case of SHG as well as the degenerate parametric mixing. For SHG, the 'pump' is the low frequency beam at frequency ω_1 and for the degenerate parametric mixing, the higher frequency beam at $2\omega_1$ is designated the 'pump'. To treat both cases together here, we use the notation shown in Table 2.5, i.e., we define $\omega \equiv \omega_1 = \omega_2$ and instead of using the three frequencies ω_1, ω_2 and ω_3, we use the frequencies ω and 2ω. Then for SHG, we will have $\omega_p = \omega$ and $\omega_s = 2\omega$ and for degenerate parametric mixing, $\omega_p = 2\omega$ and $\omega_s = \omega$

Say $\mathbf{E}^\omega$ and $\mathbf{E}^{2\omega}$ denote the field amplitudes at these two frequencies with unit vectors $\hat{e}(\omega)$ and $\hat{e}(2\omega)$ along their respective directions, i.e.,

$$
\begin{aligned}
\mathbf{E}^\omega &= \hat{e}(\omega) E^\omega \\
\mathbf{E}^{2\omega} &= \hat{e}(2\omega) E^{2\omega}.
\end{aligned}
\tag{2.35}
$$

The field $\mathbf{E}^\omega$ is the sum of the fields $\mathbf{E}_1$ and $\mathbf{E}_2$, both oscillating at frequency ω. The fields $\mathbf{E}_1$ and $\mathbf{E}_2$ can be parallel to each other or they may be orthogonal. Both options are kept open at this point and the choice of one or the other in a particular nonlinear crystal is determined by the 'phase matching' considerations, which will be discussed later.

The expression for polarization generated at frequency 2ω given in Eqn. 2.28 is then written as

$$
P_i^{2\omega} = 2\varepsilon_0 d_{il} u_l(\omega, \omega)(E^\omega)^2.
\tag{2.36}
$$

From the DFG expression given in Eqn. 2.26, the nonlinear polarization at the frequency ω is given as

$$P_i^\omega = 4\varepsilon_0 d_{il} u_l(\omega, 2\omega) E(\omega)^* E(2\omega). \tag{2.37}$$

The matrix $u(\omega, \omega)$ in Eqn. 2.36 is obtained from Eqn. 2.34 to be

$$u(\omega, \omega) = \begin{pmatrix} e_X(\omega)^2 \\ e_Y(\omega)^2 \\ e_Z(\omega)^2 \\ 2e_Y(\omega)e_Z(\omega) \\ 2e_Z(\omega)e_X(\omega) \\ 2e_X(\omega)e_Y(\omega) \end{pmatrix}. \tag{2.38}$$

2.8 Frequency Conversion And Phase Matching

Polarizations generated by nonlinear coupling of one or more incident monochromatic light beams give rise to light oscillating at new frequencies different from those initially present. This nonlinear frequency conversion has myriad practical applications and it is important to study the propagation of the incident and generated beams through the nonlinear material in detail, especially to determine the frequency conversion efficiencies that can be obtained from a nonlinear optical material. The frequency conversion processes described above, SFG, DFG, SHG, along with phenomena such as optical parametric oscillation (OPO) will be explored in detail in this book.

Nonlinear frequency mixing is also known as 'parametric frequency conversion' because different frequencies can be created by modulating the parameters of the nonlinear optical material [2]. Before the discovery of lasers and the demonstration of nonlinear optics, such parametric interactions were studied with microwaves. The nonlinearities used in microwaves exist only in junctions of devices and are localized in small regions of space compared to the wavelength. Optical nonlinearities can extend throughout the bulk of materials, i.e., over distances much larger than the wavelengths involved.

A material needs to possess nonzero value of the appropriate d coefficient for frequency conversion to take place in it, so only noncentrosymmetric crystalline solids which can have nonzero values of d coefficient in the bulk are important for efficient frequency conversion in second order materials.

However, having nonzero values of d coefficient is a necessary but not a sufficient condition for efficient frequency conversion. The nonlinear material needs to be of sufficient length and also must have sufficiently low absorption at the interacting frequencies of light. Moreover, the irradiances of the incident frequency (or frequencies) must be sufficiently high and the material must be able to withstand such high irradiances without undergoing damage or substantial changes to its optical properties.

In addition to all the above requirements, a phase matching condition needs to be satisfied for efficient frequency conversion to be achieved. It will be shown later that conversion efficiency rapidly drops off as the phase difference between the oscillating nonlinear polarization and the electric field it gives rise to becomes larger than $\pi/2$. Eqns. 2.3 and 2.4 show that the phase of the polarization wave for sum frequency generation, i.e., the term oscillating at frequency $\omega_1 + \omega_2$, is $(\mathbf{k}_1 + \mathbf{k}_2) \cdot \mathbf{r}$. If n_1 and n_2 denote the refractive indices of the medium at the frequencies ω_1 and ω_2 respectively, the magnitude of $\mathbf{k}_1 + \mathbf{k}_2$ is given by

$$k_1 + k_2 = \frac{n_1\omega_1}{c} + \frac{n_2\omega_2}{c}. \tag{2.39}$$

The phase of the generated electric field, also oscillating at frequency $\omega_1 + \omega_2$ is $k_3 \cdot \mathbf{r}$, where k_3 is the wave-vector of the generated wave, the magnitude of which is given by

$$k_3 = \frac{n_3\omega_3}{c} \tag{2.40}$$

where n_3 is the refractive index of the nonlinear medium at the frequency $\omega_3 = \omega_1 + \omega_2$.

Assuming the two incident light beams and the generated polarization (and the generated light beam) are all propagating in the same direction, say, the z direction, the three vectors k_1, k_2 and k_3 are collinear and the phase difference between the polarization and the electric field at frequency ω_3 for a distance z traveled in the nonlinear medium is given by

$$\begin{aligned}
\Delta\phi &= (k_3 - k_1 - k_2)z \tag{2.41} \\
&= \left(\frac{n_3\omega_3}{c} - \frac{n_1\omega_1}{c} - \frac{n_2\omega_2}{c} \right) z \\
&= \left(n_3 - n_1\frac{\omega_1}{\omega_3} - n_2\frac{\omega_2}{\omega_3} \right) \frac{\omega_3 z}{c} \\
&= (n_3 - n_{12})\frac{\omega_3 z}{c} \tag{2.42}
\end{aligned}$$

where

$$n_{12} \equiv n_1\frac{\omega_1}{\omega_3} + n_2\frac{\omega_2}{\omega_3} = n_1\frac{\lambda_3}{\lambda_1} + n_2\frac{\lambda_3}{\lambda_2}. \tag{2.43}$$

Since $\omega_3 = \omega_1 + \omega_2$, i.e., both frequencies ω_1 and ω_2 can range from 0 to ω_3, n_{12} takes the values from n_1 to n_2. In a normally dispersive isotropic medium, the refractive index increases with increasing frequency, so n_3 is larger than n_1 and n_2, i.e., n_3 is larger than n_{12}.

It will be shown in Chapter 5 that for plane waves and collimated beams, in material with given values of n_1, n_2 and n_3, as z increases from 0, the frequency conversion efficiency also increases, reaching a maximum at $\Delta\phi = \pi$. The value of z at which this maximum is reached is called the *coherence length* and it is denoted by l_c. Thus,

$$
\begin{aligned}
l_c &= \pi/(k_3 - k_1 - k_2) \\
&= \frac{\pi c}{\omega_3(n_3 - n_{12})} \\
&= \frac{\lambda_3}{2(n_3 - n_{12})} = \frac{1}{2\left(\dfrac{n_3}{\lambda_3} - \dfrac{n_2}{\lambda_2} - \dfrac{n_1}{\lambda_1}\right)}.
\end{aligned}
\tag{2.44}
$$

For the case of SHG, $\omega_1 = \omega_2 = \omega_p$ (say) and $\omega_3 = \omega_s = 2\,\omega_p$, so that $n_{12} = n(\omega_p)$ and $n_3 = n(2\omega_p)$. The coherence length for the SHG case is therefore given by

$$
l_c(\text{SHG}) = \frac{\lambda_p}{4\{n(2\omega_p) - n(\omega_p)\}}
\tag{2.45}
$$

since for SHG $\lambda_3 = \lambda_s = \lambda_p/2$.

For frequencies of light in the visible through infrared spectrum, and for usual optical and infrared materials with normal dispersion, l_c ranges from a few to at most a few hundreds of micrometers. The available material lengths in nonlinear crystals are typically a centimeter or more. Thus only a small fraction of the available length of a material is utilized for frequency conversion in such materials - unless special care is taken to match the phases of the generating polarization and the generated beam.

To utilize the whole length of such crystals to generate efficient frequency conversion the technique of 'phase matching' is used. One way to achieve phase matching is using the birefringence in anisotropic crystals to chose appropriate refractive indices of the interacting beams by variation of the propagation direction and/or temperature, to make the $n_3 - n_{12}$ term in the denominator of Eqn. 2.44 very small, thereby increasing the value of l_c by many orders of magnitude. Another method, known as quasi phase matching can be used for materials having no birefringence as will be discussed in detail in Chapter 5. Here we describe the technique of birefringent phase matching in anisotropic crystals.

2.8.1   Phase matching in birefringent crystals

For three collinearly propagating waves, the phase matching condition i.e., $n_3 - n_{12} = 0$, can be achieved by appropriate choice of the polarizations. Each of the three waves in a frequency mixing process can propagate in the anisotropic crystal as a *slow* wave or a *fast* wave. However, in a medium with normal dispersion, the phase matching condition cannot be satisfied if the highest frequency wave is a *slow* wave. This can be seen from Figs. 2.1 and 2.2 where n_{s1}, n_{s2} and n_{s3} denote the values of n_s at the frequencies ω_1, ω_2, and ω_3, respectively and n_{f1}, n_{f2} and n_{f3} are the corresponding values of n_f, all with the **k** vectors along some propagation direction. Since n_1 takes the values from n_{f1} to n_{s1}, and n_2 takes the values from n_{f2} to n_{s2}, n_{12} takes the values from n_{f1} to n_{s2}. In a medium with normal dispersion, for a given propagation direction, n_{s3} is always greater than n_{s2}, making it impossible to achieve the phase matching condition if the highest frequency wave is a *slow* wave. However, if the highest frequency wave is traveling as a *fast* wave, phase matching *can* be obtained with the two lower frequency waves being both *slow* waves, or with one the *slow* and the other the *fast* wave as shown in Figs. 2.1 and 2.2.

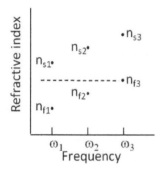 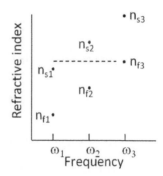

FIGURE 2.1: Ordering of the refractive indices of the slow and fast waves at three interacting frequencies.

FIGURE 2.2: Another possible ordering of the refractive indices of the slow and fast waves at three interacting frequencies.

Following the definition given in Dmitriev [2] (page 10), when both the lower frequency waves are *slow* waves, the interaction is defined to be of Type I. When the polarization directions of the two lower frequency waves are such that one is a slow wave and the other is a fast wave, the interaction is defined to be of Type II. This is summarized in Table 2.6

Type	ω_1	ω_2	ω_3
I (ssf)	slow	slow	fast
II (sff)	slow	fast	fast
II (fsf)	fast	slow	fast

TABLE 2.6. The three different phase matching conditions ssf, sff and fsf, with s and f indicating the slow and the fast waves and with ordering of the three interacting waves arranged from the lowest to highest frequencies. In a normally birefringent homogeneous medium, phase matching cannot be obtained if the highest frequency is a slow wave.

In terms of the wavelengths of light λ_1, λ_2 and λ_3 at the three interacting frequencies (in descending order of wavelength values), the phase matching condition, $n_3 - n_{12} = 0$ can be rewritten as

$$\frac{n_3}{\lambda_3} - \frac{n_2}{\lambda_2} - \frac{n_1}{\lambda_1} = 0. \qquad (2.46)$$

The phase matching angles in a given material can be determined using the following procedure: for a given material and for given values of λ_1 and λ_2 (with $\lambda_1 > \lambda_2 > \lambda_3$ and $\lambda_3 = \lambda_1\lambda_2/(\lambda_1 + \lambda_2)$) the values of n_X, n_Y and n_Z are found at all three wavelengths (usually from the Sellmeier equations). Expressions for $\mathcal{A}$, $\mathcal{B}$, $\mathcal{C}$ for arbitrary values of θ, ϕ and λ are then obtained from Eqns. 1.55 as

$$
\begin{aligned}
\mathcal{A}(\theta,\phi,\lambda) &= m_X^2 n_X(\lambda)^2 + m_Y^2 n_Y(\lambda)^2 + m_Z^2 n_Z(\lambda)^2 \\
\mathcal{B}(\theta,\phi,\lambda) &= m_X^2 n_X(\lambda)^2 (n_Y(\lambda)^2 + n_Z(\lambda)^2) + m_Y^2 n_Y(\lambda)^2 (n_Z(\lambda)^2 + n_X(\lambda)^2) \\
&\quad + m_Z^2 n_Z(\lambda)^2 (n_X(\lambda)^2 + n_Y(\lambda)^2) \\
\mathcal{C}(\lambda) &= n_X(\lambda)^2 \, n_Y(\lambda)^2 \, n_Z(\lambda)^2
\end{aligned}
$$

$$(2.47)$$

from which the expression for $\mathcal{D}$, defined as positive and real, are obtained:

$$\mathcal{D}(\theta,\phi,\lambda) = \{\mathcal{B}(\theta,\phi,\lambda)^2 - 4\mathcal{A}(\theta,\phi,\lambda)\,\mathcal{C}(\lambda)\}^{1/2}. \qquad (2.48)$$

The expressions for m_X, m_Y and m_Z in terms of the angles θ and ϕ are given in Eqns. 1.64.

From Eqns. 1.57, the expressions for the slow and the fast components of the refractive index are then

$$n_s(\theta,\phi,\lambda) = \left(\frac{\mathcal{B}(\theta,\phi,\lambda) + \mathcal{D}(\theta,\phi,\lambda)}{2\mathcal{A}(\theta,\phi,\lambda)}\right)^{1/2}$$

$$n_f(\theta,\phi,\lambda) = \left(\frac{\mathcal{B}(\theta,\phi,\lambda) - \mathcal{D}(\theta,\phi,\lambda)}{2\mathcal{A}(\theta,\phi,\lambda)}\right)^{1/2}. \qquad (2.49)$$

With these values of n_s and n_f, we define a function $f(\theta,\phi)$ with the expressions given in Table 2.7 for the three possible cases with the two types of phase matching:

For each type of three wave mixing interaction, and each set of λ_1, λ_2 and λ_3, the phase matching angle corresponds to the value of θ and ϕ at which the function $f(\theta,\phi)$ is zero. For a given type (say ssf) if $f(\theta,\phi)$ is not equal to 0 for any value of θ or ϕ, then there is no phase matching for that type of interaction. In cases where phase matching is possible, the root of the function $f(\theta,\phi)$ with respect to θ can be obtained using a computer program for each value of ϕ varying from 0 to 90°. The value of the root, designated θ_{pm} is the phase matching angle which depends on the angle ϕ. A plot of θ_{pm} vs ϕ provides the phase matching angle curve. A computer program that calculates the value of θ_{pm} for specific nonlinear optical crystals with available refractive index values can be easily written. The program SNLO available on the internet provides the phase matching angles for a large number of crystals,

Type	$f(\theta, \phi)$
I (ssf)	$\dfrac{n_s(\theta, \phi, \lambda_1)}{\lambda_1} + \dfrac{n_s(\theta, \phi, \lambda_2)}{\lambda_2} - \dfrac{n_f(\theta, \phi, \lambda_3)}{\lambda_3}$
II (sff)	$\dfrac{n_s(\theta, \phi, \lambda_1)}{\lambda_1} + \dfrac{n_f(\theta, \phi, \lambda_2)}{\lambda_2} - \dfrac{n_f(\theta, \phi, \lambda_3)}{\lambda_3}$
II (fsf)	$\dfrac{n_f(\theta, \phi, \lambda_1)}{\lambda_1} + \dfrac{n_s(\theta, \phi, \lambda_2)}{\lambda_2} - \dfrac{n_f(\theta, \phi, \lambda_3)}{\lambda_3}$

TABLE 2.7. The definition of the function $f(\theta, \phi)$ for the three types of interaction, ssf, sff and fsf.

2.8.2 Calculation of phase matching angles

As an example, we consider phase matching in the biaxial crystal KTiOPO$_4$, Potassium Titanyl Phosphate (KTP). For flux grown KTP crystals, the Sellmeier equations for n_X, n_Y and n_Z at room temperature are given in Ref. [2] as:

$$n_X(\lambda)^2 = 3.0065 + \frac{0.03901}{\lambda^2 - 0.04251} - 0.01327\lambda^2$$

$$n_Y(\lambda)^2 = 3.0333 + \frac{0.04154}{\lambda^2 - 0.04547} - 0.01408\lambda^2$$

$$n_Z(\lambda)^2 = 3.3134 + \frac{0.05694}{\lambda^2 - 0.05658} - 0.01682\lambda^2. \tag{2.50}$$

Inserting Eqns. 2.50 and the expressions for m_X, m_Y and m_Z in Eqns. 2.47 and 2.48 we obtain explicit expressions for $n_s(\theta, \phi, \lambda)$ and $n_s(\theta, \phi, \lambda)$ from Eqns. 2.49.

Inserting the expressions for n_s and n_f in the expressions for $f(\theta, \phi)$ given in Table 2.7 for the three cases, the phase matching angles θ_{pm}, in the cases where phase matching is possible, are determined as functions of ϕ.

To obtain numerical results, we consider a case of three wave mixing interaction with $\lambda_1 = 3.393$ μm and $\lambda_2 = 1.55$ μm, so that $\lambda_3 = 1.064$ μm. The nonlinear process could be sum frequency generation, difference frequency generation or optical parametric generation - the phase matching angles would be the same for any of these cases.

For interaction of these three wavelengths in KTP, the cases of ssf and fsf are found to be phase-matched for all values of ϕ, the case of sff is phase-matched for ϕ ranging from 40° to 90°, and the values of the phase matching angle θ_{pm} vs ϕ are shown in Fig. 2.3.

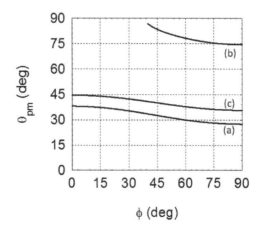

FIGURE 2.3: The values of the phase matching angle θ_{pm} as a function of ϕ for three wave interaction in KTP, with $\lambda_1 = 3.393$ μm, $\lambda_2 = 1.55$ μm and $\lambda_3 = 1.064$ μm. (a) ssf (b) fsf and (c) sff.

2.9 Walk-Off Angles

Phase matching is only a necessary and not a sufficient condition for obtaining efficient frequency conversion in crystals. Several other criteria, such as small walk-off angle between the interacting beams, low absorption at all three interacting wavelengths and sufficiently high value of the effective nonlinear optical susceptibility (d_{eff}) must also be satisfied to obtain appreciable conversion efficiency.

Walk-off angles ρ_s and ρ_f of the slow and the fast waves are given in Eqns. 1.60. Because of dispersion of the principal refractive indices, ρ_s and ρ_f are wavelength dependent, i.e., the walk-off angles at the three interacting wavelengths are dependent on the polarizations (in the *slow* or *fast* directions) as well on as the wavelengths.

2.9.1 Calculation of walk-off angles in the phase matched case in KTP

Assuming again a three wave mixing interaction in KTP, with $\lambda_1 = 3.393$ μm, $\lambda_2 = 1.55$ μm, so that $\lambda_3 = 1.064$ μm, the walk-off angles $\rho_s(\lambda)$ and $\rho_f(\lambda)$ at the three wavelengths are calculated for the three phase matched cases (ssf, fsf and sff) using Eqns. 1.60. The angles $\rho_s(\lambda)$ and $\rho_f(\lambda)$ of course depend on the angles θ and ϕ because of the dependence of n_s and n_f on these angles. We consider the three special cases of propagation along the principal planes ZX, YZ and XY and then the case of propagation along a general direction.

Propagation along the ZX plane

As discussed in Chapter 1 and as shown in Table 1.4, the optic axis separates the ZX plane into two distinct regions, $\theta < \Omega$ and $\theta > \Omega$. Since for KTP, $n_X < n_Y < n_Z$, Table 1.4 shows that for $\theta < \Omega$, $\rho_s = 0$, $\rho_f = \rho_5$ and for $\theta > \Omega$, $\rho_f = 0$, $\rho_s = \rho_5$. Figures 2.4 and 2.5 are plots of the angles ρ_s and ρ_f for the three wavelengths λ_1, λ_2 and λ_3. Because of dispersion, the values of the angle Ω for the three wavelengths are different, although the small difference is not apparent from the figure. For the three wavelengths under consideration here, Ω is 19.9°, 19.4° and 19. 6°, in order of decreasing wavelength.

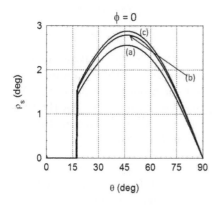

FIGURE 2.4: The walk-off angles ρ_s as a function of the polar angle θ for propagation along the ZX plane, i.e., with $\phi = 0$. (a) $\lambda = 3.393$ μm, (b) $\lambda = 1.55$ μm and (c) $\lambda = 1.06$ μm.

FIGURE 2.5: The walk-off angles ρ_f as a function of the polar angle θ for propagation along the ZX plane, i.e., with $\phi = 0$. (a) $\lambda = 3.393$ μm, (b) $\lambda = 1.55$ μm and (c) $\lambda = 1.064$ μm.

Propagation along the YZ plane

For KTP, when the propagation vector lies in the $Y-Z$ plane, i.e., for $\phi = 90°$, Table 1.3 shows that $\rho_s = \rho_4$ and $\rho_f = 0$. Plots of the angle ρ_s for the three wavelengths λ_1, λ_2 and λ_3 are shown in Fig. 2.6.

Propagation along the XY plane

When the propagation vector of light traveling in a KTP crystal lies in the $X-Y$ plane, i.e., when $\theta = 90°$, Table 1.3 shows that $\rho_s = 0$ and $\rho_f = \rho_6$. Plots of the angle ρ_f for the three wavelengths λ_1, λ_2 and λ_3 as a function of ϕ are shown in Figs. 2.7.

Propagation in a general direction

When the propagation vector of light traveling in a KTP crystal lies in a general direction which is not along any of the principal axes or principal planes, the walk-off angles ρ_s and ρ_f are determined from Eqns. 1.60. The values of ρ_s and ρ_f are plotted as functions of θ in Figs. 2.8 and Figs. 2.9 for

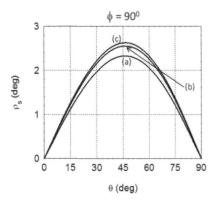

FIGURE 2.6: The walk-off angle ρ_s as a function of the polar angle θ for propagation along the YZ plane, i.e., with $\phi = 90°$ for the three wavelengths: (a) 3.393 μm, (b) 1.55 μm and (c) 1.064 μm.

FIGURE 2.7: The walk-off angle ρ_s as a function of the azimuthal angle ϕ for for propagation along the XY plane, i.e., with $\theta = 90°$ for the three wavelengths: (a) 3.393 μm, (b) 1.55 μm and (c) 1.064 μm.

ϕ equal to 15°, 45° and 75° for the wavelength of 3.393 μm. For other two wavelengths the results are similar and they have not been plotted.

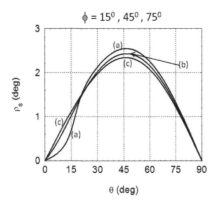

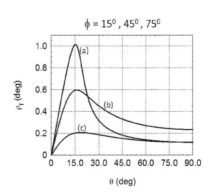

FIGURE 2.8: The walk-off angle ρ_s as a function of the polar angle θ for wavelength of 3.393 μm. (a) $\phi = 15°$, (b) $\phi = 45°$ and (c) $\phi = 75°$.

FIGURE 2.9: The walk-off angle ρ_f as a function of the polar angle θ for wavelength of 3.393 μm. (a) $\phi = 15°$, (b) $\phi = 45°$ and (c) $\phi = 75°$.

Bibliography

[1] P. A. M. Dirac, The Quantum Theory of the Emission and Absorption of Radiation, *Proc. R. Soc. Lond. A* **114**, 243–265, 1927.

[2] M. Göppert-Mayer, Über Elementaraktemit zwei Quantensprungen, *Ann. Phys.* **9**, 273–295, 1931.

[3] P. A. Franken, A. E. Hill, C. W. Peters and G. Weinreich, Generation of optical harmonics, *Phys. Rev. Lett.* **7**, 118–119, 1961.

[4] W. Kaiser and C. G. B. Garrett, Two-photon excitation in $CaF_2:Eu^{2+}$, *Phys. Rev. Lett.* **7**, 229–231, 1961.

[5] D. A. Kleinman, Nonlinear Dielectric Polarization in Optical Media, *Phys. Rev.* **126**, 1977–1979, 1962.

[6] J. A. Armstrong, N. Bloembergen, J. Ducuing and P. S. Pershan, Interactions between Light Waves in a Nonlinear Dielectric, *Phys. Rev.* **127**, 1918, 1962.

[7] W. Nernst, Über einen Versuch, von quantentheoretischen Betrachtungen zur Annahme stetiger Energieänderungen zurückzukehren, *Verhandlungen der Deutschen Physikalischen Gesellschaften* **18**, 83–116, 1916.

[8] R.W. Boyd, *Nonlinear Optics*, Academic Press, San diego, 1992.

[9] F. Zernike and J. E. Midwinter, *Applied Nonlinear Optics*, John Wiley and Sons, New York, 1973.

[10] J. F. Nye, *Physical Properties of Crystals*, Oxford University Press, London, 1964.

[11] V. G. Dmitriev, G. G. Gurzadyan, D. N. Nikogosyan, *Handbook of Nonlinear Optical Crystals*, Springer, Berlin, 1999.

3

Effective d coefficient for Three-Wave mixing Processes

> This chapter
>
> 1. Introduces the 'effective' second order nonlinear susceptibility coefficient, d_{eff}
>
> 2. Provides the coefficients of all 18 elements of the d matrix in the expression for d_{eff} for three cases of phase matched three wave interaction for propagation along principal planes of a biaxial crystal
>
> 3. Provides the coefficients of all 18 elements of the d matrix in the expression for d_{eff} for three cases of phase matched three wave interaction for propagation in an uniaxial crystal
>
> 4. Provides the expressions for d_{eff} for propagation along principal planes of some biaxial crystals
>
> 5. Provides the expressions for d_{eff} for propagation in some uniaxial crystals
>
> 6. Provides the expressions for d_{eff} for propagation in isotropic crystals

3.1 Introduction

The second order nonlinear susceptibility d introduced in Chapter 2 is a third rank tensor with 27 components. Intrinsic permutation symmetry reduces the number of independent tensor components to 18 and allows the use of a 3×6 matrix (denoted by d_{ij}), to represent the d tensor. Various material symmetry conditions further reduce the number of independent nonzero elements of the d_{ij} matrix, typically to 5 or fewer. But even after this reduction, for each of the non-zero matrix elements, the nonlinear polarization given in Eqn. 2.4 needs to be summed over the three Cartesian coordinates $(X, Y$ and $Z)$ and for all the different polarization directions of the three interacting waves, making the expression for the total polarization quite long and cumbersome.

All three interacting waves propagate through the nonlinear crystal as either a slow or a fast wave. This fact can be used to considerably simplify the polarization expression, by reducing the tensor relation between the three vectors (the generated polarization vector and the two incident electric field vectors) to a scalar relationship between the appropriate components in the polarization directions of the fast or slow waves. The summation over the coordinates X, Y and Z then allows replacement of the complicated polarization expression by a single term.

This important simplification is obtained by introducing an 'effective d coefficient', denoted by d_{eff}. It is a sum over the nonzero values of the d matrix, weighted by the applicable angular projections of the components of the unit vectors $\hat{e}_s$ and $\hat{e}_f$ in the directions of the principal dielectric axes. The d_{eff} term depends on the components of the d_{ij} matrix, the propagation direction of the **k** vector (through the angles θ and ϕ), and the walk-off angles ρ_s or ρ_f (which in turn depend on the principal refractive indices n_X, n_Y and n_Z and the angles θ and ϕ). In this chapter the d_{eff} values for biaxial and uniaxial crystals are presented for the cases of birefringent phase matched three wave mixing processes, i.e., with the highest frequency beam of light traveling as a 'fast' wave in the crystal. d_{eff} values for the biaxial crystal class with point group mm2 are given in detail in Ref. [2]. Here we present the expressions for the d_{eff} values for *all* the biaxial crystal classes, along with the effect of dispersion on the polarization directions of the electric field components. However, to avoid complication, only the case of $(X, Y, Z) \equiv (a, b, c)$ between the dielectric and crystallographic axes is considered here. Other cases of assignments can be worked out from the components of the d matrix presented in Table 2.4 (page 27) of Ref. [2].

In isotropic crystals with dispersion, birefringent phase matching is not possible, but highly efficient three-wave mixing can be accomplished through 'quasi-phase matching'. The dependence of the d_{eff} value for isotropic crystals on the angles θ and ϕ of the propagation vector are also presented in this chapter.

3.1.1 Definition of d_{eff}

The slow and the fast components of the electric field vector can be written as

$$\mathbf{E_s} = \hat{e}_s E_s \qquad \mathbf{E_f} = \hat{e}_f E_f \qquad (3.1)$$

where E_s and E_f denote the scalar amplitudes. Since the generated light also propagates as a *slow* or a *fast* wave, only the components of the generated polarization in the directions of $\hat{e}_s$ and $\hat{e}_f$ are of interest. If **P** denotes the polarization vector, its projections P_s and P_f in the directions of $\hat{e}_s$ and $\hat{e}_f$ are given by the dot products

$$P_s = \hat{e}_s \cdot \mathbf{P} = e_{si} P_i \qquad P_f = \hat{e}_f \cdot \mathbf{P} = e_{fi} P_i \qquad (3.2)$$

where the summation convention over the repeated index is assumed here and elsewhere in this chapter).

Expressions for the Cartesian components P_i of the nonlinear polarization at the three interacting frequencies were given in Eqns. 2.33, 2.36 and 2.37. As an example, $P_i(\omega_3)$ is given by

$$P_i(\omega_3) = 4\varepsilon_0 d_{il} u_l(\omega_1, \omega_2)E(\omega_1)E(\omega_2). \tag{3.3}$$

The component of $P_i(\omega_3)$ in the direction of $\hat{e}_s$ is then given by

$$P_s(\omega_3) = 4\varepsilon_0[e_{si}d_{il}u_l(\omega_1, \omega_2)]E(\omega_1)E(\omega_2) \tag{3.4}$$

using Eqn. 3.2. The expression within the square brackets in Eqn. 3.4 is called the 'effective nonlinearity' d_{eff} for this interaction, so that in this case

$$d_{\text{eff}}(\omega_3) = e_{si}d_{il}u_l(\omega_1, \omega_2). \tag{3.5}$$

Expressions for d_{eff} at the three frequencies in the non-degenerate and the degenerate three wave mixing cases are presented in the next two subsections. Lastly, coefficients of the eighteen d matrix components in the expression for d_{eff} are tabulated for the different cases of phase matching.

3.1.2 Effective nonlinearity for nondegenerate three wave mixing processes

The Cartesian components of the nonlinear polarizations at the three interacting frequencies ω_1, ω_2 and ω_3 with $\omega_1 + \omega_2 = \omega_3$ can be obtained from the sum and difference frequency generation equations (Eqns. 2.23 2.26):

$$\begin{aligned} P_i(\omega_3) &= 4\varepsilon_0 d_{il} u_l(\omega_1, \omega_2)E(\omega_1)E(\omega_2) \\ P_i(\omega_2) &= 4\varepsilon_0 d_{il} u_l(\omega_1, \omega_3)E^*(\omega_1)E(\omega_3) \\ P_i(\omega_1) &= 4\varepsilon_0 d_{il} u_l(\omega_2, \omega_3)E^*(\omega_2)E(\omega_3). \end{aligned} \tag{3.6}$$

For an ssf type interaction, the polarization component $P_f(\omega_3)$, $P_s(\omega_1)$ and $P_s(\omega_2)$ are then obtained using Eqns. 3.6 and 3.2

$$\begin{aligned} P_f(\omega_3) &= 4\varepsilon_0 e_{fi}(\omega_3)d_{il} u_l^{ss}(\omega_1, \omega_2)E(\omega_1)E(\omega_2) \\ P_s(\omega_1) &= 4\varepsilon_0 e_{si}(\omega_1)d_{il} u_l^{sf}(\omega_2, \omega_3)E(\omega_2)^* E(\omega_3) \\ P_s(\omega_2) &= 4\varepsilon_0 e_{si}(\omega_2)d_{il} u_l^{sf}(\omega_1, \omega_3)E(\omega_1)^* E(\omega_3) \end{aligned}$$

$$\tag{3.7}$$

where the appropriate superscripts are added to the column matrix u. For example,

$$u^{ss}(\omega_1, \omega_2) = \begin{pmatrix} e_{sX}(\omega_1)e_{sX}(\omega_2) \\ e_{sY}(\omega_1)e_{sY}(\omega_2) \\ e_{sZ}(\omega_1)e_{sZ}(\omega_2) \\ e_{sY}(\omega_1)e_{sZ}(\omega_2) + e_{sZ}(\omega_1)e_{sY}(\omega_2) \\ e_{sZ}(\omega_1)e_{sX}(\omega_2) + e_{sX}(\omega_1)e_{sZ}(\omega_2) \\ e_{sX}(\omega_1)e_{sY}(\omega_2) + e_{sY}(\omega_1)e_{sX}(\omega_2) \end{pmatrix} \tag{3.8}$$

and

$$u^{sf}(\omega_1,\omega_2) = \begin{pmatrix} e_{sX}(\omega_1)e_{fX}(\omega_2) \\ e_{sY}(\omega_1)e_{fY}(\omega_2) \\ e_{sZ}(\omega_1)e_{fZ}(\omega_2) \\ e_{sY}(\omega_1)e_{fZ}(\omega_2) + e_{sZ}(\omega_1)e_{fY}(\omega_2) \\ e_{sZ}(\omega_1)e_{fX}(\omega_2) + e_{sX}(\omega_1)e_{fZ}(\omega_2) \\ e_{sX}(\omega_1)e_{fY}(\omega_2) + e_{sY}(\omega_1)e_{fX}(\omega_2) \end{pmatrix}. \tag{3.9}$$

Equation 3.7 can be written in a simpler form as

$$\begin{aligned} P_f(\omega_3) &= 4\varepsilon_0 d_{\text{eff}}(\omega_3)E(\omega_1)E(\omega_2) \\ P_s(\omega_1) &= 4\varepsilon_0 d_{\text{eff}}(\omega_1)E(\omega_2)^*E(\omega_3) \\ P_s(\omega_2) &= 4\varepsilon_0 d_{\text{eff}}(\omega_2)E(\omega_1)^*E(\omega_3) \end{aligned} \tag{3.10}$$

where

$$\begin{aligned} d_{\text{eff}}(\omega_3) &= e_{fi}(\omega_3)d_{il}u_l^{ss}(\omega_1,\omega_2) \\ d_{\text{eff}}(\omega_1) &= e_{si}(\omega_1)d_{il}u_l^{sf}(\omega_2,\omega_3) \\ d_{\text{eff}}(\omega_2) &= e_{si}(\omega_2)d_{il}u_l^{sf}(\omega_1,\omega_3). \end{aligned} \tag{3.11}$$

The d_{eff} values for the other two types of birefringently phase-matched interactions (sff and fsf) can be determined similarly. Results for all three interaction types are summarized in Table 3.1

Type	$d_{\text{eff}}(\omega_1)$	$d_{\text{eff}}(\omega_2)$	$d_{\text{eff}}(\omega_3)$
I (ssf)	$e_{si}(\omega_1)d_{il}u_l^{sf}(\omega_2,\omega_3)$	$e_{si}(\omega_2)d_{il}u_l^{sf}(\omega_1,\omega_3)$	$e_{fi}(\omega_3)d_{il}u_l^{ss}(\omega_1,\omega_2)$
II (sff)	$e_{si}(\omega_1)d_{il}u_l^{ff}(\omega_2,\omega_3)$	$e_{fi}(\omega_2)d_{il}u_l^{sf}(\omega_1,\omega_3)$	$e_{fi}(\omega_3)d_{il}u_l^{sf}(\omega_1,\omega_2)$
II (fsf)	$e_{fi}(\omega_1)d_{il}u_l^{sf}(\omega_2,\omega_3)$	$e_{si}(\omega_2)d_{il}u_l^{ff}(\omega_1,\omega_3)$	$e_{fi}(\omega_3)d_{il}u_l^{fs}(\omega_1,\omega_2)$

TABLE 3.1. Expressions for d_{eff} for the three types of phase matching

3.1.3 Effective nonlinearity for the degenerate three wave mixing process

The processes of SHG and subharmonic generation constitute the degenerate cases of the three wave mixing process. In this case, the two lower frequencies ω_1 and ω_2 are equal to each other (and named ω) so that the two frequencies of interest are ω and 2ω.

As in the non-degenerate three wave mixing cases described in the last section, for birefringent phase matching to be obtained in the *degenerate* three wave mixing cases, the highest frequency beam (at frequency 2ω) must be traveling as the *fast* wave. However, the directions of oscillation of the two fields $\mathbf{E}_1$ and $\mathbf{E}_2$ can still be different - their degeneracy is only in frequency.

When the polarization directions of the two lower frequency fields $\mathbf{E}_1$ and $\mathbf{E}_2$ are the same the phase matching is said to be of Type I. When the polarization directions of the fields $\mathbf{E}_1$ and $\mathbf{E}_2$ are orthogonal, the phase matching is said to be of Type II. Below we consider these two cases separately.

3.1.4 Type I degenerate three wave mixing process

Since the higher frequency beam of light must be a fast wave for birefringent phase matching to occur, the two lower frequency beams in a type I process must both be *slow* waves. The unit vector $\hat{e}(\omega)$ in Eqn. 2.35 is equal to the unit vector for the slow wave at frequency ω, i.e.,

$$\hat{e}(\omega) = \hat{e}_s(\omega). \tag{3.12}$$

The polarization components oscillating at frequencies 2ω and ω are respectively given by

$$P_i(2\omega) = 2\varepsilon_0 d_{il} u_l^{ss}(\omega, \omega) E(\omega)^2$$
$$P_i(\omega) = 4\varepsilon_0 d_{il} u_l^{sf}(\omega, 2\omega) E(\omega)^* E(2\omega) \tag{3.13}$$

where the elements of the matrix u are given in Eqns. 3.8 and 3.9. Equations 3.13 can be written in a simpler form as

$$P_f(2\omega) = 2\varepsilon_0 d_{\text{eff}}(2\omega) E(\omega)^2$$
$$P_s(\omega) = 4\varepsilon_0 d_{\text{eff}}(\omega) E(\omega_2)^* E(2\omega) \tag{3.14}$$

where

$$d_{\text{eff}}(2\omega) = e_{fi}(2\omega) d_{il} u_l^{ss}(\omega, \omega)$$
$$d_{\text{eff}}(\omega) = e_{si}(\omega) d_{il} u_l^{sf}(\omega, 2\omega). \tag{3.15}$$

3.1.5 Type II degenerate three wave mixing process

If the two lower frequency waves in a three wave mixing process are orthogonal to each other in polarization, the phase matching process is said to be of Type II [2]. Light traveling in an arbitrary direction in an anisotropic medium can have the electric field polarization directions only along $\hat{e}_s$ or $\hat{e}_f$. We assume here that the light beam at frequency ω, with amplitude given by say $\mathbf{E}(\omega)$, is the resultant of two beams with frequency ω, but with their electric field polarization directions along $\hat{e}_s$ and $\hat{e}_f$. Suppose $\hat{e}^r(\omega)$ denotes the unit vector along the resultant $\mathbf{E}^\omega$ and that $\hat{e}^r(\omega)$ makes an angle ψ with the unit vector $\hat{d}_s$, so that

$$\hat{e}^r(\omega) = \cos\psi \, \hat{d}_s(\omega) + \sin\psi \, \hat{d}_f(\omega). \tag{3.16}$$

The polarization components oscillating at frequencies 2ω and ω are then respectively given by

$$P_i(2\omega) = 2\varepsilon_0 d_{il} u_l^{rr}(\omega, \omega) E(\omega)^2$$
$$P_i(\omega) = 4\varepsilon_0 d_{il} u_l^{rf}(\omega, 2\omega) E(\omega)^* E(2\omega) \tag{3.17}$$

where the elements of the matrix u^{rr} and u^{rf} are given in Eqns. 3.18 and 3.19.

$$u^{rr}(\omega, \omega) = \begin{pmatrix} \{e^r_X(\omega)\}^2 \\ \{e^r_Y(\omega)\}^2 \\ \{e^r_Z(\omega)\}^2 \\ 2e^r_Y(\omega)e^r_Z(\omega) \\ 2e^r_Z(\omega)e^r_X(\omega) \\ 2e^r_X(\omega)e^r_Y(\omega) \end{pmatrix} \qquad (3.18)$$

and

$$u^{rf}(\omega, 2\omega) = \begin{pmatrix} e^r_X(\omega)e_{fX}(2\omega) \\ e^r_Y(\omega)e_{fY}(2\omega) \\ e^r_Z(\omega)e_{fZ}(2\omega) \\ e^r_Y(\omega)e_{fZ}(2\omega) + e^r_Z(\omega)e_{fY}(2\omega) \\ e^r_Z(\omega)e_{fX}(2\omega) + e^r_X(\omega)e_{fZ}(2\omega) \\ e^r_X(\omega)e_{fY}(2\omega) + e^r_Y(\omega)e_{fX}(2\omega) \end{pmatrix}. \qquad (3.19)$$

The component e^r_X is given by $d_{sX}(\omega) \cos \psi + d_{fX}(\omega) \sin \psi$. Equations 3.17 can be written in a simpler form as

$$\begin{aligned} P_f(2\omega) &= 2\varepsilon_0 d_{\text{eff}}(2\omega) E(\omega)^2 \\ P_s(\omega) &= 4\varepsilon_0 d_{\text{eff}}(\omega) E(\omega_2)^* E(2\omega) \end{aligned} \qquad (3.20)$$

where

$$\begin{aligned} d_{\text{eff}}(2\omega) &= e_{fi}(2\omega) d_{il} u^{rr}_l(\omega, \omega) \\ d_{\text{eff}}(\omega) &= e_{si}(\omega) d_{il} u^{rf}_l(\omega, 2\omega). \end{aligned} \qquad (3.21)$$

Since the d_{eff} values are expressed as sums over the d coefficients in the 18 element matrix, it is important to know which of these elements are zero. For materials having a center of symmetry, like liquids and gases and centrosymmetric solids, all elements of the d matrix are zero. Among noncentrosymmetric crystals, all 18 elements of the d matrix are non-zero and in general unequal only for triclinic crystals, having point group 1. For other crystal classes, spatial symmetry relations reduce the number of independent d matrix elements to much lower value. Zernike and Midwinter ([2] [page 37, Fig. 2.1]) have summarized the properties of the d matrix for all crystal classes. In the following two sections, coefficients of d_{eff} for all 18 elements of the d matrix are tabulated for different biaxial and uniaxial crystal classes.

3.2 Expressions for d_{eff}

The d_{eff} expressions for the case of non-degenerate three wave mixing are given in Table 3.1 and the expressions for the degenerate three wave mixing case are given in Eqns. 3.15 and 3.21. For a given material, interaction type and frequency, the value of d_{eff} can be expressed in terms of the non-zero d coefficients in the d_{il} matrix by explicitly writing out the summations over the indices i and l. In Tables 3.3, 3.4 and 3.5 the coefficients for each of the 18 elements of the d matrix that need to be summed up to obtain the d_{eff} values are listed for the three interacting frequencies ω_1, ω_2 and ω_3 and for the three phase matching types ssf, sff and fsf. Although the coefficients for all 18 components of d are presented, for any particular crystal only a few of the d elements are non-zero and only those need to be considered to calculate d_{eff}. The components of the unit vectors e_s and e_f needed to calculate d_{eff} from the coefficients given in the Tables 3.3, 3.4 and 3.5 are obtained from Eqns. 1.58 for a general propagation direction of light (characterized by the angles θ, ϕ).

For propagation along the principal planes of a biaxial crystal, or in arbitrary directions in an uniaxial crystal, expressions for the components of the unit vectors e_s and e_f were summarized in Tables 1.3, 1.4, 1.5 and 1.6 in Chapter 1. Expressions for each element of the d matrix contributing to d_{eff} for these cases at the three frequencies are presented in Tables 3.23 through 3.40 at the end of this chapter. The Table 3.2 lists the different cases and the corresponding table numbers.

3.2.1 d_{eff} of biaxial crystals under Kleinman Symmetry Condition

In general, the d_{eff} value for each interaction type (ssf, sff and fsf) is frequency dependent, i.e., $d_{\text{eff}}(\omega_1)$, $d_{\text{eff}}(\omega_2)$ and $d_{\text{eff}}(\omega_3)$ are different. However, for crystals in which Kleinman symmetry holds, the d_{eff} value for each type of interaction becomes frequency independent. To show this, we use the uncontracted notation for d. For Type I (ssf) interaction,

$$\begin{aligned} d_{\text{eff}}(\omega_1) &= e_{si}(\omega_1)d_{ijk}e_{sj}(\omega_2)e_{fk}(\omega_3) \\ &= e_{sk}(\omega_1)d_{kji}e_{sj}(\omega_2)e_{fi}(\omega_3) \end{aligned} \qquad (3.22)$$

by simply interchanging the repeating indices i and k. For the same type of interaction, the $d_{\text{eff}}(\omega_3)$ is given by

$$d_{\text{eff}}(\omega_3) = e_{fi}(\omega_3)d_{ijk}e_{sj}(\omega_2)e_{sk}(\omega_1) \qquad (3.23)$$

which is identical with $d_{\text{eff}}(\omega_1)$ given in Eqn. 3.22 when the Kleinman symmetry condition $d_{ijk} = d_{kji}$ holds.

Repeating this argument for other types of interactions it can be shown that for each type of interaction, d_{eff} is frequency independent if Kleinman condition is valid.

Crystal Type	Propagation plane	Phase Matching Type	Table Number
Biaxial	YZ	ssf	3.23
		sff	3.24
		fsf	3.25
	$ZX, \theta < \Omega$	ssf	3.26
		sff	3.27
		fsf	3.28
	$ZX, \theta > \Omega$	ssf	3.29
		sff	3.30
		fsf	3.31
	XY	ssf	3.32
		sff	3.33
		fsf	3.34
Positive Uniaxial $(n_o < n_Z)$	Any	ssf	3.35
		sff	3.36
		fsf	3.37
Negative Uniaxial $(n_o > n_Z)$	Any	ssf	3.38
		sff	3.39
		fsf	3.40

TABLE 3.2. A list of the Tables at the end of this chapter which provide the d matrix coefficients that contribute to the total d_{eff} values for different phase matching types, for propagation along principal planes of a biaxial crystal or in general directions in an uniaxial crystal.

	Type I (ssf)		
	d_{eff} (ω_1)	d_{eff} (ω_2)	d_{eff} (ω_3)
d_{11}	$e_{sX}(\omega_1)e_{sX}(\omega_2)e_{fX}(\omega_3)$	$e_{sX}(\omega_2)e_{sX}(\omega_1)e_{fX}(\omega_3)$	$e_{fX}(\omega_3)e_{sX}(\omega_2)e_{sX}(\omega_1)$
d_{12}	$e_{sX}(\omega_1)e_{sY}(\omega_2)e_{fY}(\omega_3)$	$e_{sX}(\omega_2)e_{sY}(\omega_1)e_{fY}(\omega_3)$	$e_{fX}(\omega_3)e_{sY}(\omega_2)e_{sY}(\omega_1)$
d_{13}	$e_{sX}(\omega_1)e_{sZ}(\omega_2)e_{fZ}(\omega_3)$	$e_{sX}(\omega_2)e_{sZ}(\omega_1)e_{fZ}(\omega_3)$	$e_{fX}(\omega_3)e_{sZ}(\omega_2)e_{sZ}(\omega_1)$
d_{14}	$e_{sX}(\omega_1)$ $\times\{e_{sY}(\omega_2)e_{fZ}(\omega_3) + e_{sZ}(\omega_2)e_{fY}(\omega_3)\}$	$e_{sX}(\omega_2)$ $\times\{e_{sY}(\omega_1)e_{fZ}(\omega_3) + e_{sZ}(\omega_1)e_{fY}(\omega_3)\}$	$e_{fX}(\omega_3)$ $\times\{e_{sY}(\omega_2)e_{sZ}(\omega_1) + e_{sZ}(\omega_2)e_{sY}(\omega_1)\}$
d_{15}	$e_{sX}(\omega_1)$ $\times\{e_{sZ}(\omega_2)e_{fX}(\omega_3) + e_{sX}(\omega_2)e_{fZ}(\omega_3)\}$	$e_{sX}(\omega_2)$ $\times\{e_{sZ}(\omega_1)e_{fX}(\omega_3) + e_{sX}(\omega_1)e_{fZ}(\omega_3)\}$	$e_{fX}(\omega_3)$ $\times\{e_{sZ}(\omega_2)e_{sX}(\omega_1) + e_{sX}(\omega_2)e_{sZ}(\omega_1)\}$
d_{16}	$e_{sX}(\omega_1)$ $\times\{e_{sX}(\omega_2)e_{fY}(\omega_3) + e_{sY}(\omega_2)e_{fX}(\omega_3)\}$	$e_{sX}(\omega_2)$ $\times\{e_{sX}(\omega_1)e_{fY}(\omega_3) + e_{sY}(\omega_1)e_{fX}(\omega_3)\}$	$e_{fX}(\omega_3)$ $\times\{e_{sX}(\omega_2)e_{sY}(\omega_1) + e_{sY}(\omega_2)e_{sX}(\omega_1)\}$
d_{21}	$e_{sY}(\omega_1)e_{sX}(\omega_2)e_{fX}(\omega_3)$	$e_{sY}(\omega_2)e_{sX}(\omega_1)e_{fX}(\omega_3)$	$e_{fY}(\omega_3)e_{sX}(\omega_2)e_{sX}(\omega_1)$
d_{22}	$e_{sY}(\omega_1)e_{sY}(\omega_2)e_{fY}(\omega_3)$	$e_{sY}(\omega_2)e_{sY}(\omega_1)e_{fY}(\omega_3)$	$e_{fY}(\omega_3)e_{sY}(\omega_2)e_{sY}(\omega_1)$
d_{23}	$e_{sY}(\omega_1)e_{sZ}(\omega_2)e_{fZ}(\omega_3)$	$e_{sY}(\omega_2)e_{sZ}(\omega_1)e_{fZ}(\omega_3)$	$e_{fY}(\omega_3)e_{sZ}(\omega_2)e_{sZ}(\omega_1)$
d_{24}	$e_{sY}(\omega_1)$ $\times\{e_{sY}(\omega_2)e_{fZ}(\omega_3) + e_{sZ}(\omega_2)e_{fY}(\omega_3)\}$	$e_{sY}(\omega_2)$ $\times\{e_{sY}(\omega_1)e_{fZ}(\omega_3) + e_{sZ}(\omega_1)e_{fY}(\omega_3)\}$	$e_{fY}(\omega_3)$ $\times\{e_{sY}(\omega_2)e_{sZ}(\omega_1) + e_{sZ}(\omega_2)e_{sY}(\omega_1)\}$
d_{25}	$e_{sY}(\omega_1)$ $\times\{e_{sZ}(\omega_2)e_{fX}(\omega_3) + e_{sX}(\omega_2)e_{fZ}(\omega_3)\}$	$e_{sY}(\omega_2)$ $\times\{e_{sZ}(\omega_1)e_{fX}(\omega_3) + e_{sX}(\omega_1)e_{fZ}(\omega_3)\}$	$e_{fY}(\omega_3)$ $\times\{e_{sZ}(\omega_2)e_{sX}(\omega_1) + e_{sX}(\omega_2)e_{sZ}(\omega_1)\}$
d_{26}	$e_{sY}(\omega_1)$ $\times\{e_{sX}(\omega_2)e_{fY}(\omega_3) + e_{sY}(\omega_2)e_{fX}(\omega_3)\}$	$e_{sY}(\omega_2)$ $\times\{e_{sX}(\omega_1)e_{fY}(\omega_3) + e_{sY}(\omega_1)e_{fX}(\omega_3)\}$	$e_{fY}(\omega_3)$ $\times\{e_{sX}(\omega_2)e_{sY}(\omega_1) + e_{sY}(\omega_2)e_{sX}(\omega_1)\}$
d_{31}	$e_{sZ}(\omega_1)e_{sX}(\omega_2)e_{fX}(\omega_3)$	$e_{sZ}(\omega_2)e_{sX}(\omega_1)e_{fX}(\omega_3)$	$e_{fZ}(\omega_3)e_{sX}(\omega_2)e_{sX}(\omega_1)$
d_{32}	$e_{sZ}(\omega_1)e_{sY}(\omega_2)e_{fY}(\omega_3)$	$e_{sZ}(\omega_2)e_{sY}(\omega_1)e_{fY}(\omega_3)$	$e_{fZ}(\omega_3)e_{sY}(\omega_2)e_{sY}(\omega_1)$
d_{33}	$e_{sZ}(\omega_1)e_{sZ}(\omega_2)e_{fZ}(\omega_3)$	$e_{sZ}(\omega_2)e_{sZ}(\omega_1)e_{fZ}(\omega_3)$	$e_{fZ}(\omega_3)e_{sZ}(\omega_2)e_{sZ}(\omega_1)$
d_{34}	$e_{sZ}(\omega_1)$ $\times\{e_{sY}(\omega_2)e_{fZ}(\omega_3) + e_{sZ}(\omega_2)e_{fY}(\omega_3)\}$	$e_{sZ}(\omega_2)$ $\times\{e_{sY}(\omega_1)e_{fZ}(\omega_3) + e_{sZ}(\omega_1)e_{fY}(\omega_3)\}$	$e_{fZ}(\omega_3)$ $\times\{e_{sY}(\omega_2)e_{sZ}(\omega_1) + e_{sZ}(\omega_2)e_{sY}(\omega_1)\}$
d_{35}	$e_{sZ}(\omega_1)$ $\times\{e_{sZ}(\omega_2)e_{fX}(\omega_3) + e_{sX}(\omega_2)e_{fZ}(\omega_3)\}$	$e_{sZ}(\omega_2)$ $\times\{e_{sZ}(\omega_1)e_{fX}(\omega_3) + e_{sX}(\omega_1)e_{fZ}(\omega_3)\}$	$e_{fZ}(\omega_3)$ $\times\{e_{sZ}(\omega_2)e_{sX}(\omega_1) + e_{sX}(\omega_2)e_{sZ}(\omega_1)\}$
d_{36}	$e_{sZ}(\omega_1)$ $\times\{e_{sX}(\omega_2)e_{fY}(\omega_3) + e_{sY}(\omega_2)e_{fX}(\omega_3)\}$	$e_{sZ}(\omega_2)$ $\times\{e_{sX}(\omega_1)e_{fY}(\omega_3) + e_{sY}(\omega_1)e_{fX}(\omega_3)\}$	$e_{fZ}(\omega_3)$ $\times\{e_{sX}(\omega_2)e_{sY}(\omega_1) + e_{sY}(\omega_2)e_{sX}(\omega_1)\}$

TABLE 3.3. Coefficients of d_{ij} contributing to d_{eff}: ssf case

	Type II (sff)		
	$d_{\mathrm{eff}}(\omega_1)$	$d_{\mathrm{eff}}(\omega_2)$	$d_{\mathrm{eff}}(\omega_3)$
d_{11}	$e_{sX}(\omega_1)e_{fX}(\omega_2)e_{fX}(\omega_3)$	$e_{fX}(\omega_2)e_{sX}(\omega_1)e_{fX}(\omega_3)$	$e_{fX}(\omega_3)e_{fX}(\omega_2)e_{sX}(\omega_1)$
d_{12}	$e_{sX}(\omega_1)e_{fY}(\omega_2)e_{fY}(\omega_3)$	$e_{fX}(\omega_2)e_{sY}(\omega_1)e_{fY}(\omega_3)$	$e_{fX}(\omega_3)e_{fY}(\omega_2)e_{sY}(\omega_1)$
d_{13}	$e_{sX}(\omega_1)e_{fZ}(\omega_2)e_{fZ}(\omega_3)$	$e_{fX}(\omega_2)e_{sZ}(\omega_1)e_{fZ}(\omega_3)$	$e_{fX}(\omega_3)e_{fZ}(\omega_2)e_{sZ}(\omega_1)$
d_{14}	$e_{sX}(\omega_1)$ $\times \{e_{fY}(\omega_2)e_{fZ}(\omega_3) + e_{fZ}(\omega_2)e_{fY}(\omega_3)\}$	$e_{fX}(\omega_2)$ $\times \{e_{sY}(\omega_1)e_{fZ}(\omega_3) + e_{sZ}(\omega_1)e_{fY}(\omega_3)\}$	$e_{fX}(\omega_3)$ $\times \{e_{fY}(\omega_2)e_{sZ}(\omega_1) + e_{fZ}(\omega_2)e_{sY}(\omega_1)\}$
d_{15}	$e_{sX}(\omega_1)$ $\times \{e_{fZ}(\omega_2)e_{fX}(\omega_3) + e_{fX}(\omega_2)e_{fZ}(\omega_3)\}$	$e_{fX}(\omega_2)$ $\times \{e_{sZ}(\omega_1)e_{fX}(\omega_3) + e_{sX}(\omega_1)e_{fZ}(\omega_3)\}$	$e_{fX}(\omega_3)$ $\times \{e_{fZ}(\omega_2)e_{sX}(\omega_1) + e_{fX}(\omega_2)e_{sZ}(\omega_1)\}$
d_{16}	$e_{sX}(\omega_1)$ $\times \{e_{fX}(\omega_2)e_{fY}(\omega_3) + e_{fY}(\omega_2)e_{fX}(\omega_3)\}$	$e_{fX}(\omega_2)$ $\times \{e_{sX}(\omega_1)e_{fY}(\omega_3) + e_{sY}(\omega_1)e_{fX}(\omega_3)\}$	$e_{fX}(\omega_3)$ $\times \{e_{fX}(\omega_2)e_{sY}(\omega_1) + e_{fY}(\omega_2)e_{sX}(\omega_1)\}$
d_{21}	$e_{sY}(\omega_1)e_{fX}(\omega_2)e_{fX}(\omega_3)$	$e_{fY}(\omega_2)e_{sX}(\omega_1)e_{fX}(\omega_3)$	$e_{fY}(\omega_3)e_{fX}(\omega_2)e_{sX}(\omega_1)$
d_{22}	$e_{sY}(\omega_1)e_{fY}(\omega_2)e_{fY}(\omega_3)$	$e_{fY}(\omega_2)e_{sY}(\omega_1)e_{fY}(\omega_3)$	$e_{fY}(\omega_3)e_{fY}(\omega_2)e_{sY}(\omega_1)$
d_{23}	$e_{sY}(\omega_1)e_{fZ}(\omega_2)e_{fZ}(\omega_3)$	$e_{fY}(\omega_2)e_{sZ}(\omega_1)e_{fZ}(\omega_3)$	$e_{fY}(\omega_3)e_{fZ}(\omega_2)e_{sZ}(\omega_1)$
d_{24}	$e_{sY}(\omega_1)$ $\times \{e_{fY}(\omega_2)e_{fZ}(\omega_3) + e_{fZ}(\omega_2)e_{fY}(\omega_3)\}$	$e_{fY}(\omega_2)$ $\times \{e_{sY}(\omega_1)e_{fZ}(\omega_3) + e_{sZ}(\omega_1)e_{fY}(\omega_3)\}$	$e_{fY}(\omega_3)$ $\times \{e_{fY}(\omega_2)e_{sZ}(\omega_1) + e_{fZ}(\omega_2)e_{sY}(\omega_1)\}$
d_{25}	$e_{sY}(\omega_1)$ $\times \{e_{fZ}(\omega_2)e_{fX}(\omega_3) + e_{fX}(\omega_2)e_{fZ}(\omega_3)\}$	$e_{fY}(\omega_2)$ $\times \{e_{sZ}(\omega_1)e_{fX}(\omega_3) + e_{sX}(\omega_1)e_{fZ}(\omega_3)\}$	$e_{fY}(\omega_3)$ $\times \{e_{fZ}(\omega_2)e_{sX}(\omega_1) + e_{fX}(\omega_2)e_{sZ}(\omega_1)\}$
d_{26}	$e_{sY}(\omega_1)$ $\times \{e_{fX}(\omega_2)e_{fY}(\omega_3) + e_{fY}(\omega_2)e_{fX}(\omega_3)\}$	$e_{fY}(\omega_2)$ $\times \{e_{sX}(\omega_1)e_{fY}(\omega_3) + e_{sY}(\omega_1)e_{fX}(\omega_3)\}$	$e_{fY}(\omega_3)$ $\times \{e_{fX}(\omega_2)e_{sY}(\omega_1) + e_{fY}(\omega_2)e_{sX}(\omega_1)\}$
d_{31}	$e_{sZ}(\omega_1)e_{fX}(\omega_2)e_{fX}(\omega_3)$	$e_{fZ}(\omega_2)e_{sX}(\omega_1)e_{fX}(\omega_3)$	$e_{fZ}(\omega_3)e_{fX}(\omega_2)e_{sX}(\omega_1)$
d_{32}	$e_{sZ}(\omega_1)e_{fY}(\omega_2)e_{fY}(\omega_3)$	$e_{fZ}(\omega_2)e_{sY}(\omega_1)e_{fY}(\omega_3)$	$e_{fZ}(\omega_3)e_{fY}(\omega_2)e_{sY}(\omega_1)$
d_{33}	$e_{sZ}(\omega_1)e_{fZ}(\omega_2)e_{fZ}(\omega_3)$	$e_{fZ}(\omega_2)e_{sZ}(\omega_1)e_{fZ}(\omega_3)$	$e_{fZ}(\omega_3)e_{fZ}(\omega_2)e_{sZ}(\omega_1)$
d_{34}	$e_{sZ}(\omega_1)$ $\times \{e_{fY}(\omega_2)e_{fZ}(\omega_3) + e_{fZ}(\omega_2)e_{fY}(\omega_3)\}$	$e_{fZ}(\omega_2)$ $\times \{e_{sY}(\omega_1)e_{fZ}(\omega_3) + e_{sZ}(\omega_1)e_{fY}(\omega_3)\}$	$e_{fZ}(\omega_3)$ $\times \{e_{fY}(\omega_2)e_{sZ}(\omega_1) + e_{fZ}(\omega_2)e_{sY}(\omega_1)\}$
d_{35}	$e_{sZ}(\omega_1)$ $\times \{e_{fZ}(\omega_2)e_{fX}(\omega_3) + e_{fX}(\omega_2)e_{fZ}(\omega_3)\}$	$e_{fZ}(\omega_2)$ $\times \{e_{sZ}(\omega_1)e_{fX}(\omega_3) + e_{sX}(\omega_1)e_{fZ}(\omega_3)\}$	$e_{fZ}(\omega_3)$ $\times \{e_{fZ}(\omega_2)e_{sX}(\omega_1) + e_{fX}(\omega_2)e_{sZ}(\omega_1)\}$
d_{36}	$e_{sZ}(\omega_1)$ $\times \{e_{fX}(\omega_2)e_{fY}(\omega_3) + e_{fY}(\omega_2)e_{fX}(\omega_3)\}$	$e_{fZ}(\omega_2)$ $\times \{e_{sX}(\omega_1)e_{fY}(\omega_3) + e_{sY}(\omega_1)e_{fX}(\omega_3)\}$	$e_{fZ}(\omega_3)$ $\times \{e_{fX}(\omega_2)e_{sY}(\omega_1) + e_{fY}(\omega_2)e_{sX}(\omega_1)\}$

TABLE 3.4. Coefficients of d_{ij} contributing to d_{eff}: sff case

	$d_{eff}(\omega_1)$	$d_{eff}(\omega_2)$	$d_{eff}(\omega_3)$
	Type II (fsf)		
d_{11}	$e_{fX}(\omega_1)e_{sX}(\omega_2)e_{fX}(\omega_3)$	$e_{sX}(\omega_2)e_{fX}(\omega_1)e_{fX}(\omega_3)$	$e_{fX}(\omega_3)e_{sX}(\omega_2)e_{fX}(\omega_1)$
d_{12}	$e_{fX}(\omega_1)e_{sY}(\omega_2)e_{fY}(\omega_3)$	$e_{sX}(\omega_2)e_{fY}(\omega_1)e_{fY}(\omega_3)$	$e_{fX}(\omega_3)e_{sY}(\omega_2)e_{fY}(\omega_1)$
d_{13}	$e_{fX}(\omega_1)e_{sZ}(\omega_2)e_{fZ}(\omega_3)$	$e_{sX}(\omega_2)e_{fZ}(\omega_1)e_{fZ}(\omega_3)$	$e_{fX}(\omega_3)e_{sZ}(\omega_2)e_{fZ}(\omega_1)$
d_{14}	$e_{fX}(\omega_1)$ $\times\{e_{sY}(\omega_2)e_{fZ}(\omega_3)+e_{sZ}(\omega_2)e_{fY}(\omega_3)\}$	$e_{sX}(\omega_2)$ $\times\{e_{fY}(\omega_1)e_{fZ}(\omega_3)+e_{fZ}(\omega_1)e_{fY}(\omega_3)\}$	$e_{fX}(\omega_3)$ $\times\{e_{sY}(\omega_2)e_{fZ}(\omega_1)+e_{sZ}(\omega_2)e_{fY}(\omega_1)\}$
d_{15}	$e_{fX}(\omega_1)$ $\times\{e_{sZ}(\omega_2)e_{fX}(\omega_3)+e_{sX}(\omega_2)e_{fZ}(\omega_3)\}$	$e_{sX}(\omega_2)$ $\times\{e_{fZ}(\omega_1)e_{fX}(\omega_3)+e_{fX}(\omega_1)e_{fZ}(\omega_3)\}$	$e_{fX}(\omega_3)$ $\times\{e_{sZ}(\omega_2)e_{fX}(\omega_1)+e_{sX}(\omega_2)e_{fZ}(\omega_1)\}$
d_{16}	$e_{fX}(\omega_1)$ $\times\{e_{sX}(\omega_2)e_{fY}(\omega_3)+e_{sY}(\omega_2)e_{fX}(\omega_3)\}$	$e_{sX}(\omega_2)$ $\times\{e_{fX}(\omega_1)e_{fY}(\omega_3)+e_{fY}(\omega_1)e_{fX}(\omega_3)\}$	$e_{fX}(\omega_3)$ $\times\{e_{sX}(\omega_2)e_{fY}(\omega_1)+e_{sY}(\omega_2)e_{fX}(\omega_1)\}$
d_{21}	$e_{fY}(\omega_1)e_{sX}(\omega_2)e_{fX}(\omega_3)$	$e_{sY}(\omega_2)e_{fX}(\omega_1)e_{fX}(\omega_3)$	$e_{fY}(\omega_3)e_{sX}(\omega_2)e_{fX}(\omega_1)$
d_{22}	$e_{fY}(\omega_1)e_{sY}(\omega_2)e_{fY}(\omega_3)$	$e_{sY}(\omega_2)e_{fY}(\omega_1)e_{fY}(\omega_3)$	$e_{fY}(\omega_3)e_{sY}(\omega_2)e_{fY}(\omega_1)$
d_{23}	$e_{fY}(\omega_1)e_{sZ}(\omega_2)e_{fZ}(\omega_3)$	$e_{sY}(\omega_2)e_{fZ}(\omega_1)e_{fZ}(\omega_3)$	$e_{fY}(\omega_3)e_{sZ}(\omega_2)e_{fZ}(\omega_1)$
d_{24}	$e_{fY}(\omega_1)$ $\times\{e_{sY}(\omega_2)e_{fZ}(\omega_3)+e_{sZ}(\omega_2)e_{fY}(\omega_3)\}$	$e_{sY}(\omega_2)$ $\times\{e_{fY}(\omega_1)e_{fZ}(\omega_3)+e_{fZ}(\omega_1)e_{fY}(\omega_3)\}$	$e_{fY}(\omega_3)$ $\times\{e_{sY}(\omega_2)e_{fZ}(\omega_1)+e_{sZ}(\omega_2)e_{fY}(\omega_1)\}$
d_{25}	$e_{fY}(\omega_1)$ $\times\{e_{sZ}(\omega_2)e_{fX}(\omega_3)+e_{sX}(\omega_2)e_{fZ}(\omega_3)\}$	$e_{sY}(\omega_2)$ $\times\{e_{fZ}(\omega_1)e_{fX}(\omega_3)+e_{fX}(\omega_1)e_{fZ}(\omega_3)\}$	$e_{fY}(\omega_3)$ $\times\{e_{sZ}(\omega_2)e_{fX}(\omega_1)+e_{sX}(\omega_2)e_{fZ}(\omega_1)\}$
d_{26}	$e_{fY}(\omega_1)$ $\times\{e_{sX}(\omega_2)e_{fY}(\omega_3)+e_{sY}(\omega_2)e_{fX}(\omega_3)\}$	$e_{sY}(\omega_2)$ $\times\{e_{fX}(\omega_1)e_{fY}(\omega_3)+e_{fY}(\omega_1)e_{fX}(\omega_3)\}$	$e_{fY}(\omega_3)$ $\times\{e_{sX}(\omega_2)e_{fY}(\omega_1)+e_{sY}(\omega_2)e_{fX}(\omega_1)\}$
d_{31}	$e_{fZ}(\omega_1)e_{sX}(\omega_2)e_{fX}(\omega_3)$	$e_{sZ}(\omega_2)e_{fX}(\omega_1)e_{fX}(\omega_3)$	$e_{fZ}(\omega_3)e_{sX}(\omega_2)e_{fX}(\omega_1)$
d_{32}	$e_{fZ}(\omega_1)e_{sY}(\omega_2)e_{fY}(\omega_3)$	$e_{sZ}(\omega_2)e_{fY}(\omega_1)e_{fY}(\omega_3)$	$e_{fZ}(\omega_3)e_{sY}(\omega_2)e_{fY}(\omega_1)$
d_{33}	$e_{fZ}(\omega_1)e_{sZ}(\omega_2)e_{fZ}(\omega_3)$	$e_{sZ}(\omega_2)e_{fZ}(\omega_1)e_{fZ}(\omega_3)$	$e_{fZ}(\omega_3)e_{sZ}(\omega_2)e_{fZ}(\omega_1)$
d_{34}	$e_{fZ}(\omega_1)$ $\times\{e_{sY}(\omega_2)e_{fZ}(\omega_3)+e_{sZ}(\omega_2)e_{fY}(\omega_3)\}$	$e_{sZ}(\omega_2)$ $\times\{e_{fY}(\omega_1)e_{fZ}(\omega_3)+e_{fZ}(\omega_1)e_{fY}(\omega_3)\}$	$e_{fZ}(\omega_3)$ $\times\{e_{sY}(\omega_2)e_{fZ}(\omega_1)+e_{sZ}(\omega_2)e_{fY}(\omega_1)\}$
d_{35}	$e_{fZ}(\omega_1)$ $\times\{e_{sZ}(\omega_2)e_{fX}(\omega_3)+e_{sX}(\omega_2)e_{fZ}(\omega_3)\}$	$e_{sZ}(\omega_2)$ $\times\{e_{fZ}(\omega_1)e_{fX}(\omega_3)+e_{fX}(\omega_1)e_{fZ}(\omega_3)\}$	$e_{fZ}(\omega_3)$ $\times\{e_{sZ}(\omega_2)e_{fX}(\omega_1)+e_{sX}(\omega_2)e_{fZ}(\omega_1)\}$
d_{36}	$e_{fZ}(\omega_1)$ $\times\{e_{sX}(\omega_2)e_{fY}(\omega_3)+e_{sY}(\omega_2)e_{fX}(\omega_3)\}$	$e_{sZ}(\omega_2)$ $\times\{e_{fX}(\omega_1)e_{fY}(\omega_3)+e_{fY}(\omega_1)e_{fX}(\omega_3)\}$	$e_{fZ}(\omega_3)$ $\times\{e_{sX}(\omega_2)e_{fY}(\omega_1)+e_{sY}(\omega_2)e_{fX}(\omega_1)\}$

TABLE 3.5. Coefficients of d_{ij} contributing to d_{eff}: fsf case

3.2.2   Reduction of d_{eff} to expressions in the literature

According to Ref. [2] (page 27), 'the correct expressions for the $d_{\text{eff}}^{\text{ss-f}}$ and $d_{\text{eff}}^{\text{sf-f}}$ .. were first obtained by Lavrovskaya et al.' [3]). Since the explicit expressions (without walk-off) for the ssf and sff cases are given in Ref. [2], we first check the expressions above to see if we obtain the same results.

For crystals belonging to point group mm2, there are 5 non-zero d coefficients: d_{15}, d_{24}, d_{31}, d_{32} and d_{33}. For a Type I(ssf) interaction, $d_{\text{eff}}(\omega_3)$ is obtained from Table 3.3 as:

$$
\begin{aligned}
d_{\text{eff}}(\omega_3) \;=\;& e_{fi}(\omega_3)d_{il}u_l^{ss} \\
=\;& d_{15}e_{fX}(\omega_3)\{e_{sZ}(\omega_1)e_{sX}(\omega_2)+e_{sX}(\omega_1)e_{sZ}(\omega_2)\} \\
& + d_{24}e_{fY}(\omega_3)\{e_{sY}(\omega_1)e_{sZ}(\omega_2)+e_{sZ}(\omega_1)e_{sY}(\omega_2)\} \\
& + d_{31}e_{fZ}(\omega_3)e_{sX}(\omega_1)e_{sX}(\omega_2) \\
& + d_{32}e_{fZ}(\omega_3)e_{sY}(\omega_1)e_{sY}(\omega_2) \\
& + d_{33}e_{fZ}(\omega_3)e_{sZ}(\omega_1)e_{sZ}(\omega_2).
\end{aligned}
\tag{3.24}
$$

The components of $\hat{e}_s$ and $\hat{e}_f$ are given in Chapter 1 (Eqns. 1.96 and 1.98). With the walk-off angles ρ_s and ρ_f set equal to 0, we obtain for the case of $n_X < n_Y < n_Z$

$$
\begin{aligned}
e_{sX} &= \cos\theta\;\cos\phi\;\cos\delta - \sin\phi\;\sin\delta \\
e_{sY} &= \cos\theta\;\sin\phi\;\cos\delta + \cos\phi\;\sin\delta \\
e_{sZ} &= -\sin\theta\;\cos\delta
\end{aligned}
\tag{3.25}
$$

$$
\begin{aligned}
e_{fX} &= \cos\theta\;\cos\phi\;\sin\delta + \sin\phi\;\cos\delta \\
e_{fY} &= \cos\theta\;\sin\phi\;\sin\delta - \cos\phi\;\cos\delta \\
e_{fZ} &= -\sin\theta\;\sin\delta.
\end{aligned}
\tag{3.26}
$$

Assuming dispersion of the angle δ to be small, and inserting the unit vector components in Eqn. 3.24, we find for example the coefficient of the d_{15} term to be

$$
\begin{aligned}
& e_{fX}(e_{sZ}e_{sX}+e_{sX}e_{sZ}) \\
=\;& 2e_{fX}e_{sZ}e_{sX} \\
=\;& -2(\cos\theta\;\cos\phi\;\sin\delta + \sin\phi\;\cos\delta)(\cos\theta\;\cos\phi\;\cos\delta - \sin\phi\;\sin\delta) \\
& \times \sin\theta\;\cos\delta.
\end{aligned}
\tag{3.27}
$$

It is a bit laborious although quite straightforward to show that this coefficient matches the expression for the coefficient of d_{15} in Ref. [2] (page 27). By similar procedure, all the other terms in Eqn. 3.24 can be shown to match the Lavrovskaya form. It may be worth reiterating here that the expressions for $d_{\text{eff}}^{\text{ss-f}}$ and $d_{\text{eff}}^{\text{sf-f}}$ given in Ref. [2] (page 27) are valid only for small walk-off angles, i.e., with $\cos\rho_s$ and $\cos\rho_f \approx 1$ and $\sin\rho_s$ and $\sin\rho_f \approx 0$.

3.3 d_{eff} Values for Some Biaxial and Uniaxial Crystals of Different Classes

For light propagating with **k** vector in a general direction (characterized by the angles θ, ϕ, say) in a biaxial or uniaxial crystal, the d_{eff} values for the three frequencies ω_1, ω_2 and ω_3 for the three types of phase matching (ssf, sff and fsf) can be written out explicitly using the equations presented earlier in Chapter 1 and in Table 2.2.

For the special cases of propagation along the principal planes or along the principal axes, the expressions for d_{eff} are obtained from Tables 1.3 and 1.4 in Chapter 1.

Before getting to specific crystals, we collect here a set of relevant equations from Chapter 1, with the parameters expressed in terms of the angles θ, ϕ and the wavelength λ (instead of the frequency ω), with λ taking the value of λ_1, λ_2 or λ_3.

Since the principal dielectric permittivities are wavelength dependent, so are the principal dielectric indices n_X, n_Y and n_Z, to reflect which the indices are written as $n_X(\lambda)$, $n_Y(\lambda)$ and $n_Z(\lambda)$. Empirical relationships between the refractive indices and wavelength are usually known as Sellmeier equations and

The expressions for the slow and fast components of the refractive index denoted by $n_s(\theta, \phi, \lambda)$ and $n_f(\theta, \phi, \lambda)$ are provided in Chapter 2 (Eqns. 2.49). Using Eqns. 1.60 the walk-off angles ρ_s and ρ_f as functions of θ, ϕ and λ can then be obtained from the relations

$$\sin \rho_s(\theta, \phi, \lambda) = \frac{1}{n_s(\theta, \phi, \lambda)\mathcal{E}_s(\theta, \phi, \lambda)}$$

$$\sin \rho_f(\theta, \phi, \lambda) = \frac{1}{n_f(\theta, \phi, \lambda)\mathcal{E}_f(\theta, \phi, \lambda)} \tag{3.28}$$

where

$$\mathcal{E}_s(\theta, \phi, \lambda) \equiv$$

$$\sqrt{\left(\frac{\sin\theta\cos\phi}{n_s^2(\theta,\phi,\lambda) - n_X^2(\lambda)}\right)^2 + \left(\frac{\sin\theta\sin\phi}{n_s^2(\theta,\phi,\lambda) - n_Y^2(\lambda)}\right)^2 + \left(\frac{\cos\theta}{n_s^2(\theta,\phi,\lambda) - n_Z^2(\lambda)}\right)^2} \tag{3.29}$$

and

$$\mathcal{E}_f(\theta, \phi, \lambda) \equiv$$

$$\sqrt{\left(\frac{\sin\theta\cos\phi}{n_f^2(\theta,\phi,\lambda) - n_X^2(\lambda)}\right)^2 + \left(\frac{\sin\theta\sin\phi}{n_f^2(\theta,\phi,\lambda) - n_Y^2(\lambda)}\right)^2 + \left(\frac{\cos\theta}{n_f^2(\theta,\phi,\lambda) - n_Z^2(\lambda)}\right)^2}. \tag{3.30}$$

Using Eqns. 1.58 the Cartesian components of the unit vectors $\hat{e}_s$ and $\hat{e}_f$ can then be written as

$$e_{sX}(\theta,\phi,\lambda) = \frac{\sin\theta\cos\phi}{1-\dfrac{n_X^2(\lambda)}{n_s^2(\theta,\phi,\lambda)}}\sin\rho_s = \frac{n_s(\theta,\phi,\lambda)\sin\theta\cos\phi}{\mathcal{E}_s(\theta,\phi,\lambda)\{n_s^2(\theta,\phi,\lambda)-n_X^2(\lambda)\}}$$

$$e_{sY}(\theta,\phi,\lambda) = \frac{\sin\theta\sin\phi}{1-\dfrac{n_Y^2(\lambda)}{n_s^2(\theta,\phi,\lambda)}}\sin\rho_s = \frac{n_s(\theta,\phi,\lambda)\sin\theta\cos\phi}{\mathcal{E}_s(\theta,\phi,\lambda)\{n_s^2(\theta,\phi,\lambda)-n_Y^2(\lambda)\}}$$

$$e_{sZ}(\theta,\phi,\lambda) = \frac{\cos\theta}{1-\dfrac{n_Z^2(\lambda)}{n_s^2(\theta,\phi,\lambda)}}\sin\rho_s = \frac{n_s(\theta,\phi,\lambda)\sin\theta\cos\phi}{\mathcal{E}_s(\theta,\phi,\lambda)\{n_s^2(\theta,\phi,\lambda)-n_Z^2(\lambda)\}}$$

$$(3.31)$$

and

$$e_{fX}(\theta,\phi,\lambda) = \frac{\sin\theta\cos\phi}{1-\dfrac{n_X^2(\lambda)}{n_f^2(\theta,\phi,\lambda)}}\sin\rho_f = \frac{n_f(\theta,\phi,\lambda)\sin\theta\cos\phi}{\mathcal{E}_f(\theta,\phi,\lambda)\{n_f^2(\theta,\phi,\lambda)-n_X^2(\lambda)\}}$$

$$e_{fY}(\theta,\phi,\lambda) = \frac{\sin\theta\sin\phi}{1-\dfrac{n_Y^2(\lambda)}{n_f^2(\theta,\phi,\lambda)}}\sin\rho_f = \frac{n_f(\theta,\phi,\lambda)\sin\theta\cos\phi}{\mathcal{E}_f(\theta,\phi,\lambda)\{n_f^2(\theta,\phi,\lambda)-n_Y^2(\lambda)\}}$$

$$e_{fZ}(\theta,\phi,\lambda) = \frac{\cos\theta}{1-\dfrac{n_Z^2(\lambda)}{n_f^2(\theta,\phi,\lambda)}}\sin\rho_f = \frac{n_f(\theta,\phi,\lambda)\sin\theta\cos\phi}{\mathcal{E}_f(\theta,\phi,\lambda)\{n_f^2(\theta,\phi,\lambda)-n_Z^2(\lambda)\}}.$$

$$(3.32)$$

Using these values of the field components along with the appropriate d coefficients, the d_{eff} values can be determined from Table 3.1.

3.3.1 d_{eff} for KTP for propagation in a general direction

KTP (Potassium Titanyl Phosphate) is a positive biaxial crystal with point group mm2, so the nonzero elements of its d matrix are d_{15}, d_{24}, d_{31}, d_{32} and d_{33} from Table 2.2. Under Kleinman symmetry, $d_{31} = d_{15}$, and $d_{24} = d_{32}$, so that there are only three independent components, d_{15}, d_{24} and d_{33}. To evaluate the d_{eff} values, the coefficients of all five nonzero components need to be determined from the tables.

Because of the importance of KTP in nonlinear optical applications, we present here the explicit expressions for d_{eff} for the ssf, sff and the fsf type processes and provide the expressions for d_{eff} first for a general propagation direction and then for four distinct propagation directions: $\mathbf{k}$ along the YZ plane, the XY plane, the ZX plane with $\theta < \Omega$ and the ZX plane with $\theta > \Omega$.

Boulanger et al [4] presented the expression for d_{eff} of KTP in the case of second harmonic generation in terms of field factors F_{ij}, the details of which for general three wave mixing were given in Ref. [5]. For ease of computer programming, we denote here the field factors for the cases of Type I (ssf), Type II (sff) and Type II(fsf) interactions by F_{ij}, G_{ij} and H_{ij}, respectively, .

3.3.1.1 d_{eff} for KTP for a Type I (ssf) mixing process

For Type I, ssf type of interaction we have from Table 3.3

$$
\begin{aligned}
d_{\text{eff}}^{\text{I(ssf)}}(\omega_3) &= e_{fi}(\omega_3)d_{il}\mathrm{u}_l^{ss}(\omega_1,\omega_2)\\
&\equiv F_{15}^{\lambda_3}(\theta,\phi)d_{15}+F_{24}^{\lambda_3}(\theta,\phi)d_{24}+F_{31}^{\lambda_3}(\theta,\phi)d_{31}\\
&\quad +F_{32}^{\lambda_3}(\theta,\phi)d_{32}+F_{33}^{\lambda_3}(\theta,\phi)d_{33}
\end{aligned}
\tag{3.33}
$$

where

$$
F_{15}^{\lambda_3}(\theta,\phi)\equiv e_{fX}(\theta,\phi,\lambda_3)\{e_{sZ}(\theta,\phi,\lambda_1)e_{sX}(\theta,\phi,\lambda_2)+e_{sX}(\theta,\phi,\lambda_1)e_{sZ}(\theta,\phi,\lambda_2)\}
$$
$$
F_{24}^{\lambda_3}(\theta,\phi)= e_{fY}(\theta,\phi,\lambda_3)\{e_{sY}(\theta,\phi,\lambda_1)e_{sZ}(\theta,\phi,\lambda_2)+e_{sZ}(\theta,\phi,\lambda_1)e_{sY}(\theta,\phi,\lambda_2)\}
$$
$$
F_{31}^{\lambda_3}(\theta,\phi)= e_{fZ}(\theta,\phi,\lambda_3)e_{sX}(\theta,\phi,\lambda_1)e_{sX}(\theta,\phi,\lambda_2)
$$
$$
F_{32}^{\lambda_3}(\theta,\phi)= e_{fZ}(\theta,\phi,\lambda_3)e_{sY}(\theta,\phi,\lambda_1)e_{sY}(\theta,\phi,\lambda_2)
$$
$$
F_{33}^{\lambda_3}(\theta,\phi)= e_{fZ}(\theta,\phi,\lambda_3)e_{sZ}(\theta,\phi,\lambda_1)e_{sZ}(\theta,\phi,\lambda_2).
\tag{3.34}
$$

Similarly,

$$
\begin{aligned}
d_{\text{eff}}^{\text{I(ssf)}}(\omega_2) &= e_{si}(\omega_2)d_{il}\mathrm{u}_l^{sf}(\omega_1,\omega_3)\\
&\equiv F_{15}^{\lambda_2}(\theta,\phi)d_{15}+F_{24}^{\lambda_2}(\theta,\phi)d_{24}+F_{31}^{\lambda_2}(\theta,\phi)d_{31}\\
&\quad +F_{32}^{\lambda_2}(\theta,\phi)d_{32}+F_{33}^{\lambda_2}(\theta,\phi)d_{33}
\end{aligned}
\tag{3.35}
$$

where

$$
F_{15}^{\lambda_2}(\theta,\phi)\equiv e_{sX}(\theta,\phi,\lambda_2)\{e_{sZ}(\theta,\phi,\lambda_1)e_{fX}(\theta,\phi,\lambda_3)+e_{sX}(\theta,\phi,\lambda_1)e_{fZ}(\theta,\phi,\lambda_3)\}
$$
$$
F_{24}^{\lambda_2}(\theta,\phi)= e_{sY}(\theta,\phi,\lambda_2)\{e_{sY}(\theta,\phi,\lambda_1)e_{fZ}(\theta,\phi,\lambda_3)+e_{sZ}(\theta,\phi,\lambda_1)e_{fY}(\theta,\phi,\lambda_3)\}
$$
$$
F_{31}^{\lambda_2}(\theta,\phi)= e_{sZ}(\theta,\phi,\lambda_2)e_{sX}(\theta,\phi,\lambda_1)e_{fX}(\theta,\phi,\lambda_3)
$$
$$
F_{32}^{\lambda_2}(\theta,\phi)= e_{sZ}(\theta,\phi,\lambda_2)e_{sY}(\theta,\phi,\lambda_1)e_{fY}(\theta,\phi,\lambda_3)
$$
$$
F_{33}^{\lambda_2}(\theta,\phi)= e_{sZ}(\theta,\phi,\lambda_2)e_{sZ}(\theta,\phi,\lambda_1)e_{fZ}(\theta,\phi,\lambda_3).
\tag{3.36}
$$

and

$$
\begin{aligned}
d_{\text{eff}}^{\text{I(ssf)}}(\omega_1) &= e_{si}(\omega_1)d_{il}\mathrm{u}_l^{sf}(\omega_2,\omega_3)\\
&\equiv F_{15}^{\lambda_1}(\theta,\phi)d_{15}+F_{24}^{\lambda_1}(\theta,\phi)d_{24}+F_{31}^{\lambda_1}(\theta,\phi)d_{31}\\
&\quad +F_{32}^{\lambda_1}(\theta,\phi)d_{32}+F_{33}^{\lambda_1}(\theta,\phi)d_{33}
\end{aligned}
\tag{3.37}
$$

where

$$F_{15}^{\lambda_1}(\theta,\phi) \equiv e_{sX}(\theta,\phi,\lambda_1)\{e_{sZ}(\theta,\phi,\lambda_2)e_{fX}(\theta,\phi,\lambda_3)+e_{sX}(\theta,\phi,\lambda_2)e_{fZ}(\theta,\phi,\lambda_3)\}$$

$$F_{24}^{\lambda_1}(\theta,\phi) = e_{sY}(\theta,\phi,\lambda_1)\{e_{sY}(\theta,\phi,\lambda_2)e_{fZ}(\theta,\phi,\lambda_3)+e_{sZ}(\theta,\phi,\lambda_2)e_{fY}(\theta,\phi,\lambda_3)\}$$

$$F_{31}^{\lambda_1}(\theta,\phi) = e_{sZ}(\theta,\phi,\lambda_1)e_{sX}(\theta,\phi,\lambda_2)e_{fX}(\theta,\phi,\lambda_3)$$

$$F_{32}^{\lambda_1}(\theta,\phi) = e_{sZ}(\theta,\phi,\lambda_1)e_{sY}(\theta,\phi,\lambda_2)e_{fY}(\theta,\phi,\lambda_3)$$

$$F_{33}^{\lambda_1}(\theta,\phi) = e_{sZ}(\theta,\phi,\lambda_1)e_{sZ}(\theta,\phi,\lambda_2)e_{fZ}(\theta,\phi,\lambda_3). \tag{3.38}$$

3.3.1.2 d_{eff} for KTP for a Type II(sff) mixing process

For Type II, sff type of interaction we have,

$$\begin{aligned}
d_{\text{eff}}^{\text{II(sff)}}(\omega_3) &= e_{fi}(\omega_3)d_{il}\mathsf{u}_l^{sf}(\omega_1,\omega_2)\\
&\equiv G_{15}^{\lambda_3}(\theta,\phi)d_{15}+G_{24}^{\lambda_3}(\theta,\phi)d_{24}+G_{31}^{\lambda_3}(\theta,\phi)d_{31}\\
&\quad +G_{32}^{\lambda_3}(\theta,\phi)d_{32}+G_{33}^{\lambda_3}(\theta,\phi)d_{33}
\end{aligned} \tag{3.39}$$

where

$$G_{15}^{\lambda_3}(\theta,\phi) \equiv e_{fX}(\theta,\phi,\lambda_3)\{e_{sZ}(\theta,\phi,\lambda_1)e_{fX}(\theta,\phi,\lambda_2)+e_{sX}(\theta,\phi,\lambda_1)e_{fZ}(\theta,\phi,\lambda_2)\}$$

$$G_{24}^{\lambda_3}(\theta,\phi) = e_{fY}(\theta,\phi,\lambda_3)\{e_{sY}(\theta,\phi,\lambda_1)e_{fZ}(\theta,\phi,\lambda_2)+e_{sZ}(\theta,\phi,\lambda_1)e_{fY}(\theta,\phi,\lambda_2)\}$$

$$G_{31}^{\lambda_3}(\theta,\phi) = e_{fZ}(\theta,\phi,\lambda_3)e_{sX}(\theta,\phi,\lambda_1)e_{fX}(\theta,\phi,\lambda_2)$$

$$G_{32}^{\lambda_3}(\theta,\phi) = e_{fZ}(\theta,\phi,\lambda_3)e_{sY}(\theta,\phi,\lambda_1)e_{fY}(\theta,\phi,\lambda_2)$$

$$G_{33}^{\lambda_3}(\theta,\phi) = e_{fZ}(\theta,\phi,\lambda_3)e_{sZ}(\theta,\phi,\lambda_1)e_{fZ}(\theta,\phi,\lambda_2). \tag{3.40}$$

Similarly,

$$\begin{aligned}
d_{\text{eff}}^{\text{II(sff)}}(\omega_2) &= e_{fi}(\omega_2)d_{il}\mathsf{u}_l^{sf}(\omega_1,\omega_3)\\
&\equiv G_{15}^{\lambda_2}(\theta,\phi)d_{15}+G_{24}^{\lambda_2}(\theta,\phi)d_{24}+G_{31}^{\lambda_2}(\theta,\phi)d_{31}\\
&\quad +G_{32}^{\lambda_2}(\theta,\phi)d_{32}+G_{33}^{\lambda_2}(\theta,\phi)d_{33}
\end{aligned} \tag{3.41}$$

where

$$G_{15}^{\lambda_2}(\theta,\phi) \equiv e_{fX}(\theta,\phi,\lambda_2)\{e_{sZ}(\theta,\phi,\lambda_1)e_{fX}(\theta,\phi,\lambda_3)+e_{sX}(\theta,\phi,\lambda_1)e_{fZ}(\theta,\phi,\lambda_3)\}$$

$$G_{24}^{\lambda_2}(\theta,\phi) = e_{fY}(\theta,\phi,\lambda_2)\{e_{sY}(\theta,\phi,\lambda_1)e_{fZ}(\theta,\phi,\lambda_3)+e_{sZ}(\theta,\phi,\lambda_1)e_{fY}(\theta,\phi,\lambda_3)\}$$

$$G_{31}^{\lambda_2}(\theta,\phi) = e_{fZ}(\theta,\phi,\lambda_2)e_{sX}(\theta,\phi,\lambda_1)e_{fX}(\theta,\phi,\lambda_3)$$

$$G_{32}^{\lambda_2}(\theta,\phi) = e_{fZ}(\theta,\phi,\lambda_2)e_{sY}(\theta,\phi,\lambda_1)e_{fY}(\theta,\phi,\lambda_3)$$

$$G_{33}^{\lambda_2}(\theta,\phi) = e_{fZ}(\theta,\phi,\lambda_2)e_{sZ}(\theta,\phi,\lambda_1)e_{fZ}(\theta,\phi,\lambda_3). \tag{3.42}$$

and

$$\begin{aligned}
d_{\text{eff}}^{\text{II(sff)}}(\omega_1) &= e_{si}(\omega_1)d_{il}\mathsf{u}_l^{ff}(\omega_2,\omega_3)\\
&\equiv G_{15}^{\lambda_1}(\theta,\phi)d_{15}+G_{24}^{\lambda_1}(\theta,\phi)d_{24}+G_{31}^{\lambda_1}(\theta,\phi)\\
&\quad +G_{32}^{\lambda_1}(\theta,\phi)d_{32}+G_{33}^{\lambda_1}(\theta,\phi)d_{33}
\end{aligned} \tag{3.43}$$

where

$$G_{15}^{\lambda_1}(\theta,\phi) \equiv e_{sX}(\theta,\phi,\lambda_1)\{e_{fZ}(\theta,\phi,\lambda_2)e_{fX}(\theta,\phi,\lambda_3)+e_{fX}(\theta,\phi,\lambda_2)e_{fZ}(\theta,\phi,\lambda_3)\}$$

$$G_{24}^{\lambda_1}(\theta,\phi) = e_{sY}(\theta,\phi,\lambda_1)\{e_{fY}(\theta,\phi,\lambda_2)e_{fZ}(\theta,\phi,\lambda_3)+e_{fZ}(\theta,\phi,\lambda_2)e_{fY}(\theta,\phi,\lambda_3)\}$$

$$G_{31}^{\lambda_1}(\theta,\phi) = e_{sZ}(\theta,\phi,\lambda_1)e_{fX}(\theta,\phi,\lambda_2)e_{fX}(\theta,\phi,\lambda_3)$$

$$G_{32}^{\lambda_1}(\theta,\phi) = e_{sZ}(\theta,\phi,\lambda_1)e_{fY}(\theta,\phi,\lambda_2)e_{fY}(\theta,\phi,\lambda_3)$$

$$G_{33}^{\lambda_1}(\theta,\phi) = e_{sZ}(\theta,\phi,\lambda_1)e_{fZ}(\theta,\phi,\lambda_2)e_{fZ}(\theta,\phi,\lambda_3). \tag{3.44}$$

3.3.1.3 d_{eff} for KTP for a Type II(fsf) mixing process

For Type II, fsf type of interaction we have,

$$
\begin{aligned}
d_{\text{eff}}^{\text{II(fsf)}}(\omega_3) &= e_{fi}(\omega_3)d_{il}u_l^{fs}(\omega_1,\omega_2) \\
&\equiv H_{15}^{\lambda_3}(\theta,\phi)d_{15} + H_{24}^{\lambda_3}(\theta,\phi)d_{24} + H_{31}^{\lambda_3}(\theta,\phi)d_{31} \\
&\quad + H_{32}^{\lambda_3}(\theta,\phi)d_{32} + H_{33}^{\lambda_3}(\theta,\phi)d_{33}
\end{aligned} \tag{3.45}
$$

where

$$H_{15}^{\lambda_3}(\theta,\phi) \equiv e_{fX}(\theta,\phi,\lambda_3)\{e_{fZ}(\theta,\phi,\lambda_1)e_{sX}(\theta,\phi,\lambda_2)+e_{fX}(\theta,\phi,\lambda_1)e_{sZ}(\theta,\phi,\lambda_2)\}$$

$$H_{24}^{\lambda_3}(\theta,\phi) = e_{fY}(\theta,\phi,\lambda_3)\{e_{fY}(\theta,\phi,\lambda_1)e_{sZ}(\theta,\phi,\lambda_2)+e_{fZ}(\theta,\phi,\lambda_1)e_{sY}(\theta,\phi,\lambda_2)\}$$

$$H_{31}^{\lambda_3}(\theta,\phi) = e_{fZ}(\theta,\phi,\lambda_3)e_{fX}(\theta,\phi,\lambda_1)e_{sX}(\theta,\phi,\lambda_2)$$

$$H_{32}^{\lambda_3}(\theta,\phi) = e_{fZ}(\theta,\phi,\lambda_3)e_{fY}(\theta,\phi,\lambda_1)e_{sY}(\theta,\phi,\lambda_2)$$

$$H_{33}^{\lambda_3}(\theta,\phi) = e_{fZ}(\theta,\phi,\lambda_3)e_{fZ}(\theta,\phi,\lambda_1)e_{sZ}(\theta,\phi,\lambda_2). \tag{3.46}$$

Similarly,

$$
\begin{aligned}
d_{\text{eff}}^{\text{II(fsf)}}(\omega_2) &= e_{si}(\omega_2)d_{il}u_l^{ff}(\omega_1,\omega_3) \\
&\equiv H_{15}^{\lambda_2}(\theta,\phi)d_{15} + H_{24}^{\lambda_2}(\theta,\phi)d_{24} + H_{31}^{\lambda_2}(\theta,\phi)d_{31} \\
&\quad + H_{32}^{\lambda_2}(\theta,\phi)d_{32} + H_{33}^{\lambda_2}(\theta,\phi)d_{33}
\end{aligned} \tag{3.47}
$$

where

$$H_{15}^{\lambda_2}(\theta,\phi) \equiv e_{sX}(\theta,\phi,\lambda_2)\{e_{fZ}(\theta,\phi,\lambda_1)e_{fX}(\theta,\phi,\lambda_3)+e_{fX}(\theta,\phi,\lambda_1)e_{fZ}(\theta,\phi,\lambda_3)\}$$

$$H_{24}^{\lambda_2}(\theta,\phi) = e_{sY}(\theta,\phi,\lambda_2)\{e_{fY}(\theta,\phi,\lambda_1)e_{fZ}(\theta,\phi,\lambda_3)+e_{fZ}(\theta,\phi,\lambda_1)e_{fY}(\theta,\phi,\lambda_3)\}$$

$$H_{31}^{\lambda_2}(\theta,\phi) = e_{sZ}(\theta,\phi,\lambda_2)e_{fX}(\theta,\phi,\lambda_1)e_{fX}(\theta,\phi,\lambda_3)$$

$$H_{32}^{\lambda_2}(\theta,\phi) = e_{sZ}(\theta,\phi,\lambda_2)e_{fY}(\theta,\phi,\lambda_1)e_{fY}(\theta,\phi,\lambda_3)$$

$$H_{33}^{\lambda_2}(\theta,\phi) = e_{sZ}(\theta,\phi,\lambda_2)e_{fZ}(\theta,\phi,\lambda_1)e_{fZ}(\theta,\phi,\lambda_3). \tag{3.48}$$

and

$$
\begin{aligned}
d_{\text{eff}}^{\text{II(fsf)}}(\omega_1) &= e_{fi}(\omega_1)d_{il}u_l^{sf}(\omega_2,\omega_3) \\
&\equiv H_{15}^{\lambda_1}(\theta,\phi)d_{15} + H_{24}^{\lambda_1}(\theta,\phi)d_{24} + H_{31}^{\lambda_1}(\theta,\phi)d_{31} \\
&\quad + H_{32}^{\lambda_1}(\theta,\phi)d_{32} + H_{33}^{\lambda_1}(\theta,\phi)d_{33}
\end{aligned} \tag{3.49}
$$

where

$$H_{15}^{\lambda_1}(\theta,\phi) \equiv e_{fX}(\theta,\phi,\lambda_1)\{e_{sZ}(\theta,\phi,\lambda_2)e_{fX}(\theta,\phi,\lambda_3) + e_{sX}(\theta,\phi,\lambda_2)e_{fZ}(\theta,\phi,\lambda_3)\}$$
$$H_{24}^{\lambda_1}(\theta,\phi) = e_{fY}(\theta,\phi,\lambda_1)\{e_{sY}(\theta,\phi,\lambda_2)e_{fZ}(\theta,\phi,\lambda_3) + e_{sZ}(\theta,\phi,\lambda_2)e_{fY}(\theta,\phi,\lambda_3)\}$$
$$H_{31}^{\lambda_1}(\theta,\phi) = e_{fZ}(\theta,\phi,\lambda_1)e_{sX}(\theta,\phi,\lambda_2)e_{fX}(\theta,\phi,\lambda_3)$$
$$H_{32}^{\lambda_1}(\theta,\phi) = e_{fZ}(\theta,\phi,\lambda_1)e_{sY}(\theta,\phi,\lambda_2)e_{fY}(\theta,\phi,\lambda_3)$$
$$H_{33}^{\lambda_1}(\theta,\phi) = e_{fZ}(\theta,\phi,\lambda_1)e_{sZ}(\theta,\phi,\lambda_2)e_{fZ}(\theta,\phi,\lambda_3). \tag{3.50}$$

3.3.2 d_{eff} for KTP for propagation along principal planes

For KTP we have $n_X < n_Y < n_Z$ at each wavelength. Using the walk-off angles from Sec. 2.9.1 and the field directions from Eqns. 1.96 and 1.98 in Sec. 1.6.3, the results for propagation along the principal planes are summarized in the following tables:

Propagation along the ZX plane $(\phi = 0)$, for $\theta < \Omega$

Type	ssf	sff	fsf
$d_{\text{eff}}(\omega_1)$	$-d_{24}\sin(\theta - \rho_{53})$	0	0
$d_{\text{eff}}(\omega_2)$	$-d_{24}\sin(\theta - \rho_{53})$	0	0
$d_{\text{eff}}(\omega_3)$	$-d_{32}\sin(\theta - \rho_{53})$	0	0

$\rho_{53} \equiv \rho_5(\omega_3)$, with ρ_5 given in Eqn 1.133

TABLE 3.6. d_{eff} values for KTP for **k** along the ZX plane, for $\theta < \Omega$

Propagation along the ZX plane $(\phi = 0)$, $\theta > \Omega$

Type	ssf	sff	fsf
$d_{\text{eff}}(\omega_1)$	0	$-d_{32}\sin(\theta - \rho_{51})$	$-d_{24}\sin(\theta - \rho_{52})$
$d_{\text{eff}}(\omega_2)$	0	$-d_{24}\sin(\theta - \rho_{51})$	$-d_{32}\sin(\theta - \rho_{52})$
$d_{\text{eff}}(\omega_3)$	0	$-d_{24}\sin(\theta - \rho_{51})$	$-d_{24}\sin(\theta - \rho_{52})$

$\rho_{51} \equiv \rho_5(\omega_1)$, $\rho_{52} \equiv \rho_5(\omega_2)$, $\rho_{53} \equiv \rho_5(\omega_3)$, , with ρ_5 given in Eqn. 1.133

TABLE 3.7. d_{eff} values for KTP for **k** along the ZX plane, for $\theta > \Omega$

For KTP, $d_{15} = 1.4$, $d_{24} = 2.65$, $d_{33} = 10.7$ (all in pm/V) [4]. Assuming Kleinman's symmetry condition to be valid, $d_{31} = d_{15} = 1.4$ pm/V and $d_{32} = d_{24} = 2.65$ pm/V. The magnitude of the d_{eff} values of KTP for the three cases of phase matching (ssf, sff and fsf) are plotted in Figs. 3.1, 3.2 and 3.3 as functions of the angle θ for various values of ϕ ranging from 0 to 90°. Since

Propagation along the YZ plane, ($\phi = \pi/2$)

Type	ssf	sff	fsf
$d_{\text{eff}}(\omega_1)$	0	$-d_{31}\sin(\theta - \rho_{41})$	$-d_{15}\sin(\theta - \rho_{42})$
$d_{\text{eff}}(\omega_2)$	0	$-d_{15}\sin(\theta - \rho_{41})$	$-d_{31}\sin(\theta - \rho_{42})$
$d_{\text{eff}}(\omega_3)$	0	$-d_{15}\sin(\theta - \rho_{41})$	$-d_{15}\sin(\theta - \rho_{42})$

$\rho_{41} \equiv \rho_4(\omega_1)$, $\rho_{42} \equiv \rho_4(\omega_2)$, with ρ_4 given in Eqn. 1.120

TABLE 3.8. d_{eff} values for KTP for **k** along the YZ plane

Propagation along the XY plane, ($\theta = \pi/2$)

	ssf	sff	fsf
$d_{\text{eff}}(\omega_1)$	0	$-d_{31}\sin(\phi + \rho_{62})\sin(\phi + \rho_{63})$ $-d_{32}\cos(\phi + \rho_{62})\cos(\phi + \rho_{63})$	$-d_{15}\sin(\phi + \rho_{61})\sin(\phi + \rho_{63})$ $-d_{24}\cos(\phi + \rho_{61})\cos(\phi + \rho_{63})$
$d_{\text{eff}}(\omega_2)$	0	$-d_{15}\sin(\phi + \rho_{62})\sin(\phi + \rho_{63})$ $-d_{24}\cos(\phi + \rho_{62})\cos(\phi + \rho_{63})$	$-d_{31}\sin(\phi + \rho_{61})\sin(\phi + \rho_{63})$ $-d_{32}\cos(\phi + \rho_{61})\cos(\phi + \rho_{63})$
$d_{\text{eff}}(\omega_3)$	0	$-d_{15}\sin(\phi + \rho_{62})\sin(\phi + \rho_{63})$ $-d_{24}\cos(\phi + \rho_{62})\cos(\phi + \rho_{63})$	$-d_{15}\sin(\phi + \rho_{61})\sin(\phi + \rho_{63})$ $-d_{24}\cos(\phi + \rho_{61})\cos(\phi + \rho_{63})$

$\rho_{61} \equiv \rho_6(\omega_1)$, $\rho_{62} \equiv \rho_6(\omega_2)$, $\rho_{63} \equiv \rho_6(\omega_3)$, with ρ_6 given in Eqn. 1.145

TABLE 3.9. d_{eff} values for KTP for **k** along the XY plane

Kleinman's symmetry is assumed, the d_{eff} values for a given phase matching type are the same for all three interacting wavelengths, which assumed here to be 1.064 μm, 1.55 μm and 3.39 μm.

The $|d_{\text{eff}}|$ values for $\phi = 0$ correspond to propagation on the ZX plane - and shows discontinuous behavior at $\theta = \Omega$, near 18°. For the fsf phase matching, there is a spike in the $|d_{\text{eff}}|$ for θ approximately 17.5°, as shown in Fig. 3.3 and in greater detail in Fig. 3.4. This behavior is due to the dispersion in the value of Ω (as discussed in Chapter 1).

The case of propagation along the XY plane ($\theta = 90°$) is shown in Fig. 3.5. For the ssf phasematching case, d_{eff} is zero for $\theta = 90°$. For the sff and the fsf phasematching cases, the values of d_{eff} are the same for $\theta = 90°$.

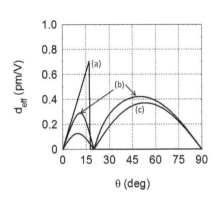

FIGURE 3.1: Magnitude of the d_{eff} values for ssf type phasematching. (a) $\phi = 0$, (b) $\phi = 30°$, (c) $\phi = 60°$.

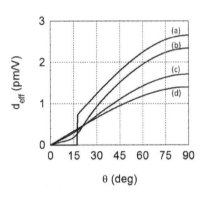

FIGURE 3.2: Magnitude of the d_{eff} values for sff type phasematching. (a) $\phi = 0$, (b) $\phi = 30°$, (c) $\phi = 60°$, (d) $\phi = 90°$.

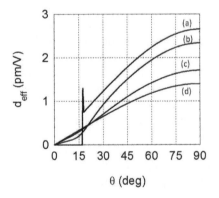

FIGURE 3.3: Magnitude of the d_{eff} values for fsf type phasematching. (a) $\phi = 0$, (b) $\phi = 30°$, (c) $\phi = 60°$, (d) $\phi = 90°$.

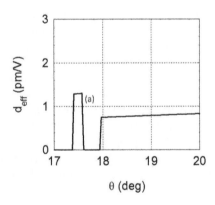

FIGURE 3.4: Magnitude of the d_{eff} values for fsf type phasematching at $\phi = 0$.

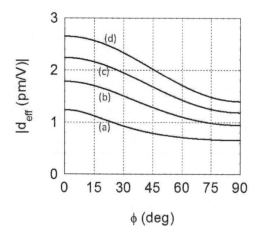

FIGURE 3.5: Magnitude of the d_{eff} values for the sff and fsf type phasematching cases plotted as a function of ϕ, (a) $\theta = 30°$, (b) $\theta = 45°$, (c) $\phi = 60°$, (d) $\phi = 90°$.

3.4 d_{eff} for Uniaxial Crystals

Geometric factors for each of the eighteen d coefficients for the three types of birefringent phase matched interactions (ssf, sfs and fsf) in uniaxial crystals are given in Tables 3.35 - 3.40. In any given crystal, only a few of the eighteen coefficients are nonzero and unique, and the d_{eff} value can be obtained by summing over just these nonzero coefficients. In this section, examples of some commonly used uniaxial nonlinear optical crystals belonging to different point groups are considered.

Negative uniaxial crystals with point group 3m

Crystals such as lithium niobate (LiNbO$_3$), beta-barium borate (β-BaB$_2$O$_4$, BBO), proustite (Ag$_3$AsS$_3$), pyragyrite (Ag$_3$SbS$_3$) and thallium arsenide selenide (Tl$_3$AsSe$_3$, TAS) are negative uniaxial belonging to class 3m.

Table 2.3 shows that for crystals belonging to the class 3m there are eight nonzero values of the d matrix: d_{15}, d_{16}, d_{21}, d_{22}, d_{24}, d_{31}, d_{32} and d_{33}, of which only d_{15}, d_{16} and d_{33} are independent (when crystal symmetry and Kleinman symmetry conditions are applied). The geometric factors for each of the nonzero d coefficients contributing to the d_{eff} at frequency ω_3 are listed in Table 3.10.

Under Kleinman symmetry conditions, using Table 2.3 we have

$$d_{15} = d_{24} = d_{31} = d_{32}$$
$$d_{16} = d_{21} = -d_{22} \tag{3.51}$$

and we obtain the expressions shown in Table 3.11 from Table 3.10:

	ssf (ooe)	sff (oee)	fsf (eoe)
d_{15}	0	$-\sin\phi\cos\phi\sin\theta_2''\cos\theta_3''$	$-\sin\phi\cos\phi\sin\theta_1''\cos\theta_3''$
d_{16}	$\cos\phi\sin 2\phi\cos\theta_3''$	$-\cos\phi\cos 2\phi\cos\theta_2''\cos\theta_3''$	$-\cos\phi\cos 2\phi\cos\theta_1''\cos\theta_3''$
d_{21}	$-\sin^3\phi\cos\theta_3''$	$\sin^2\phi\cos\phi\cos\theta_2''\cos\theta_3''$	$\sin^2\phi\cos\phi\cos\theta_1''\cos\theta_3''$
d_{22}	$-\sin\phi\cos^2\phi\cos\theta_3''$	$-\sin^2\phi\cos\phi\cos\theta_2''\cos\theta_3''$	$-\sin^2\phi\cos\phi\cos\theta_1''\cos\theta_3''$
d_{24}	0	$\sin\phi\cos\phi\sin\theta_2''\cos\theta_3''$	$\sin\phi\cos\phi\sin\theta_1''\cos\theta_3''$
d_{31}	$\sin^2\phi\sin\theta_3''$	$-\sin\phi\cos\phi\cos\theta_2''\sin\theta_3''$	$-\sin\phi\cos\phi\cos\theta_1''\sin\theta_3''$
d_{32}	$\cos^2\phi\sin\theta_3''$	$\sin\phi\cos\phi\cos\theta_2''\sin\theta_3''$	$\sin\phi\cos\phi\cos\theta_1''\sin\theta_3''$
d_{33}	0	0	0

$\theta_1'' \equiv \theta + \rho(\omega_1)$, $\theta_2'' \equiv \theta + \rho(\omega_2)$, $\theta_3'' \equiv \theta + \rho(\omega_3)$, with ρ given in Table 1.6.

TABLE 3.10. Geometric factors contributing to $d_{\text{eff}}(\omega_3)$ for each of the non-zero elements of the d matrix, for negative uniaxial crystals with point group $3m$, such as LiNbO$_3$

	ssf (ooe)	sff (oee)	fsf (eoe)
d_{eff}	$d_{15}\sin\theta_3'' + d_{16}\cos\theta_3''\sin 3\phi$	$-d_{16}\cos\theta_2''\cos\theta_3''\cos 3\phi$	$-d_{16}\cos\theta_2''\cos\theta_3''\cos 3\phi$

TABLE 3.11. d_{eff} values for negative uniaxial crystals with point group $3m$, such as LiNbO$_3$

Negative uniaxial crystals with point group $\bar{4}$2m

Crystals such as silver thiogallate (AgGaS$_2$), silver gallium selenide (AgGaSe$_2$), cadmium germanium arsenide (CdGeAs$_2$), potassium dihydrogen phosphate (KH$_2$PO$_4$, KDP), potassium dihydrogen arsenate (KH$_2$AsO$_4$, KDA) are negative uniaxial crystals in class $\bar{4}$2m.

For crystals belonging to the point group $\bar{4}$2m, Table 2.3 shows that the non-zero d coefficients are d_{14}, d_{25} and d_{36}, of which $d_{25} = d_{14}$. Further, under Kleinman symmetry condition, $d_{36} = d_{14}$, i.e., all three d coefficients are equal.

The geometric factors for the three nonzero d coefficients contributing to the d_{eff} at frequency ω_3 are listed in table 3.12.

	ssf (ooe)	sff (oee)	fsf (eoe)
d_{14}	0	$\cos^2\phi\sin\theta_2''\cos\theta_3''$	$\cos^2\phi\sin\theta_1''\cos\theta_3''$
d_{25}	0	$-\sin^2\phi\sin\theta_2''\cos\theta_3''$	$-\sin^2\phi\sin\theta_1''\cos\theta_3''$
d_{36}	$-\sin 2\phi\sin\theta_3''$	$\cos 2\phi\cos\theta_2''\sin\theta_3''$	$\cos 2\phi\cos\theta_1''\sin\theta_3''$

$\theta_1'' \equiv \theta + \rho(\omega_1)$, $\theta_2'' \equiv \theta + \rho(\omega_2)$, $\theta_3'' \equiv \theta + \rho(\omega_3)$, with ρ given in Table 1.6.

TABLE 3.12. Geometric factors contributing to $d_{\text{eff}}(\omega_3)$ for each of the non-zero elements of the d matrix, for negative uniaxial crystals with point group $\bar{4}$2m, such as AgGaSe$_2$

The expressions for d_{eff} for negative uniaxial crystals such as AgGaSe$_2$ (when Kleinman symmetry conditions hold) are shown in Table 3.13

	ssf (ooe)	sff (oee)	fsf (eoe)
d_{eff}	$-d_{14}\sin 2\phi \sin\theta_3''$	$d_{14}\cos 2\phi \sin(\theta_2'' + \theta_3'')$	$d_{14}\cos 2\phi \sin(\theta_1'' + \theta_3'')$

TABLE 3.13. d_{eff} values for negative uniaxial crystals with point group $\bar{4}$2m, such as AgGaSe$_2$

Positive uniaxial crystals with point group $\bar{4}$2m

Crystals such as zinc germanium diphosphide (ZnGeP$_2$, ZGP), cadmium germanium arsenide (CdGeAs$_2$, CGA) and urea (CO(NH$_2$)) are examples of positive uniaxial crystals in class $\bar{4}$2m.

The geometric factors for the three nonzero d coefficients contributing to the d_{eff} at frequency ω_3 for positive uniaxial crystals with point group $\bar{4}$2m are listed in Table 3.14.

	ssf (eeo)	sff (eoo)	fsf (oeo)
d_{14}	$-\sin^2 \phi \sin(\theta_1' + \theta_2')$	$\sin\phi\cos\phi\sin\theta_1'$	$\sin\phi\cos\phi\sin\theta_2'$
d_{25}	$\cos^2 \phi \sin(\theta_1' + \theta_2')$	$\sin\phi\cos\phi\sin\theta_1'$	$\sin\phi\cos\phi\sin\theta_2'$
d_{36}	0	0	0

$\theta_1' \equiv \theta - \rho(\omega_1)$, $\theta_2' \equiv \theta - \rho(\omega_2)$, with ρ given in Table 1.6.

TABLE 3.14. Geometric factors contributing to $d_{\text{eff}}(\omega_3)$ for each of the nonzero elements of the d matrix, for positive uniaxial crystals with point group $\bar{4}$2m, such as ZnGeP$_2$

The expressions for d_{eff} for negative uniaxial crystals such as ZnGeP$_2$ are shown in Table 3.15

	ssf (eeo)	sff (eoo)	fsf (oeo)
d_{eff}	$d_{14}\cos 2\phi \sin(\theta_1' + \theta_2')$	$d_{14}\sin 2\phi \sin\theta_1'$	$d_{14}\sin 2\phi \sin\theta_2'$

TABLE 3.15. d_{eff} values for positive uniaxial crystals with point group $\bar{4}$2m, such as ZnGeP$_2$

Positive uniaxial crystal with point group 6mm

For positive uniaxial crystals such as cadmium selenide (CdSe) belonging to the point group 6mm, Table 2.3 shows that the non-zero d coefficients are d_{15}, d_{24}, d_{31}, d_{32} and d_{33}, of which $d_{24} = d_{15}$ and $d_{32} = d_{31}$. Further, under Kleinman symmetry condition, $d_{31} = d_{15}$.

	ssf (eeo)	sff (eoo)	fsf (oeo)
d_{15}	$-\sin\phi\cos\phi\sin(\theta_1'+\theta_2')$	$-\sin^2\phi\sin\theta_1'$	$-\sin^2\phi\sin\theta_2'$
d_{24}	$\sin\phi\cos\phi\sin(\theta_1'+\theta_2')$	$-\cos^2\phi\sin\theta_1'$	$-\cos^2\phi\sin\theta_2'$
d_{31}	0	0	0
d_{32}	0	0	0
d_{33}	0	0	0

$\theta_1' \equiv \theta - \rho(\omega_1)$, $\theta_2' \equiv \theta - \rho(\omega_2)$, with ρ given in Table 1.6.

TABLE 3.16. Geometric factors contributing to $d_{\text{eff}}(\omega_3)$ for each of the non-zero elements of the d matrix, for positive uniaxial crystals with point group 6mm, such as CdSe

The geometric factors for the three nonzero d coefficients contributing to the d_{eff} at frequency ω_3 are listed in Table 3.16.

The expressions for d_{eff} for positive uniaxial crystals such as CdSe are shown in Table 3.17

	ssf (eeo)	sff (eoo)	fsf (oeo)
d_{eff}	0	$-d_{15}\sin\theta_1'$	$-d_{15}\sin\theta_2'$

TABLE 3.17. d_{eff} for positive uniaxial crystals with point group 6mm, such as CdSe

Negative uniaxial crystals with point group $\bar{6}$m2

For negative uniaxial crystals belonging to the point group $\bar{6}$m2, such as gallium selenide (GaSe), the non-zero d coefficients are d_{16}, d_{21} and d_{22}, of which $d_{21} = d_{16}$ and $d_{22} = -d_{16}$.

The geometric factors for the three nonzero d coefficients contributing to the d_{eff} at frequency ω_3 are listed in Table 3.18.

	ssf (ooe)	sff (oee)	fsf (eoe)
d_{16}	$\cos\phi\sin 2\phi\cos\theta_3''$	$-\cos\phi\cos 2\phi\cos\theta_2''\cos\theta_3''$	$-\cos\phi\cos 2\phi\cos\theta_1''\cos\theta_3''$
d_{21}	$-\sin^3\phi\cos\theta_3''$	$\sin^2\phi\cos\phi\cos\theta_2''\cos\theta_3''$	$\sin^2\phi\cos\phi\cos\theta_1''\cos\theta_3''$
d_{22}	$-\sin\phi\cos^2\phi\cos\theta_3''$	$-\sin^2\phi\cos\phi\cos\theta_2''\cos\theta_3''$	$-\sin^2\phi\cos\phi\cos\theta_1''\cos\theta_3''$

$\theta_1'' \equiv \theta + \rho(\omega_1)$, $\theta_2'' \equiv \theta + \rho(\omega_2)$, $\theta_3'' \equiv \theta + \rho(\omega_3)$, with ρ given in Table 1.6.

TABLE 3.18. Geometric factors contributing to $d_{\text{eff}}(\omega_3)$ for each of the non-zero elements of the d matrix, for negative uniaxial crystals with point group $\bar{6}$m2, such as GaSe

The expressions for d_{eff} for negative uniaxial crystals with point group $\bar{6}$m2, such as GaSe, are shown in Table 3.19

	ssf (*ooe*)	sff (*oee*)	fsf (*eoe*)
d_{eff}	$d_{16} \sin 3\phi \cos \theta_3''$	$-d_{16} \cos 3\phi \cos \theta_2'' \cos \theta_3''$	$-d_{16} \cos 3\phi \cos \theta_1'' \cos \theta_3''$

TABLE 3.19. d_{eff} value for negative uniaxial crystals with point group $\bar{6}$m2, such as GaSe

Positive uniaxial crystals with point group 32

For positive uniaxial crystals belonging to the point group 32, such as selenium (Se), tellurium (Te), or cinnabar (HgS), , the non-zero d coefficients are d_{11}, d_{12}, d_{14}, d_{25}, and d_{26}, of which $d_{12} = -d_{11}$, $d_{25} = -d_{14}$ and $d_{26} = -d_{11}$. Since Kleinman symmetry requires $d_{25} = d_{14}$, the coefficients d_{14} and d_{25} are zero.

The geometric factors for the three nonzero d coefficients contributing to the d_{eff} at frequency ω_3 are listed in table 3.18.

	ssf (*eeo*)	sff (*eoo*)	fsf (*oeo*)
d_{11}	$\sin \phi \cos^2 \phi \cos \theta_1' \cos \theta_2'$	$\sin^2 \phi \cos \phi \cos \theta_1'$	$\sin^2 \phi \cos \phi \cos \theta_2'$
d_{12}	$\sin^3 \phi \cos \theta_1' \cos \theta_2'$	$- \sin^2 \phi \cos \phi \cos \theta_1'$	$- \sin^2 \phi \cos \phi \cos \theta_2'$
d_{26}	$- \cos \phi \sin 2\phi \cos \theta_1' \cos \theta_2'$	$\cos \phi \cos 2\phi \cos \theta_1'$	$\cos \phi \cos 2\phi \cos \theta_2'$

$\theta_1' \equiv \theta - \rho(\omega_1)$, $\theta_2' \equiv \theta - \rho(\omega_2)$, with ρ given in Table 1.6.

TABLE 3.20. Geometric factors contributing to $d_{\text{eff}}(\omega_3)$ for each of the non-zero elements of the d matrix, for positive uniaxial crystal with point group 32, such as Se, Te, HgS

The expressions for d_{eff} for positive uniaxial crystals such as Se, Te or HgS are shown in Table 3.21

	ssf (*eeo*)	sff (*eoo*)	fsf (*oeo*)
d_{eff}	$d_{11} \sin 3\phi \cos \theta_1' \cos \theta_2'$	$-d_{11} \cos 3\phi \cos \theta_1'$	$-d_{11} \cos 3\phi \cos \theta_2'$

TABLE 3.21. d_{eff} value for positive uniaxial crystal with point group 32, such as Se, Te, HgS

3.5 d_{eff} for Isotropic Crystals

The magnitude of the d coefficients of many cubic crystals such as GaAs or CdTe can be substantial, exceeding 50 pm/V. Although birefringent phase matching is not possible in them because they are optically isotropic, the technique of quasi phase-matching (QPM), which will be discussed in detail in Chapter 5, provides a way to obtain large frequency conversion efficiency from such materials. To increase the efficiency of the QPM process, the propagation direction and polarization directions of the incident beams of light need to be chosen to maximize the value of d_{eff}, the effective nonlinear coefficient. Here we derive the expressions for d_{eff} in isotropic crystals with the aim of determining the optimum angles for the orientations of the propagation vector ($\mathbf{k}$) and the polarization directions of the electric fields with respect to the crystal axes.

In isotropic materials, the electric field $\mathbf{E}$ (with unit vector $\hat{e}$) and the displacement vector $\mathbf{D}$ are parallel to each other, so $\mathbf{k}$ and $\hat{e}$ are perpendicular. For a given direction of $\mathbf{k}$, $\hat{e}$ must lie on a plane perpendicular to $\mathbf{k}$, but is allowed to point in any direction on that plane.

For $\mathbf{k}$ pointing along the direction characterized by the angles θ, ϕ with respect to the dielectric axes X,Y,Z (as shown in Fig. 1.3), unit vectors $\hat{u}_1$ and $\hat{u}_2$ lying on the plane perpendicular to $\mathbf{k}$ were defined in Chapter 1, (Eqns. 1.75 and 1.76) as

$$\begin{aligned}
\hat{u}_1 &= \hat{X}\,\cos\theta\cos\phi + \hat{Y}\,\cos\theta\sin\phi - \hat{Z}\,\sin\theta \\
\hat{u}_2 &= \hat{X}\,\sin\phi - \hat{Y}\,\cos\phi.
\end{aligned} \tag{3.52}$$

For light traveling with propagation vector $\mathbf{k}$, $\hat{e}$ lies on the plane containing $\hat{u}_1$ and $\hat{u}_2$. Since $\hat{u}_1$ and $\hat{u}_2$ are perpendicular to each other, they can be used as the axes of a two dimensional coordinate system on the plane perpendicular to $\mathbf{k}$, and the mutually perpendicular vectors ($\hat{u}_1$, $\hat{u}_2$, $\mathbf{k}$) can be used as the axes of a three dimensional coordinate system as the alternate to the principal dielectric axes (X, Y, Z).

We choose here the process of sum-frequency generation as a representative three wave mixing process. If $\hat{e}_1$ and $\hat{e}_2$ denote the unit vectors at frequencies ω_1 and ω_2 which make angles ψ_1 and ψ_2 with respect to $\hat{u}_2$, as shown in Fig. 3.6, we obtain from Eqn. 3.52

$$\begin{aligned}
\hat{e}(\omega_1) &= \hat{u}_1\,\sin\psi_1 + \hat{u}_2\,\cos\psi_1 \\
&= \hat{X}(\cos\theta\cos\phi\sin\psi_1 + \sin\phi\cos\psi_1) + \hat{Y}(\cos\theta\sin\phi\sin\psi_1 - \cos\phi\cos\psi_1) \\
&\quad - \hat{Z}(\sin\theta\sin\psi_1)
\end{aligned}$$

$$\tag{3.53}$$

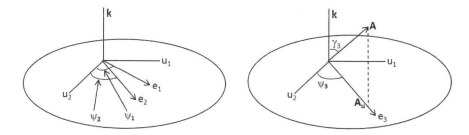

FIGURE 3.6: The unit vectors $\hat{e}_1$ and $\hat{e}_2$ of the electric fields at frequencies ω_1 and ω_2 with respect to the u_1-u_2 axes.

FIGURE 3.7: The unit vector $\hat{e}_3$ of the electric field at frequency ω_3 with respect to the u_1-u_2 axes.

and similarly,

$$\hat{e}(\omega_2) = \hat{u}_1 \, \sin\psi_2 + \hat{u}_2 \, \cos\psi_2$$
$$= \hat{X}(\cos\theta\cos\phi\sin\psi_2 + \sin\phi\cos\psi_2) + \hat{Y}(\cos\theta\sin\phi\sin\psi_2 - \cos\phi\cos\psi_2)$$
$$- \hat{Z}(\sin\theta\sin\psi_2).$$

$$(3.54)$$

In isotropic crystals, there are only three nonzero d coefficients, d_{14}, d_{25} and d_{36}, all of which are equal to each other. From Eqns. 3.6 the Cartesian components of the nonlinear polarization, say at frequency ω_3, are given by

$$
\begin{aligned}
P_X(\omega_3) &= 4\varepsilon_0 d_{14} u_4(\omega_1,\omega_2) E_1 E_2 \\
&= 4\varepsilon_0 d_{14}\{e_Y(\omega_1)e_Z(\omega_2) + e_Z(\omega_1)e_Y(\omega_2)\}E_1 E_2 \\
P_Y(\omega_3) &= 4\varepsilon_0 d_{25} u_5(\omega_1,\omega_2) E_1 E_2 \\
&= 4\varepsilon_0 d_{25}\{e_Z(\omega_1)e_X(\omega_2) + e_X(\omega_1)e_Z(\omega_2)\}E_1 E_2 \\
P_Z(\omega_3) &= 4\varepsilon_0 d_{36} u_6(\omega_1,\omega_2) E_1 E_2 \\
&= 4\varepsilon_0 d_{36}\{e_X(\omega_1)e_Y(\omega_2) + e_Y(\omega_1)e_X(\omega_2)\}E_1 E_2. \quad (3.55)
\end{aligned}
$$

Substituting the values of e_X, e_Y and e_Z at the two frequencies from Eqns. 3.53 and 3.54 and using the equality of the three nonzero d coefficients, we find

$$
\begin{aligned}
P_X(\omega_3) &= 4\varepsilon_0 d_{14} A_X E_1 E_2 \\
P_Y(\omega_3) &= 4\varepsilon_0 d_{14} A_Y E_1 E_2 \\
P_Z(\omega_3) &= 4\varepsilon_0 d_{14} A_Z E_1 E_2 \quad (3.56)
\end{aligned}
$$

where E_1 and E_2 denote the field amplitudes at frequencies ω_1 and ω_2 respectively, and

$$A_X = \sin\theta\cos\phi\sin(\psi_1 + \psi_2) - \sin 2\theta\sin\phi\sin\psi_1\sin\psi_2$$
$$A_Y = -\sin\theta\sin\phi\sin(\psi_1 + \psi_2) - \sin 2\theta\cos\phi\sin\psi_1\sin\psi_2$$
$$A_Z = -\cos\theta\cos 2\phi\sin(\psi_1 + \psi_2) - \sin 2\phi(\cos\psi_1\cos\psi_2 - \cos^2\theta\sin\psi_1\sin\psi_2).$$
$$(3.57)$$

Defining

$$p \equiv \sin(\psi_1 + \psi_2), \quad q \equiv \sin\psi_1\sin\psi_2, \quad r \equiv \cos\psi_1\cos\psi_2 \qquad (3.58)$$

we can rewrite Eqns. 3.57 as

$$A_X = p\,\sin\theta\cos\phi - q\,\sin 2\theta\sin\phi$$
$$A_Y = -p\,\sin\theta\sin\phi - q\,\sin 2\theta\cos\phi$$
$$A_Z = -p\,\cos\theta\cos 2\phi + q\,\cos^2\theta\sin 2\phi - r\,\sin 2\phi. \qquad (3.59)$$

The generated nonlinear polarization vector at frequency ω_3 is therefore given by

$$\mathbf{P}(\omega_3) = 4\varepsilon_0 d_{14}\mathbf{A}E_1E_2 \qquad (3.60)$$

where the vector $\mathbf{A}$ with components given in Eqns. 3.57 is along $\mathbf{P}(\omega_3)$ with magnitude

$$A = (A_X^2 + A_Y^2 + A_Z^2)^{1/2}. \qquad (3.61)$$

For light generated at frequency ω_3 and traveling with propagation vector along $\mathbf{k}$, the electric field (denoted by $\mathbf{E}_3$) lies on the u_1-u_2 plane. The 'effective' nonlinear polarization component that gives rise to the light beam traveling collinearly with the two incident beams with wave vector $\mathbf{k}$ is the component of $\mathbf{P}(\omega_3)$ on the u_1-u_2 plane. Denoting this component as $\mathbf{P}_u(\omega_3)$, we have

$$\mathbf{P}_u(\omega_3) = 4\varepsilon_0 d_{14}\mathbf{A}_\mathbf{u}E_1E_2$$
$$\equiv 4\varepsilon_0 d_{\text{eff}}E_1E_2 \qquad (3.62)$$

where

$$d_{\text{eff}} = d_{14}\mathbf{A}_\mathbf{u} \qquad (3.63)$$

the vector $\mathbf{A}_u$ being the projection of the vector $\mathbf{A}$ on the u_1-u_2 plane.

$\mathbf{A}$ can be expressed in the mutually perpendicular $\hat{u}_1$, $\hat{u}_2$ and $\hat{m}$ coordinate system as

$$\mathbf{A} = A_{u1}\,\hat{u}_1 + A_{u2}\,\hat{u}_2 + A_m\,\hat{m} \qquad (3.64)$$

where A_{u1}, A_{u2} and A_m are the components of $\mathbf{A}$ along the three new coordinate axes, given by

$$
\begin{aligned}
A_{u1} &\equiv \mathbf{A} \cdot \hat{u}_1 &&= \cos\theta\cos\phi A_X + \cos\theta\sin\phi A_Y - \sin\theta A_Z \\
A_{u2} &\equiv \mathbf{A} \cdot \hat{u}_2 &&= \sin\phi A_X - \cos\phi A_Y \\
A_m &\equiv \mathbf{A} \cdot \hat{m} &&= \sin\theta\cos\phi A_X + \sin\theta\sin\phi A_Y + \sin\theta A_Z.
\end{aligned} \tag{3.65}
$$

Using the values of A_X, A_Y and A_Z from Eqns. 3.59 we obtain

$$
\begin{aligned}
A_{u1} &= p\sin 2\theta\cos 2\phi - \frac{3}{2}q\sin 2\theta\cos\theta\sin 2\phi + r\sin\theta\sin 2\phi \\
A_{u2} &= p\sin\theta\sin 2\phi + q\sin 2\theta\cos 2\phi
\end{aligned} \tag{3.66}
$$

and

$$
\mathbf{A}_u = A_{u1}\,\hat{u}_1 + A_{u2}\,\hat{u}_2. \tag{3.67}
$$

From Eqn. 3.63, we than have

$$
d_{\text{eff}} = d_{14}\sqrt{A_{u1}^2 + A_{u2}^2}. \tag{3.68}
$$

Calculating d_{eff} for different values of θ, ϕ, ψ_1 and ψ_2, it is found that for any value of θ and ϕ, the maximum of d_{eff} is obtained with $\psi_1 = \psi_2 \equiv \psi$, say, and that the maximum possible value of d_{eff}/d_{14} is $2/\sqrt{3}$, i.e., 1.1547. This maximum is obtained for the angle θ equal to 45°, the angle ϕ equal to either 0 or 90° and the angle ψ given by

$$
\begin{aligned}
\psi &= \tan^{-1}\frac{1}{\cos\theta} \\
&= \tan^{-1}\sqrt{2} = 54.735°.
\end{aligned} \tag{3.69}
$$

This combination of θ, ϕ and ψ corresponds to the electric field direction along $< 111 >$ crystal direction, i.e., along the body diagonals, for which $\hat{e}_3$ has components of equal magnitudes along all three principal dielectric axes.

Figures 3.8 - 3.11 show the dependence of d_{eff} on the angle ψ (assuming again that $\psi_1 = \psi_2 \equiv \psi$) for various directions of $\mathbf{k}$ ($\phi = 0$, 30°, 45° and 60°, and $\theta = 30°$, 45° and 60° and 90°). When $\theta = 0$, $d_{\text{eff}} = 0$ for all values ϕ and ψ. For $\phi = 0$, $d_{\text{eff}} = 0$ for both $\theta = 0°$ and $\theta = 90°$ cases. Also, the results for $\phi = 90°$ are identical with those for $\phi = 0°$.

In Type II interactions, the two incident beams in a three wave mixing process are distinct by polarization. Denoting by ψ_1 and ψ_2 the angles between the electric field vectors of the two incident beams with the unit vector $\hat{u}_2$, the values of $(d_{\text{eff}}/d_{14})^2$ (for $\theta = 45°$ and $\phi = 0$) are plotted against ψ_1 for different values of ψ_2 in Fig. 3.12. We find that at $\psi_2 = 0$, the maximum value of $(d_{\text{eff}}/d_{14})^2$ is 1, which is obtained at ψ_1 equal to 90° and 270°. As ψ_2 increases, the maximum value of $(d_{\text{eff}}/d_{14})^2$ also increases, reaching the value of 4/3 at $\psi_1 = \psi_2 = \tan^{-1}\sqrt{2}$, i.e., 54.7356°. When ψ_2 is equal to 90°, d_{eff} is equal to d_{14}, irrespective of the value of the angle ψ_1 as shown by the dashed line for case (f) in Fig. 3.12.

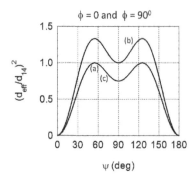

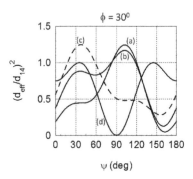

FIGURE 3.8: Values of $(d_{\text{eff}}/d_{14})^2$ plotted against ψ for $\phi = 0°$ and $90°$ (a) $\theta = 30°$, (b) $\theta = 45°$, (c) $\theta = 60°$

FIGURE 3.9: Values of $(d_{\text{eff}}/d_{14})^2$ plotted against ψ for $\phi = 30°$.(a) $\theta = 30°$, (b) $\theta = 45°$, (c) $\theta = 60°$, (d) $\theta = 90°$

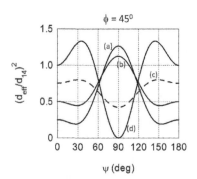

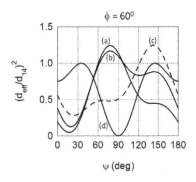

FIGURE 3.10: Values of $(d_{\text{eff}}/d_{14})^2$ plotted against ψ for $\phi = 45°$ (a) $\theta = 30°$, (b) $\theta = 45°$, (c) $\theta = 60°$, (d) $\theta = 90°$

FIGURE 3.11: Values of $(d_{\text{eff}}/d_{14})^2$ plotted against ψ for $\phi = 60°$.(a) $\theta = 30°$, (b) $\theta = 45°$, (c) $\theta = 60°$, (d) $\theta = 90°$

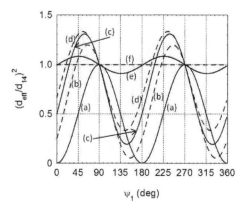

FIGURE 3.12: Values of $(d_{\text{eff}}/d_{14})^2$ plotted against ψ for $\theta = 45°$ and $\phi = 60°$, for different values of ψ_2:(a) $\psi_2 = 0°$, (b) $\psi_2 = 30°$, (c) $\psi_2 = 45°$, (d) $\psi_2 = 54.7356°$, (e) $\psi_2 = 85°$ and (f) $\psi_2 = 90°$.

3.5.1 The direction of the nonlinear polarization

The angle ψ_3 between $\mathbf{A}_u$ and $\hat{u}_2$ as shown in Fig. 3.7 is given by

$$\cos \psi_3 = \frac{A_{u2}}{\sqrt{A_{u1}^2 + A_{u2}^2}}. \tag{3.70}$$

Since ψ_3 can vary between 0 and 360° depending on the sign of the components A_{u1} and A_{u2}, we define the acute angle

$$\psi_{30} = \cos^{-1} \frac{|A_{u2}|}{\sqrt{A_{u1}^2 + A_{u2}^2}} \tag{3.71}$$

and use the Table 3.22 below to determine the value of ψ_3.

A_{u1}	A_{u2}	ψ_3
positive	positive	ψ_{30}
positive	negative	$180° - \psi_{30}$
negative	positive	$360° - \psi_{30}$
negative	negative	$180° + \psi_{30}$

TABLE 3.22. The values of angle ψ_3 for different signs of A_{u1} and A_{u2}

As an example, let us consider propagation of light beams along the [011] direction, i.e., with $\mathbf{k}$ vector on the YZ plane at an angle of 45° with the Z axis as shown in Fig. 3.14. Thus for this case $\theta = 45°$ and $\phi = 90°$, for which we get from Eqns. 3.66

$$A_{u1} = -p \quad \text{and} \quad A_{u2} = -q. \tag{3.72}$$

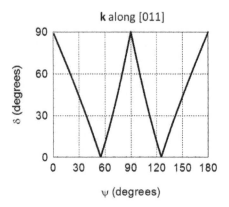

FIGURE 3.13: Values of the angle δ plotted against ψ.

With $\psi_1 = \psi_2 = \psi$ (that maximizes d_{eff}) we have $p = \sin 2\psi$ and $q = \sin^2 \psi$ and the angle ψ_3 can be calculated using Eqn. 3.71 and Table 3.22, with

$$
\begin{aligned}
\psi_{30} &= \cos^{-1} \frac{q}{\sqrt{p^2 + q^2}} \\
&= \cos^{-1} \frac{\sin^2 \psi}{\sqrt{\sin^2 2\psi + \sin^4 \psi}} \\
&= \cos^{-1} \frac{\sin \psi}{\sqrt{4 - 3\sin^2 \psi}}.
\end{aligned} \tag{3.73}
$$

As ψ increases from 0 to 90°, we find that ψ_{30} decreases from 90° to 0, while A_{u1} and A_{u2} are both negative, i.e., $\mathbf{A}_u$ lies in the third quadrant of the u_1-u_2 coordinate system. When ψ increases from 90° to 180°, ψ_{30} increases from 0 to 90°, while A_{u1} is positive and A_{u2} is negative, i.e., $\mathbf{A}_u$ lies in the second quadrant of the u_1-u_2 coordinate system.

Since polarizers can measure only the direction of orientation (and not the sense) of the polarization vectors, the angle between the directions of polarization of the incident fields and the generated field is the smaller of the two supplementary angles formed by the crossing of the two vectors, and for this case ($\mathbf{k}$ along [011]) it can be shown to be given by

$$
\begin{aligned}
\delta &= |\psi_{30} - \psi| & \text{for } \psi \text{ between 0 and 90°} \\
&= |\psi_{30} - (\pi - \psi)| & \text{for } \psi \text{ between 90° and 180°.}
\end{aligned} \tag{3.74}
$$

Figure 3.13 shows the dependence of δ on ψ for a three wave mixing process with $\mathbf{k}$ along the [011] direction of an isotropic crystal.

3.5.2 Propagation along principal planes

Since semiconductor wafers are commonly available with faces normal to the dielectric principal axes and they cleave easily along diagonal planes such as (110), propagation with **k** vector along the $< 110 >$ set of directions is of interest. For these **k** directions, d_{eff} takes simple forms. The angles θ and ϕ for different diagonal directions in these planes are shown in Fig. 3.14. The directions of the electric fields and the projection of the nonlinear polarization on the u_1-u_2 plane for propagation along YZ, ZX and XY planes are shown below.

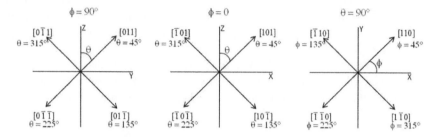

FIGURE 3.14: The diagonal directions along the principal planes YZ, ZX and XY.

Propagation along the YZ plane ($\phi = 90°$)

The directions of the **k** vector, the unit vectors $\hat{u}_1$ and $\hat{u}_2$ defined in Eqn. 3.52, the electric field (from Eqn. 3.53) and the component of the nonlinear polarization on the u_1-u_2 plane for a (100) face wafer are shown in Fig. 3.15. For $\phi = 90°$, we have

$$
\begin{aligned}
\hat{m} &= \hat{Y}\sin\theta + \hat{Z}\cos\theta \\
\hat{u}_1 &= \hat{Y}\cos\theta - \hat{Z}\sin\theta \\
\hat{u}_2 &= \hat{X} \\
\hat{e} &= \hat{X}\cos\psi + \hat{Y}\cos\theta\sin\psi - \hat{Z}\sin\theta\sin\psi \\
\mathbf{A_u} &= \hat{u}_1\sin\gamma + \hat{u}_2\cos\gamma
\end{aligned}
\tag{3.75}
$$

where $\hat{m}$ is the unit vector along **k**. From Eqn. 3.66 we get

$$
A_{u1} = -p\,\sin 2\theta, \qquad A_{u2} = -q\,\sin 2\theta
\tag{3.76}
$$

so that

$$
d_{\text{eff}} = d_{14}\,\sqrt{p^2 + q^2}\,\sin 2\theta
\tag{3.77}
$$

and the angle γ between the polarization component on the u_1-u_2 plane and the u_2 axis is given by

$$\begin{aligned}
\tan\gamma &= \frac{A_{u1}}{A_{u2}} \\
&= \frac{p}{q} \\
&= \frac{\sin(\psi_1 + \psi_2)}{\sin\psi_1 \sin\psi_2}.
\end{aligned} \qquad (3.78)$$

Thus for given values of ψ_1 and ψ_2, d_{eff} is maximized with $\sin 2\theta$ equal to 1, i.e., with θ equal to $45°$, which corresponds to $\mathbf{k}$ along $[011]$ as shown in Fig. 3.14. It is also straightforward to show analytically that d_{eff}^2 maximizes when $\psi_1 = \psi_2 = \psi$, where $\tan\psi = \sqrt{2}$, i.e., $\psi \approx 54.736°$ with maximum value of $4/3$ for $(d_{\text{eff}}/d_{14})^2$.

Since in a cubic crystal, X, Y and Z are interchangeable, similar results are obtained for propagation along the ZX and the XY planes (with the angle ϕ replacing θ for the propagation in the XY plane), as shown below.

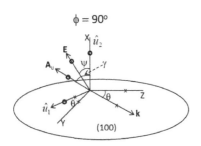

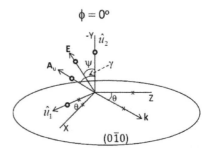

FIGURE 3.15: Directions of the different vectors with respect to the principal dielectric axes for propagation along the YZ plane. Multiple vectors marked with the same symbol are coplanar.

FIGURE 3.16: Directions of the different vectors with respect to the principal dielectric axes for propagation along the ZX plane. Multiple vectors marked with the same symbol are coplanar.

Propagation along the ZX plane ($\phi = 0°$)

The directions of the $\mathbf{k}$ vector, the unit vectors $\hat{u}_1$ and $\hat{u}_2$ defined in Eqn. 3.52, the electric field (from Eqn. 3.53) and the component of the nonlinear polarization on the u_1-u_2 plane for a $(0\bar{1}1)$ face wafer are shown in Fig. 3.16. For

$\phi = 0°$, we have

$$
\begin{aligned}
\hat{m} &= \hat{X}\sin\theta + \hat{Z}\cos\theta \\
\hat{u}_1 &= \hat{X}\cos\theta - \hat{Z}\sin\theta \\
\hat{u}_2 &= -\hat{Y} \\
\hat{e} &= \hat{X}\cos\theta\sin\psi - \hat{Y}\cos\psi - \hat{Z}\sin\theta\sin\psi \\
\mathbf{A_u} &= \hat{u}_1\sin\gamma + \hat{u}_2\cos\gamma.
\end{aligned}
\tag{3.79}
$$

From Eqn. 3.66 we get

$$
A_{u1} = p\ \sin 2\theta, \qquad A_{u2} = q\ \sin 2\theta \tag{3.80}
$$

so that

$$
d_{\text{eff}} = d_{14}\ \sqrt{p^2 + q^2}\sin 2\theta \tag{3.81}
$$

and the angle γ between the polarization component on the u_1-u_2 plane and the u_2 axis is the same as that in Eqn. 3.78.

Propagation along the XY plane ($\theta = 90°$)

The directions of the $\mathbf{k}$ vector, the unit vectors $\hat{u}_1$ and $\hat{u}_2$ defined in Eqn. 3.52, the electric field (from Eqn. 3.53) and the component of the nonlinear polarization on the u_1-u_2 plane for a $(00\bar{1})$ face wafer are shown in Fig. 3.17. For $\theta = 90°$, we have

$$
\begin{aligned}
\hat{m} &= \hat{X}\cos\phi + \hat{Y}\sin\phi \\
\hat{u}_1 &= -\hat{Z} \\
\hat{u}_2 &= \hat{X}\sin\theta - \hat{Y}\cos\phi \\
\hat{e} &= \hat{X}\sin\phi\cos\psi - \hat{Y}\cos\phi\cos\psi - \hat{Z}\sin\psi \\
\mathbf{A_u} &= \hat{u}_1\sin\gamma + \hat{u}_2\cos\gamma.
\end{aligned}
\tag{3.82}
$$

From Eqn. 3.66 we get

$$
A_{u1} = r\ \sin 2\phi, \qquad A_{u2} = p\ \sin 2\phi \tag{3.83}
$$

so that

$$
d_{\text{eff}} = d_{14}\ \sqrt{r^2 + p^2}\sin 2\phi \tag{3.84}
$$

and the angle γ between the polarization component on the u_1-u_2 plane and the u_2 axis is given by

$$
\begin{aligned}
\tan\gamma &= \frac{A_{u1}}{A_{u2}} \\
&= \frac{r}{p} \\
&= \frac{\cos\psi_1\cos\psi_2}{\sin(\psi_1 + \psi_2)}.
\end{aligned}
\tag{3.85}
$$

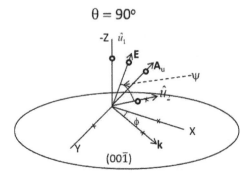

FIGURE 3.17: Directions of the different vectors with respect to the principal dielectric axes for propagation along the XY plane. Multiple vectors marked with the same symbol are coplanar.

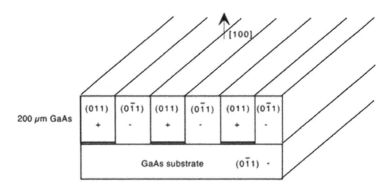

FIGURE 3.18: Orientation Patterning in GaAs, modified figure from [6].

3.5.3 Propagation through orientation patterned material

As will be discussed in detail in Chapter 5, a way to obtain substantial frequency conversion efficiency in cubic crystals is to orientation pattern them in a layered structure in which the alternate layers have d coefficients with opposite signs. If the layer thicknesses are odd multiples of the coherence length l_c (defined in Eqn. 2.44), the generated field builds up through each layer. If there are N such layers, the field generated at the end of N layers is proportional to N, and the generated irradiance is proportional to N^2. Here we discuss how the d coefficients with alternating signs can be obtained from alternating layer orientation.

Figure 3.18 shows an example of orientation patterned GaAs, in which layers of GaAs are epitaxially grown on a bulk GaAs substrate, with the crystallographic orientation of each layer oriented 90° with respect to that of its nearest neighbor layers [6].

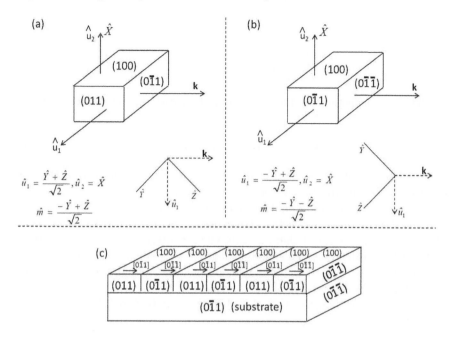

FIGURE 3.19: Orientation Patterning, (a) $\mathbf{k}$ along $[0\bar{1}1]$, (b) $\mathbf{k}$ along $[0\bar{1}\bar{1}]$, (c) arrangement of the orientation pattern on top of a (100) face substrate. The orientations of $\mathbf{k}$ with respect to the principal axes in each segment are shown by the directions indicated by the square brackets and arrows.

To determine the direction of the nonlinear polarization for incident light propagating in a given direction through such layers, we consider here propagation through just 2 of the layers as shown in Fig. 3.19(a) and 3.19(b). The details of the orientation of the multiple layers, with the same orientations as in [6] are shown in Fig. 3.19(c).

For a cubic crystal shaped as a rectangular prism with faces (100), (011) and $(0\bar{1}1)$ and with the propagation direction along $[0\bar{1}1]$, as shown in Fig. 3.19(a), we have from Eqn. 3.65

$$\text{for } \mathbf{k} \text{ along } [0\bar{1}1], \quad \theta = 315°, \quad \phi = 90°$$
$$A_{u1} = p, \quad A_{u2} = q$$
$$\hat{e}_3 = \hat{u}_1 \sin \psi_3 + \hat{u}_2 \cos \psi_3 \tag{3.86}$$

so that the projection of the nonlinear polarization in the direction of the electric field is

$$\mathbf{A} \cdot \hat{e}_3 = p \sin \psi_3 + q \cos \psi_3. \tag{3.87}$$

When the propagation direction is along $[0\bar{1}\bar{1}]$, as shown in Fig. 3.19(b), we have

$$\text{for } \mathbf{k} \text{ along } [0\bar{1}\bar{1}], \quad \theta = 225°, \quad \phi = 90°$$
$$A_{u1} = -p, \quad A_{u2} = -q \quad (3.88)$$

so that the projection of the nonlinear polarization in the direction of the electric field is

$$\mathbf{A} \cdot \hat{e}_3 = -p \sin \psi_3 - q \cos \psi_3 \quad (3.89)$$

if the direction of the electric field indicated by the angle ψ_3 remains unchanged under the rotation of the dielectric axes.

Comparing Eqns. 3.87 and 3.89 we find that for any polarization angles ψ_1 and ψ_2 of the two incident fields, the projection of the vector $\mathbf{A}$ on the u_1-u_2 plane (shown in Fig. 3.7) changes sign when $\mathbf{k}$ alternates between the $[0\bar{1}1]$ and the $[0\bar{1}\bar{1}]$ directions. This is equivalent to the value of d_{eff} changing its sign from one layer to the next as light propagates through the alternatively oriented series of layers.

Eqn. 3.77 directly shows that as θ changes from 315° to 225°, d_{eff} changes its sign.

Bibliography

[1] V.G. Dmitriev, G.G. Gurzadyan, D.N. Nikogosyan, *Handbook of Nonlinear Optical Crystals*, Springer, Berlin, 1999.

[2] F. Zernike and J. E. Midwinter, *Applied Nonlinear Optics*, John Wiley and Sons, New York, 1973.

[3] O. I. Lavrovskaya, N. I. Pavlova and A. V. Tarasov, Second harmonic generation of light from an YAG:Nd^{3+} laser in an optically biaxial crystal KTiOPO$_4$, *Sov. Phys. Crystallogr.* **31**, 678–681, 1986.

[4] B. Boulanger, J. P. Fève, G. Marnier, B. Ménaert, X. Cabirol, P. Villeval and C. Bonnin, Relative sign and absolute magnitude of $d^{(2)}$ nonlinear coefficients of KTP from second-harmonic-generation measurements, *J. Opt. Soc. Am. B* **11**, 750, 1994.

[5] B. Boulanger and G. Marnier, Field factor calculation for the study of the relationships between all the 3-wave non-linear optical interactions in uniaxial and biaxial crystals, *J. Phys:Condensed Matter*, 3, 8327, 1991.

[6] L. A. Eyres, P. J. Tourreau, T. J. Pinguet, C. B. Ebert, J. S. Harris, M. M. Fejer, L. Becouarn, B. Gerard and E. Lallier, All-epitaxial fabrication of thick, orientation-patterned GaAs films for nonlinear optical frequency conversion, *Appl. Phys. Lett.* **79**, 904, 2001.

	Case 1 ($n_X < n_Y < n_Z$)			Case 2 ($n_X > n_Y > n_Z$)		
	ω_1	ω_2	ω_3	ω_1	ω_2	ω_3
d_{11}	0	0	0	0	0	0
d_{12}	0	0	$\cos\theta'_{41}\cos\theta'_{42}$	0	0	0
d_{13}	0	0	$\sin\theta'_{41}\sin\theta'_{42}$	0	0	0
d_{14}	0	0	$-\sin(\theta'_{41}+\theta'_{42})$	0	0	0
d_{15}	0	0	0	$\sin\theta''_{43}$	$\sin\theta''_{43}$	0
d_{16}	0	0	0	$-\cos\theta''_{43}$	$-\cos\theta''_{43}$	0
d_{21}	0	0	0	0	0	$-\cos\theta''_{43}$
d_{22}	0	0	0	0	0	0
d_{23}	0	0	0	0	0	0
d_{24}	0	0	0	0	0	0
d_{25}	$-\cos\theta'_{41}\sin\theta'_{42}$	$-\sin\theta'_{41}\cos\theta'_{42}$	0	0	0	0
d_{26}	$\cos\theta'_{41}\cos\theta'_{42}$	$\cos\theta'_{41}\cos\theta'_{42}$	0	0	0	0
d_{31}	0	0	0	0	0	$\sin\theta''_{43}$
d_{32}	0	0	0	0	0	0
d_{33}	0	0	0	0	0	0
d_{34}	0	0	0	0	0	0
d_{35}	$\sin\theta'_{41}\sin\theta'_{42}$	$\sin\theta'_{41}\sin\theta'_{42}$	0	0	0	0
d_{36}	$-\sin\theta'_{41}\cos\theta'_{42}$	$-\cos\theta'_{41}\sin\theta'_{42}$	0	0	0	0

$\theta''_{41} \equiv \theta + \rho_4(\omega_1)$, $\theta'_{42} \equiv \theta - \rho_4(\omega_2)$, $\theta''_{43} \equiv \theta + \rho_4(\omega_3)$, with ρ_4 given in Eqn. 1.120.

TABLE 3.23. Propagation in a biaxial crystal: **k** on YZ plane, ssf case

	Case 1 ($n_X < n_Y < n_Z$)			Case 2 ($n_X > n_Y > n_Z$)		
	ω_1	ω_2	ω_3	ω_1	ω_2	ω_3
d_{11}	0	0	0	0	0	0
d_{12}	0	0	0	$-\cos\theta''_{42}\cos\theta''_{43}$	0	0
d_{13}	0	0	0	$-\sin\theta''_{42}\sin\theta''_{43}$	0	0
d_{14}	0	0	0	$-\sin(\theta''_{42}+\theta''_{43})$	0	0
d_{15}	0	$-\sin\theta'_{41}$	$-\sin\theta'_{41}$	0	0	0
d_{16}	0	$\cos\theta'_{41}$	$\cos\theta'_{41}$	0	0	0
d_{21}	$\cos\theta'_{41}$	0	0	0	0	0
d_{22}	0	0	0	0	0	0
d_{23}	0	0	0	0	0	0
d_{24}	0	0	0	0	0	0
d_{25}	0	0	0	0	$\cos\theta''_{42}\sin\theta''_{43}$	$\sin\theta''_{42}\cos\theta''_{43}$
d_{26}	0	0	0	0	$-\cos\theta''_{42}\cos\theta''_{43}$	$-\cos\theta''_{42}\cos\theta''_{43}$
d_{31}	$-\sin\theta'_{41}$	0	0	0	0	0
d_{32}	0	0	0	0	0	0
d_{33}	0	0	0	0	0	0
d_{34}	0	0	0	0	0	0
d_{35}	0	0	0	0	$-\sin\theta''_{42}\sin\theta''_{43}$	$-\sin\theta''_{42}\sin\theta''_{43}$
d_{36}	0	0	0	0	$\sin\theta''_{42}\cos\theta''_{43}$	$\cos\theta''_{42}\sin\theta''_{43}$

$\theta'_{41} \equiv \theta - \rho_4(\omega_1)$, $\theta''_{42} \equiv \theta - \rho_4(\omega_2)$, $\theta''_{43} \equiv \theta + \rho_4(\omega_3)$, with ρ_4 given in Eqn. 1.120.

TABLE 3.24. Propagation in a biaxial crystal: **k** on YZ plane, sff case

	Case 1 $(n_X < n_Y < n_Z)$			Case 2 $(n_X > n_Y > n_Z)$		
	ω_1	ω_2	ω_3	ω_1	ω_2	ω_3
d_{11}	0	0	0	0	0	0
d_{12}	0	0	0	0	$-\cos\theta''_{41}\cos\theta''_{43}$	0
d_{13}	0	0	0	0	$-\sin\theta''_{41}\sin\theta''_{43}$	0
d_{14}	0	0	0	0	$\sin(\theta''_{41}+\theta''_{43})$	0
d_{15}	$-\sin\theta'_{42}$	0	$-\sin\theta'_{42}$	0	0	0
d_{16}	$\cos\theta'_{42}$	0	$\cos\theta'_{42}$	0	0	0
d_{21}	0	$\cos\theta'_{42}$	0	0	0	0
d_{22}	0	0	0	0	0	0
d_{23}	0	0	0	0	0	0
d_{24}	0	0	0	0	0	0
d_{25}	0	0	0	$\cos\theta''_{41}\sin\theta''_{43}$	0	$\sin\theta''_{41}\cos\theta''_{43}$
d_{26}	0	0	0	$-\cos\theta''_{41}\cos\theta''_{43}$	0	$-\cos\theta''_{41}\cos\theta''_{43}$
d_{31}	0	$-\sin\theta'_{42}$	0	0	0	0
d_{32}	0	0	0	0	0	0
d_{33}	0	0	0	0	0	0
d_{34}	0	0	0	0	0	0
d_{35}	0	0	0	$-\sin\theta''_{41}\sin\theta''_{43}$	0	$-\sin\theta''_{41}\sin\theta''_{43}$
d_{36}	0	0	0	$\sin\theta''_{41}\cos\theta''_{43}$	0	$\cos\theta''_{41}\sin\theta''_{43}$

$\theta''_{41} \equiv \theta - \rho_4(\omega_1)$, $\theta'_{42} \equiv \theta + \rho_4(\omega_2)$, $\theta''_{43} \equiv \theta + \rho_4(\omega_3)$, with ρ_4 given in Eqn. 1.120.

TABLE 3.25. Propagation in a biaxial crystal: **k** on YZ plane, fsf case

	Case 1 $(n_X < n_Y < n_Z)$			Case 2 $(n_X > n_Y > n_Z)$		
	ω_1	ω_2	ω_3	ω_1	ω_2	ω_3
d_{11}	0	0	0	0	0	0
d_{12}	0	0	$\cos\theta'_{53}$	0	0	0
d_{13}	0	0	0	0	0	0
d_{14}	0	0	0	$\cos\theta''_{51}\sin\theta''_{52}$	$\sin\theta''_{51}\cos\theta''_{52}$	0
d_{15}	0	0	0	0	0	0
d_{16}	0	0	0	$-\cos\theta''_{51}\cos\theta''_{52}$	$-\cos\theta''_{51}\cos\theta''_{52}$	0
d_{21}	0	0	0	0	0	$-\cos\theta''_{51}\cos\theta''_{52}$
d_{22}	0	0	0	0	0	0
d_{23}	0	0	0	0	0	$-\sin\theta''_{51}\sin\theta''_{52}$
d_{24}	$-\sin\theta'_{53}$	$-\sin\theta'_{53}$	0	0	0	0
d_{25}	0	0	0	0	0	$\sin(\theta''_{51}+\theta''_{52})$
d_{26}	$\cos\theta'_{53}$	$\cos\theta'_{53}$	0	0	0	0
d_{31}	0	0	0	0	0	0
d_{32}	0	0	$-\sin\theta'_{53}$	0	0	0
d_{33}	0	0	0	0	0	0
d_{34}	0	0	0	$-\sin\theta''_{51}\sin\theta''_{52}$	$-\sin\theta''_{51}\sin\theta''_{52}$	0
d_{35}	0	0	0	0	0	0
d_{36}	0	0	0	$\sin\theta''_{51}\cos\theta''_{52}$	$\sin\theta''_{52}\cos\theta''_{51}$	0

$\theta''_{51} \equiv \theta + \rho_5(\omega_1)$, $\theta''_{52} \equiv \theta + \rho_5(\omega_2)$, $\theta'_{53} \equiv \theta - \rho_5(\omega_3)$, with ρ_5 given in Eqn. 1.133.

TABLE 3.26. Propagation in a biaxial crystal: **k** on ZX plane, $\theta < \Omega$, ssf case

	Case 1 $(n_X < n_Y < n_Z)$			Case 2 $(n_X > n_Y > n_Z)$		
	ω_1	ω_2	ω_3	ω_1	ω_2	ω_3
d_{11}	0	0	0	0	0	0
d_{12}	0	0	0	$-\cos\theta''_{51}$	0	0
d_{13}	0	0	0	0	0	0
d_{14}	0	$-\cos\theta'_{52}\sin\theta'_{53}$	$-\cos\theta'_{53}\sin\theta'_{52}$	0	0	0
d_{15}	0	0	0	0	0	0
d_{16}	0	$\cos\theta'_{52}\cos\theta'_{53}$	$\cos\theta'_{52}\cos\theta'_{53}$	0	0	0
d_{21}	$\cos\theta'_{52}\cos\theta'_{53}$	0	0	0	0	
d_{22}	0	0	0	0	0	0
d_{23}	$\sin\theta'_{52}\sin\theta'_{53}$	0	0	0	0	0
d_{24}	0	0	0	0	$\sin\theta''_{51}$	$\sin\theta''_{51}$
d_{25}	$-\sin(\theta'_{52}+\theta'_{53})$	0	0	0	0	0
d_{26}	0	0	0	0	$-\cos\theta''_{51}$	$-\cos\theta''_{51}$
d_{31}	0	0	0	0	0	0
d_{32}	0	0	0	$\sin\theta''_{51}$	0	0
d_{33}	0	0	0	0	0	0
d_{34}	0	$\sin\theta'_{52}\sin\theta'_{53}$	$\sin\theta'_{52}\sin\theta'_{53}$	0	0	0
d_{35}	0	0	0	0	0	0
d_{36}	0	$-\sin\theta'_{52}\cos\theta'_{53}$	$-\cos\theta'_{52}\sin\theta'_{53}$	0	0	0

$\theta''_{51} \equiv \theta + \rho_5(\omega_1)$, $\theta'_{52} \equiv \theta - \rho_5(\omega_2)$, $\theta'_{53} \equiv \theta - \rho_5(\omega_3)$, with ρ_5 given in Eqn. 1.133.

TABLE 3.27. Propagation in a biaxial crystal: $\mathbf{k}$ on ZX plane, $\theta < \Omega$, sff case

	Case 1 $(n_X < n_Y < n_Z)$			Case 2 $(n_X > n_Y > n_Z)$		
	ω_1	ω_2	ω_3	ω_1	ω_2	ω_3
d_{11}	0	0	0	0	0	0
d_{12}	0	0	0	0	$-\cos\theta''_{52}$	0
d_{13}	0	0	0	0	0	0
d_{14}	$-\cos\theta'_{51}\sin\theta'_{53}$	0	$-\sin\theta'_{51}\cos\theta'_{53}$	0	0	0
d_{15}	0	0	0	0	0	0
d_{16}	$\cos\theta'_{51}\cos\theta'_{53}$	0	$\cos\theta'_{51}\cos\theta'_{53}$	0	0	0
d_{21}	0	$\cos\theta'_{51}\cos\theta'_{53}$	0	0	0	0
d_{22}	0	0	0	0	0	0
d_{23}	0	$\sin\theta'_{51}\sin\theta'_{53}$	0	0	0	0
d_{24}	0	0	0	$\sin\theta''_{52}$	0	$\sin\theta''_{52}$
d_{25}	0	$-\sin(\theta'_{51}+\theta'_{53})$	0	0	0	0
d_{24}	0	0	0	$-\cos\theta''_{52}$	0	$-\cos\theta''_{52}$
d_{31}	0	0	0	0	0	0
d_{32}	0	0	0	0	$\sin\theta''_{52}$	0
d_{33}	0	0	0	0	0	0
d_{34}	$\sin\theta'_{51}\sin\theta'_{53}$	0	$\sin\theta'_{51}\sin\theta'_{53}$	0	0	0
d_{35}	0	0	0	0	0	0
d_{36}	$-\sin\theta'_{51}\cos\theta'_{53}$	0	$-\sin\theta'_{53}\cos\theta'_{51}$	0	0	0

$\theta'_{51} \equiv \theta - \rho_5(\omega_1)$, $\theta''_{52} \equiv \theta + \rho_5(\omega_2)$, $\theta'_{53} \equiv \theta - \rho_5(\omega_3)$, with ρ_5 given in Eqn. 1.133.

TABLE 3.28. Propagation in a biaxial crystal: $\mathbf{k}$ on ZX plane, $\theta < \Omega$, fsf case

	Case 1 ($n_X < n_Y < n_Z$)			Case 2 ($n_X > n_Y > n_Z$)		
	ω_1	ω_2	ω_3	ω_1	ω_2	ω_3
d_{11}	0	0	0	0	0	0
d_{12}	0	0	0	0	0	$-\cos\theta''_{53}$
d_{13}	0	0	0	0	0	0
d_{14}	$\cos\theta'_{51}\sin\theta'_{52}$	$\sin\theta'_{51}\cos\theta'_{52}$	0	0	0	0
d_{15}	0	0	0	0	0	0
d_{16}	$-\cos\theta'_{51}\cos\theta'_{52}$	$-\cos\theta'_{51}\cos\theta'_{52}$	0	0	0	0
d_{21}	0	0	$-\cos\theta'_{52}\cos\theta'_{51}$	0	0	0
d_{22}	0	0	0	0	0	0
d_{23}	0	0	$-\sin\theta'_{52}\sin\theta'_{51}$	0	0	0
d_{24}	0	0	0	$\sin\theta''_{53}$	$\sin\theta''_{53}$	0
d_{25}	0	0	$\sin(\theta'_{51}+\theta'_{52})$	0	0	0
d_{26}	0	0	0	$-\cos\theta''_{53}$	$-\cos\theta''_{53}$	0
d_{31}	0	0	0	0	0	0
d_{32}	0	0	0	0	0	$\sin\theta''_{53}$
d_{33}	0	0	0	0	0	0
d_{34}	$\sin\theta'_{51}\sin\theta'_{52}$	$\sin\theta'_{51}\sin\theta'_{52}$	0	0	0	0
d_{35}	0	0	0	0	0	0
d_{36}	$\sin\theta'_{51}\cos\theta'_{52}$	$\cos\theta'_{51}\sin\theta'_{52}$	0	0	0	0

$\theta'_{51} \equiv \theta - \rho_5(\omega_1)$, $\theta'_{52} \equiv \theta - \rho_5(\omega_2)$, $\theta''_{53} \equiv \theta + \rho_5(\omega_3)$, with ρ_5 given in Eqn. 1.133.

TABLE 3.29. Propagation in a biaxial crystal: **k** on ZX plane, $\theta > \Omega$, ssf case

	Case 1 ($n_X < n_Y < n_Z$)			Case 2 ($n_X > n_Y > n_Z$)		
	ω_1	ω_2	ω_3	ω_1	ω_2	ω_3
d_{11}	0	0	0	0	0	0
d_{12}	$\cos\theta'_{51}$	0	0	0	0	0
d_{13}	0	0	0	0	0	0
d_{14}	0	0	0	0	$-\cos\theta''_{52}\sin\theta''_{53}$	$-\sin\theta''_{52}\cos\theta''_{53}$
d_{15}	0	0	0	0	0	0
d_{16}	0	0	0	0	$\cos\theta''_{52}\cos\theta''_{53}$	$\cos\theta''_{52}\cos\theta''_{53}$
d_{21}	0	0	0	$\cos\theta''_{52}\cos\theta''_{53}$	0	0
d_{22}	0	0	0	0	0	0
d_{23}	0	0	0	$\sin\theta''_{52}\sin\theta''_{53}$	0	0
d_{24}	0	$-\sin\theta'_{51}$	$-\sin\theta'_{51}$	0	0	0
d_{25}	0	0	0	$-\sin(\theta''_{52}+\theta''_{53})$	0	0
d_{26}	0	$\cos\theta'_{51}$	$\cos\theta'_{51}$	0	0	0
d_{31}	0	0	0	0	0	0
d_{32}	$-\sin\theta'_{51}$	0	0	0	0	0
d_{33}	0	0	0	0	0	0
d_{34}	0	0	0	0	$\sin\theta''_{52}\sin\theta''_{53}$	$\sin\theta''_{52}\sin\theta''_{53}$
d_{35}	0	0	0	0	0	0
d_{36}	0	0	0	0	$-\sin\theta''_{52}\cos\theta''_{53}$	$-\cos\theta''_{52}\sin\theta''_{53}$

$\theta'_{51} \equiv \theta - \rho_5(\omega_1)$, $\theta''_{52} \equiv \theta + \rho_5(\omega_2)$, $\theta''_{53} \equiv \theta + \rho_5(\omega_3)$, with ρ_5 given in Eqn. 1.133.

TABLE 3.30. Propagation in a biaxial crystal: **k** on ZX plane, $\theta > \Omega$, sff case

	Case 1 $(n_X < n_Y < n_Z)$			Case 2 $(n_X > n_Y > n_Z)$		
	ω_1	ω_2	ω_3	ω_1	ω_2	ω_3
d_{11}	0	0	0	0	0	0
d_{12}	0	$\cos\theta_{52}'$	0	0	0	0
d_{13}	0	0	0	0	0	0
d_{14}	0	0	0	$-\cos\theta_{51}''\sin\theta_{53}''$	0	$-\sin\theta_{51}''\cos\theta_{53}''$
d_{15}	0	0	0	0	0	0
d_{16}	0	0	0	$\cos\theta_{51}''\cos\theta_{53}''$	0	$\cos\theta_{51}''\cos\theta_{53}''$
d_{21}	0	0	0	0	$\cos\theta_{51}''\cos\theta_{53}''$	0
d_{22}	0	0	0	0	0	0
d_{23}	0	0	0	0	$\sin\theta_{51}''\sin\theta_{53}''$	0
d_{24}	$-\sin\theta_{52}'$	0	$-\sin\theta_{52}'$	0	0	0
d_{25}	0	0	0	0	$-\sin(\theta_{51}''+\theta_{53}'')$	0
d_{26}	$\cos\theta_{52}'$	0	$\cos\theta_{52}'$	0	0	0
d_{31}	0	0	0	0	0	0
d_{32}	0	$-\sin\theta_{52}'$	0	0	0	0
d_{33}	0	0	0	0	0	0
d_{34}	0	0	0	$\sin\theta_{51}''\sin\theta_{53}''$	0	$\sin\theta_{51}''\sin\theta_{53}''$
d_{35}	0	0	0	0	0	0
d_{36}	0	0	0	$-\sin\theta_{51}''\cos\theta_{53}''$	0	$-\cos\theta_{51}''\sin\theta_{53}''$

$\theta_{51}'' \equiv \theta + \rho_5(\omega_1), \theta_{52}' \equiv \theta - \rho_5(\omega_2), \theta_{53}'' \equiv \theta + \rho_5(\omega_3)$, with ρ_5 given in Eqn. 1.133.

TABLE 3.31. Propagation in a biaxial crystal: $\mathbf{k}$ on ZX plane, $\theta > \Omega$, fsf case

	Case 1 $(n_X < n_Y < n_Z)$			Case 2 $(n_X > n_Y > n_Z)$		
	ω_1	ω_2	ω_3	ω_1	ω_2	ω_3
d_{11}	0	0	0	0	0	0
d_{12}	0	0	0	0	0	0
d_{13}	0	0	$\sin\phi_{63}''$	0	0	0
d_{14}	0	0	0	$-\sin\phi_{61}'\cos\phi_{62}'$	$-\sin\phi_{62}'\cos\phi_{61}'$	0
d_{15}	0	0	0	$\sin\phi_{61}'\sin\phi_{62}'$	$\sin\phi_{61}'\sin\phi_{62}'$	0
d_{16}	0	0	0	0	0	0
d_{21}	0	0	0	0	0	0
d_{22}	0	0	0	0	0	0
d_{23}	0	0	$-\cos\phi_{63}''$	0	0	0
d_{24}	0	0	0	$\cos\phi_{61}'\cos\phi_{62}'$	$\cos\phi_{61}'\cos\phi_{62}'$	0
d_{25}	0	0	0	$-\sin\phi_{62}'\cos\phi_{61}'$	$-\cos\phi_{62}'\sin\phi_{61}'$	0
d_{26}	0	0	0	0	0	0
d_{31}	0	0	0	0	0	$\sin\phi_{61}'\sin\phi_{62}'$
d_{32}	0	0	0	0	0	$\cos\phi_{61}'\cos\phi_{62}'$
d_{33}	0	0	0	0	0	0
d_{34}	$-\cos\phi_{63}''$	$-\cos\phi_{63}''$	0	0	0	0
d_{35}	$\sin\phi_{63}''$	$\sin\phi_{63}''$	0	0	0	0
d_{36}	0	0	0	0	0	$-\sin(\phi_{61}'+\phi_{62}')$

$\phi_{61}' \equiv \phi - \rho_6(\omega_1)$, $\phi_{62}' \equiv \phi - \rho_6(\omega_2)$, $\phi_{63}'' \equiv \phi + \rho_6(\omega_3)$, with ρ_6 given in Eqn. 1.145.

TABLE 3.32. Propagation in a biaxial crystal: $\mathbf{k}$ on XY plane, ssf case

	Case 1 $(n_X < n_Y < n_Z)$			Case 2 $(n_X > n_Y > n_Z)$		
	ω_1	ω_2	ω_3	ω_1	ω_2	ω_3
d_{11}	0	0	0	0	0	0
d_{12}	0	0	0	0	0	0
d_{13}	0	0	0	$-\sin\phi'_{61}$	0	0
d_{14}	0	$\sin\phi''_{62}\cos\phi''_{63}$	$\cos\phi''_{62}\sin\phi''_{63}$	0	0	0
d_{15}	0	$-\sin\phi''_{62}\sin\phi''_{63}$	$-\sin\phi''_{62}\sin\phi''_{63}$	0	0	0
d_{16}	0	0	0	0	0	0
d_{21}	0	0	0	0	0	0
d_{22}	0	0	0	0	0	0
d_{23}	0	0	0	$\cos\phi'_{61}$	0	0
d_{24}	0	$-\cos\phi''_{62}\cos\phi''_{63}$	$-\cos\phi''_{62}\cos\phi''_{63}$	0	0	0
d_{25}	0	$\cos\phi''_{62}\sin\phi''_{63}$	$\sin\phi''_{62}\cos\phi''_{63}$	0	0	0
d_{26}	0	0	0	0	0	0
d_{31}	$-\sin\phi''_{62}\sin\phi''_{63}$	0	0	0	0	0
d_{32}	$-\cos\phi''_{62}\cos\phi''_{63}$	0	0	0	0	0
d_{33}	0	0	0	0	0	0
d_{34}	0	0	0	0	$\cos\phi'_{61}$	$\cos\phi'_{61}$
d_{35}	0	0	0	0	$-\sin\phi'_{61}$	$-\sin\phi'_{61}$
d_{36}	$\sin(\phi''_{62}+\phi''_{63})$	0	0	0	0	0

$\phi'_{61} \equiv \phi - \rho_6(\omega_1)$, $\phi''_{62} \equiv \phi + \rho_6(\omega_2)$, $\phi''_{63} \equiv \phi + \rho_6(\omega_3)$, with ρ_6 given in Eqn. 1.145.

TABLE 3.33. Propagation in a biaxial crystal: **k** on XY plane, sff case

	Case 1 $(n_X < n_Y < n_Z)$			Case 2 $(n_X > n_Y > n_Z)$		
	ω_1	ω_2	ω_3	ω_1	ω_2	ω_3
d_{11}	0	0	0	0	0	0
d_{12}	0	0	0	0	0	0
d_{13}	0	0	0	0	$-\sin\phi'_{62}$	0
d_{14}	$\sin\phi''_{61}\cos\phi''_{63}$	0	$\sin\phi''_{63}\cos\phi''_{61}$	0	0	0
d_{15}	$-\sin\phi''_{61}\sin\phi''_{63}$	0	$-\sin\phi''_{61}\sin\phi''_{63}$	0	0	0
d_{16}	0	0	0	0	0	0
d_{21}	0	0	0	0	0	0
d_{22}	0	0	0	0	0	0
d_{23}	0	0	0	0	$\cos\phi'_{62}$	0
d_{24}	$-\cos\phi''_{61}\cos\phi''_{63}$	0	$-\cos\phi''_{61}\cos\phi''_{63}$	0	0	0
d_{25}	$\cos\phi''_{61}\sin\phi''_{63}$	0	$\sin\phi''_{61}\cos\phi''_{63}$	0	0	0
d_{26}	0	0	0	0	0	0
d_{31}	0	$-\sin\phi''_{61}\sin\phi''_{63}$	0	0	0	0
d_{32}	0	$-\cos\phi''_{61}\cos\phi''_{63}$	0	0	0	0
d_{33}	0	0	0	0	0	0
d_{34}	0	0	0	$\cos\phi'_{62}$	0	$\cos\phi'_{62}$
d_{35}	0	0	0	$-\sin\phi'_{62}$	0	$-\sin\phi'_{62}$
d_{36}	0	$\sin(\phi''_{61}+\phi''_{63})$	0	0	0	0

$\phi''_{61} \equiv \phi + \rho_6(\omega_1)$, $\phi'_{62} \equiv \phi - \rho_6(\omega_2)$, $\phi''_{63} \equiv \phi + \rho_6(\omega_3)$, with ρ_6 given in Eqn. 1.145.

TABLE 3.34. Propagation in a biaxial crystal: **k** on XY plane, fsf case

	Positive uniaxial: ssf = *eeo*		
	ω_1	ω_2	ω_3
d_{11}	$\sin\phi\cos^2\phi\cos\theta'_1\cos\theta'_2$	$\sin\phi\cos^2\phi\cos\theta'_1\cos\theta'_2$	$\sin\phi\cos^2\phi\cos\theta'_1\cos\theta'_2$
d_{12}	$-\sin\phi\cos^2\phi\cos\theta'_1\cos\theta'_2$	$-\sin\phi\cos^2\phi\cos\theta'_1\cos\theta'_2$	$\sin^3\phi\cos\theta'_1\cos\theta'_2$
d_{13}	0	0	$\sin\phi\sin\theta'_1\sin\theta'_2$
d_{14}	$\cos^2\phi\cos\theta'_1\sin\theta'_2$	$\cos^2\phi\sin\theta'_1\cos\theta'_2$	$-\sin^2\phi\sin(\theta'_1+\theta'_2)$
d_{15}	$-\sin\phi\cos\phi\cos\theta'_1\sin\theta'_2$	$-\sin\phi\cos\phi\sin\theta'_1\cos\theta'_2$	$-\sin\phi\cos\phi\sin(\theta'_1+\theta'_2)$
d_{16}	$-\cos\phi\cos 2\phi\cos\theta'_1\cos\theta'_2$	$-\cos\phi\cos 2\phi\cos\theta'_1\cos\theta'_2$	$\sin\phi\sin 2\phi\cos\theta'_1\cos\theta'_2$
d_{21}	$\sin^2\phi\cos\phi\cos\theta'_1\cos\theta'_2$	$\sin^2\phi\cos\phi\cos\theta'_1\cos\theta'_2$	$-\cos^3\phi\cos\theta'_1\cos\theta'_2$
d_{22}	$-\sin^2\phi\cos\phi\cos\theta'_1\cos\theta'_2$	$-\sin^2\phi\cos\phi\cos\theta'_1\cos\theta'_2$	$-\cos\phi\sin^2\phi\cos\theta'_1\cos\theta'_2$
d_{23}	0	0	$-\cos\phi\sin\theta'_1\sin\theta'_2$
d_{24}	$\sin\phi\cos\phi\cos\theta'_1\sin\theta'_2$	$\sin\phi\cos\phi\sin\theta'_1\cos\theta'_2$	$\cos\phi\sin\phi\sin(\theta'_1+\theta'_2)$
d_{25}	$-\sin^2\phi\cos\theta'_1\sin\theta'_2$	$-\sin^2\phi\sin\theta'_1\cos\theta'_2$	$\cos^2\phi\sin(\theta'_1+\theta'_2)$
d_{26}	$-\sin\phi\cos 2\phi\cos\theta'_1\cos\theta'_2$	$-\sin\phi\cos 2\phi\cos\theta'_1\cos\theta'_2$	$-\cos\phi\sin 2\phi\cos\theta'_1\cos\theta'_2$
d_{31}	$-\sin\phi\cos\phi\sin\theta'_1\cos\theta'_2$	$-\sin\phi\cos\phi\cos\theta'_1\sin\theta'_2$	0
d_{32}	$\sin\phi\cos\phi\sin\theta'_1\cos\theta'_2$	$\sin\phi\cos\phi\cos\theta'_1\sin\theta'_2$	0
d_{33}	0	0	0
d_{34}	$-\cos\phi\sin\theta'_1\sin\theta'_2$	$-\cos\phi\sin\theta'_1\sin\theta'_2$	0
d_{35}	$\sin\phi\sin\theta'_1\sin\theta'_2$	$\sin\phi\sin\theta'_1\sin\theta'_2$	0
d_{36}	$\cos 2\phi\sin\theta'_1\cos\theta'_2$	$\cos 2\phi\cos\theta'_1\sin\theta'_2$	0

$\theta'_1 \equiv \theta - \rho(\omega_1)$, $\theta'_2 \equiv \theta - \rho(\omega_2)$, with ρ given in Table 1.6.

TABLE 3.35. Propagation in a positive uniaxial crystal: $n_o < n_Z$, ssf case

	Positive uniaxial: *eoo* ($n_o < n_Z$)		
	ω_1	ω_2	ω_3
d_{11}	$\sin^2\phi\cos\phi\cos\theta'_1$	$\sin^2\phi\cos\phi\cos\theta'_1$	$\sin^2\phi\cos\phi\cos\theta'_1$
d_{12}	$\cos^3\phi\cos\theta'_1$	$-\sin^2\phi\cos\phi\cos\theta'_1$	$-\sin^2\phi\cos\phi\cos\theta'_1$
d_{13}	0	0	0
d_{14}	0	$\sin\phi\cos\phi\sin\theta'_1$	$\sin\phi\cos\phi\sin\theta'_1$
d_{15}	0	$-\sin^2\phi\sin\theta'_1$	$-\sin^2\phi\sin\theta'_1$
d_{16}	$-\cos\phi\sin 2\phi\cos\theta'_1$	$-\sin\phi\cos 2\phi\cos\theta'_1$	$-\sin\phi\cos 2\phi\cos\theta'_1$
d_{21}	$\sin^3\phi\cos\theta'_1$	$-\cos^2\phi\sin\phi\cos\theta'_1$	$-\sin\phi\cos^2\phi\sin\theta'_1$
d_{22}	$\sin\phi\cos^2\phi\cos\theta'_1$	$\sin\phi\cos^2\phi\cos\theta'_1$	$\sin\phi\cos^2\phi\cos\theta'_1$
d_{23}	0	0	0
d_{24}	0	$-\cos^2\phi\sin\theta'_1$	$-\cos^2\phi\sin\theta'_1$
d_{25}	0	$\sin\phi\cos\phi\sin\theta'_1$	$\sin\phi\cos\phi\sin\theta'_1$
d_{26}	$-\sin\phi\sin 2\phi\cos\theta'_1$	$\cos\phi\cos 2\phi\cos\theta'_1$	$\cos\phi\cos 2\phi\cos\theta'_1$
d_{31}	$-\sin^2\phi\sin\theta'_1$	0	0
d_{32}	$-\cos^2\phi\sin\theta'_1$	0	0
d_{33}	0	0	0
d_{34}	0	0	0
d_{35}	0	0	0
d_{36}	$\sin 2\phi\sin\theta'_1$	0	0

$\theta'_1 \equiv \theta - \rho(\omega_1)$, with ρ given in Table 1.6.

TABLE 3.36. Propagation in a positive uniaxial crystal: $n_o < n_Z$, sff case

	ω_1	ω_2	ω_3
	Positive uniaxial: fsf $= oeo$		
d_{11}	$\sin^2\phi\cos\phi\cos\theta'_2$	$\sin^2\phi\cos\phi\cos\theta'_2$	$\sin^2\phi\cos\phi\cos\theta'_2$
d_{12}	$-\sin^2\phi\cos\phi\cos\theta'_2$	$\cos^3\phi\cos\theta'_2$	$-\sin^2\phi\cos\phi\cos\theta'_2$
d_{13}	0	0	0
d_{14}	$\sin\phi\cos\phi\sin\theta'_2$	0	$\sin\phi\cos\phi\sin\theta'_2$
d_{15}	$-\sin^2\phi\sin\theta'_2$	0	$-\sin^2\phi\sin\theta'_2$
d_{16}	$-\sin\phi\cos2\phi\cos\theta'_2$	$-\sin2\phi\cos\phi\cos\theta'_2$	$-\sin\phi\cos2\phi\cos\theta'_2$
d_{21}	$-\sin\phi\cos^2\phi\cos\theta'_2$	$\sin^3\phi\cos\theta'_2$	$-\sin\phi\cos^2\phi\cos\theta'_2$
d_{22}	$\sin\phi\cos^2\phi\cos\theta'_2$	$\sin\phi\cos^2\phi\cos\theta'_2$	$\cos^2\phi\sin\phi\sin\theta'_2$
d_{23}	0	0	0
d_{24}	$-\cos^2\phi\sin\theta'_2$	0	$-\cos^2\phi\sin\theta'_2$
d_{25}	$\sin\phi\cos\phi\sin\theta'_2$	0	$\sin\phi\cos\phi\sin\theta'_2$
d_{26}	$\cos\phi\cos2\phi\cos\theta'_2$	$-\sin\phi\sin2\phi\cos\theta'_2$	$\cos\phi\cos2\phi\cos\theta'_2$
d_{31}	0	$-\sin^2\phi\sin\theta'_2$	0
d_{32}	0	$-\cos^2\phi\sin\theta'_2$	0
d_{33}	0	0	0
d_{34}	0	0	0
d_{35}	0	0	0
d_{36}	0	$\sin2\phi\sin\theta'_2$	0

$\theta'_2 \equiv \theta - \rho(\omega_2)$, with ρ given in Table 1.6.

TABLE 3.37. Propagation in a positive uniaxial crystal: $n_o < n_Z$, fsf case

	ω_1	ω_2	ω_3
	Negative uniaxial: ssf $= ooe$		
d_{11}	$-\sin^2\phi\cos\phi\cos\theta''_3$	$-\sin^2\phi\cos\phi\cos\theta''_3$	$-\sin^2\phi\cos\phi\cos\theta''_3$
d_{12}	$\sin^2\phi\cos\phi\cos\theta''_3$	$\sin^2\phi\cos\phi\cos\theta''_3$	$-\cos^3\phi\cos\theta''_3$
d_{13}	0	0	0
d_{14}	$-\sin\phi\cos\phi\sin\theta''_3$	$-\sin\phi\cos\phi\sin\theta''_3$	0
d_{15}	$\sin^2\phi\sin\theta''_3$	$\sin^2\phi\sin\theta''_3$	0
d_{16}	$\sin\phi\cos2\phi\cos\theta''_3$	$\sin\phi\cos2\phi\cos\theta''_3$	$\cos\phi\sin2\phi\cos\theta''_3$
d_{21}	$\cos^2\phi\sin\phi\cos\theta''_3$	$\cos^2\phi\sin\phi\cos\theta''_3$	$-\sin^3\phi\cos\theta''_3$
d_{22}	$-\cos^2\phi\sin\phi\cos\theta''_3$	$-\cos^2\phi\sin\phi\cos\theta''_3$	$-\sin\phi\cos^2\phi\cos\theta''_3$
d_{23}	0	0	0
d_{24}	$\cos^2\phi\sin\theta''_3$	$\cos^2\phi\sin\theta''_3$	0
d_{25}	$-\sin\phi\cos\phi\sin\theta''_3$	$-\sin\phi\cos\phi\sin\theta''_3$	0
d_{26}	$-\cos\phi\cos2\phi\cos\theta''_3$	$-\cos\phi\cos2\phi\cos\theta''_3$	$\sin\phi\sin2\phi\cos\theta''_3$
d_{31}	0	0	$\sin^2\phi\sin\theta''_3$
d_{32}	0	0	$\cos^2\phi\sin\theta''_3$
d_{33}	0	0	0
d_{34}	0	0	0
d_{35}	0	0	0
d_{36}	0	0	$-\sin2\phi\sin\theta''_3$

$\theta''_3 \equiv \theta + \rho(\omega_3)$, with ρ given in Table 1.6.

TABLE 3.38. Propagation in a negative uniaxial crystal: $n_o > n_Z$, ssf case

	ω_1	ω_2	ω_3
		Negative uniaxial: sff $= oee$	
d_{11}	$\sin\phi\cos^2\phi\cos\theta_2''\cos\theta_3''$	$\sin\phi\cos^2\phi\cos\theta_2''\cos\theta_3''$	$\sin\phi\cos^2\phi\cos\theta_2''\cos\theta_3''$
d_{12}	$\sin^3\phi\cos\theta_2''\cos\theta_3''$	$-\cos^2\phi\sin\phi\cos\theta_2''\cos\theta_3''$	$-\cos^2\phi\sin\phi\cos\theta_2''\cos\theta_3''$
d_{13}	$\sin\phi\sin\theta_2''\sin\theta_3''$	0	0
d_{14}	$-\sin^2\phi\sin(\theta_2''+\theta_3'')$	$\cos^2\phi\cos\theta_2''\sin\theta_3''$	$\cos^2\phi\sin\theta_2''\cos\theta_3''$
d_{15}	$-\sin\phi\cos\phi\sin(\theta_2''+\theta_3'')$	$-\sin\phi\cos\phi\cos\theta_2''\sin\theta_3''$	$-\sin\phi\cos\phi\sin\theta_2''\cos\theta_3''$
d_{16}	$\sin\phi\sin2\phi\cos\theta_2''\cos\theta_3''$	$-\cos\phi\cos2\phi\cos\theta_2''\cos\theta_3''$	$-\cos\phi\cos2\phi\cos\theta_2''\cos\theta_3''$
d_{21}	$-\cos^3\phi\cos\theta_2''\cos\theta_3''$	$\sin^2\phi\cos\phi\cos\theta_2''\cos\theta_3''$	$\sin^2\phi\cos\phi\cos\theta_2''\cos\theta_3''$
d_{22}	$-\sin^2\phi\cos\phi\cos\theta_2''\cos\theta_3''$	$-\sin^2\phi\cos\phi\cos\theta_2''\cos\theta_3''$	$-\sin^2\phi\cos\phi\cos\theta_2''\cos\theta_3''$
d_{23}	$-\cos\phi\sin\theta_2''\sin\theta_3''$	0	0
d_{24}	$\sin\phi\cos\phi\sin(\theta_2''+\theta_3'')$	$\sin\phi\cos\phi\cos\theta_2''\sin\theta_3''$	$\sin\phi\cos\phi\sin\theta_2''\cos\theta_3''$
d_{25}	$\cos^2\phi\sin(\theta_2''+\theta_3'')$	$-\sin^2\phi\cos\theta_2''\sin\theta_3''$	$-\sin^2\phi\sin\theta_2''\cos\theta_3''$
d_{26}	$-\cos\phi\sin2\phi\cos\theta_2''\cos\theta_3''$	$-\sin\phi\cos2\phi\cos\theta_2''\cos\theta_3''$	$-\sin\phi\cos2\phi\cos\theta_2''\cos\theta_3''$
d_{31}	0	$-\sin\phi\cos\phi\sin\theta_2''\cos\theta_3''$	$-\sin\phi\cos\phi\cos\theta_2''\sin\theta_3''$
d_{32}	0	$\sin\phi\cos\phi\sin\theta_2''\cos\theta_3''$	$\sin\phi\cos\phi\cos\theta_2''\sin\theta_3''$
d_{33}	0	0	0
d_{34}	0	$\cos\phi\sin\theta_2''\sin\theta_3''$	$-\cos\phi\sin\theta_2''\sin\theta_3''$
d_{35}	0	$\sin\phi\sin\theta_2''\sin\theta_3''$	$\sin\phi\sin\theta_2''\sin\theta_3''$
d_{36}	0	$\cos2\phi\sin\theta_2''\cos\theta_3''$	$\cos2\phi\cos\theta_2''\sin\theta_3''$

$\theta_2'' \equiv \theta + \rho(\omega_3), \theta_3'' \equiv \theta + \rho(\omega_3)$, with ρ given in Table 1.6.

TABLE 3.39. Propagation in a negative uniaxial crystal: $n_o > n_Z$, sff case

	ω_1	ω_2	ω_3
		Negative uniaxial: fsf $= eoe$	
d_{11}	$\sin\phi\cos^2\phi\cos\theta_1''\cos\theta_3''$	$\sin\phi\cos^2\phi\cos\theta_1''\cos\theta_3''$	$\sin\phi\cos^2\phi\cos\theta_1''\cos\theta_3''$
d_{12}	$-\sin\phi\cos^2\phi\cos\theta_1''\cos\theta_3''$	$\sin^3\phi\cos\theta_1''\cos\theta_3''$	$-\sin\phi\cos^2\phi\cos\theta_1''\cos\theta_3''$
d_{13}	0	$\sin\phi\sin\theta_1''\sin\theta_3''$	0
d_{14}	$\cos^2\phi\cos\theta_1''\sin\theta_3''$	$-\sin^2\phi\sin(\theta_3''+\theta_1'')$	$\cos^2\phi\sin\theta_1''\cos\theta_3''$
d_{15}	$-\sin\phi\cos\phi\cos\theta_1''\sin\theta_3''$	$-\sin\phi\cos\phi\sin(\theta_1''+\theta_3'')$	$-\sin\phi\cos\phi\sin\theta_1''\cos\theta_3''$
d_{16}	$-\cos\phi\cos2\phi\cos\theta_1''\cos\theta_3''$	$\sin\phi\sin2\phi\cos\theta_1''\cos\theta_3''$	$-\cos\phi\cos2\phi\cos\theta_1''\cos\theta_3''$
d_{21}	$\sin^2\phi\cos\phi\cos\theta_1''\cos\theta_3''$	$-\cos^3\phi\cos\theta_1''\cos\theta_3''$	$\sin^2\phi\cos\phi\cos\theta_1''\cos\theta_3''$
d_{22}	$-\sin^2\phi\cos\phi\cos\theta_1''\cos\theta_3''$	$-\sin^2\phi\cos\phi\cos\theta_1''\cos\theta_3''$	$-\sin^2\phi\cos\phi\cos\theta_1''\cos\theta_3''$
d_{23}	0	$-\cos\phi\sin\theta_1''\sin\theta_3''$	0
d_{24}	$\sin\phi\cos\phi\cos\theta_1''\sin\theta_3''$	$\sin\phi\cos\phi\sin(\theta_1''+\theta_3'')$	$\sin\phi\cos\phi\sin\theta_1''\cos\theta_3''$
d_{25}	$-\sin^2\phi\cos\theta_1''\sin\theta_3''$	$\cos^2\phi\sin(\theta_1''+\theta_3'')$	$-\sin^2\phi\sin\theta_1''\cos\theta_3''$
d_{26}	$-\sin\phi\cos2\phi\cos\theta_1''\cos\theta_3''$	$-\cos\phi\sin2\phi\cos\theta_1''\cos\theta_3''$	$-\sin\phi\cos2\phi\cos\theta_1''\cos\theta_3''$
d_{31}	$-\sin\phi\cos\phi\sin\theta_1''\cos\theta_3''$	0	$-\sin\phi\cos\phi\cos\theta_1''\sin\theta_3''$
d_{32}	$\sin\phi\cos\phi\sin\theta_1''\cos\theta_3''$	0	$\sin\phi\cos\phi\cos\theta_1''\sin\theta_3''$
d_{33}	0	0	0
d_{34}	$-\cos\phi\sin\theta_1''\sin\theta_3''$	0	$-\cos\phi\sin\theta_1''\sin\theta_3''$
d_{35}	$\sin\phi\sin\theta_1''\sin\theta_3''$	0	$\sin\phi\sin\theta_1''\sin\theta_3''$
d_{36}	$\cos2\phi\sin\theta_1''\cos\theta_3''$	0	$\cos2\phi\cos\theta_1''\sin\theta_3''$

$\theta_1'' \equiv \theta + \rho(\omega_1), \theta_3'' \equiv \theta + \rho(\omega_3)$, with ρ given in Table 1.6.

TABLE 3.40. Propagation in a negative uniaxial crystal: $n_o > n_Z$, fsf case

4

Nonlinear Propagation Equations and Solutions

In this chapter the light propagation equations for three wave mixing in second order nonlinear optical media are established. The cases of sum and difference frequency generation as well as second harmonic generation are considered. Analytical solutions of these equations are presented for special cases. The numerical procedure used to solve the equations in the general case is described. Results for the optimum frequency conversion efficiency including effects of phase mismatch, beam focusing and walk-off, are presented. Polynomial fits are presented for certain ranges of the parameters.

4.1 Nonlinear Propagation Equations

The equation describing propagation of light through a linear anisotropic medium was derived in Chapter 1 (Eqn. 1.200). To derive the propagation equations in a *nonlinear* medium, the nonlinear terms in the expression for polarization given in Eqn. 2.2 in Chapter 2 need to be included in Eqn. 1.20, i.e., the displacement vector $\tilde{\mathbf{D}}$ in Eqn. 1.19 needs to be replaced by the total displacement vector

$$
\begin{aligned}
\tilde{\mathbf{D}}_{\text{tot}} &= \epsilon_0 \tilde{\mathbf{E}} + \tilde{\mathbf{P}} \\
&= (\epsilon_0 \tilde{\mathbf{E}} + \tilde{\mathbf{P}}_L) + \tilde{\mathbf{P}}_{NL} \\
&= \tilde{\mathbf{D}} + \tilde{\mathbf{P}}_{NL}
\end{aligned}
\tag{4.1}
$$

where $\tilde{\mathbf{D}}$ is the linear portion of the total displacement vector. Since $\tilde{\mathbf{D}}$ and $\tilde{\mathbf{P}}_{NL}$ are both oscillating at the same frequency (say ω)

$$
\begin{aligned}
\tilde{\mathbf{D}} &= e^{-i\omega t}\mathbf{D} \qquad \text{and} \\
\tilde{\mathbf{P}}_{NL} &= e^{-i\omega t}\mathbf{P}_{\mathbf{NL}}
\end{aligned}
\tag{4.2}
$$

so that

$$
\begin{aligned}
\tilde{\mathbf{D}}_{\text{tot}} &= e^{-i\omega t}(\mathbf{D} + \mathbf{P}_{\mathbf{NL}}) \\
&\equiv e^{-i\omega t}\mathbf{D}_{\text{tot}}.
\end{aligned}
\tag{4.3}
$$

137

Substituting $\mathbf{D}$ by $\mathbf{D}_{\text{tot}}$ in the right hand side of Eqn. 1.186 in Chapter 1, we obtain

$$
\begin{aligned}
\nabla(\nabla \cdot \mathbf{E}) - \nabla^2 \mathbf{E} &= \mu_0 \omega^2 \mathbf{D}_{\text{tot}} \\
&= \mu_0 \omega^2 (\mathbf{D} + \mathbf{P_{NL}}) \\
&= \mu_0 \omega^2 \psi (\mathfrak{D} + \mathbf{P_{NL}} \psi^{-1})
\end{aligned}
\tag{4.4}
$$

using the relation $\mathbf{D} = \mathfrak{D} \psi$ from Eqn. 1.184 in Chapter 1.

In the presence of nonlinearity, the vector $\mathfrak{D}$ in the right hand side of Eqn. 1.199 in Chapter 1 gets replaced by

$$
\mathfrak{D}_{\text{tot}} \equiv \mathfrak{D} + \mathbf{P}_{NL} \psi^{-1}.
\tag{4.5}
$$

Replacing $\hat{e} \cdot \vec{\mathfrak{D}}$ in Eqn. 1.195 by

$$
\hat{e} \cdot \mathfrak{D}_{\text{tot}} = \mathfrak{D} \cos \rho + (\mathbf{P}_{NL} \cdot \hat{e}) \psi^{-1}
\tag{4.6}
$$

we obtain from Eqn. 1.199

$$
\begin{aligned}
2ik \sin \rho \cos \rho \frac{\partial A}{\partial x} &- 2ik \cos^2 \rho \frac{\partial A}{\partial z} - \nabla_T^2 A \\
&= \mu_0 \omega^2 \hat{e} \cdot \vec{\mathfrak{D}}_{\text{tot}} - k^2 A \cos^2 \rho \\
&= \mu_0 \omega^2 (\mathfrak{D} - k^2 A \cos \rho) \cos \rho + \mu_0 \omega^2 (\mathbf{P}_{NL} \cdot \hat{e}) \psi^{-1} \\
&= \mu_0 \omega^2 (\mathbf{P}_{NL} \cdot \hat{e}) \psi^{-1}
\end{aligned}
\tag{4.7}
$$

and using Eqn. 1.50 with $k = n\omega/c$, Eqn. 4.7 can be re-written as

$$
\frac{\partial A}{\partial z} = \frac{i}{2k \cos^2 \rho} \nabla_T^2 A + \tan \rho \frac{\partial A}{\partial x} + \frac{i}{2k \cos^2 \rho} \mu_0 \omega^2 (\mathbf{P}_{NL} \cdot \hat{e}) \psi^{-1}.
\tag{4.8}
$$

In this treatment of light propagation through crystals, linear absorption of light has so far been ignored. Denoting the linear absorption coefficient by α, a linear absorption term is introduced here phenomenologically, so that Eqn. 4.8 is modified to

$$
\frac{\partial A}{\partial z} = \frac{i}{2k \cos^2 \rho} \nabla_T^2 A + \tan \rho \frac{\partial A}{\partial x} - \frac{\alpha}{2} A + \frac{i}{2k \cos^2 \rho} \mu_0 \omega^2 (\mathbf{P}_{NL} \cdot \hat{e}) \psi^{-1}.
\tag{4.9}
$$

The propagation equation including the linear absorption term has been derived in [1] starting from the Maxwell's equations, with the assumption that absorption arises from free charges in the medium. It can also be derived assuming the linear dielectric constant to be complex instead of real. Here we add the term just phenomenologically - instead of repeating the well known derivations.

Since walk-off angles are not more than a few degrees for most cases of interest, the $\cos^2 \rho$ terms in equations derived from Eqn. 4.8 will be set to unity.

The factor $\mathbf{P}_{NL} \cdot \hat{e}$ in the last term of Eqn. 4.9 depends upon the type of nonlinear interaction. For example, for an ssf type of interaction in the case of non-degenerate three wave mixing, the values of $\mathbf{P}_{NL} \cdot \hat{e}$ are given at the three interacting frequencies by Eqn. 3.10. If A_1, A_2 and A_3 denote the slowly varying amplitudes of the oscillating electric fields at the frequencies ω_1, ω_2 and ω_3, respectively, the propagation equations at the three frequencies are

$$\frac{\partial A_1}{\partial z} = \frac{i}{2k_1} \nabla_T^2 A_1 + \tan \rho_1 \frac{\partial A_1}{\partial x} - \frac{\alpha_1}{2} A_1$$

$$+ \frac{i}{2k_1} \mu_0 \omega_1^2 \, 4\varepsilon_0 d_{\text{eff}}(\omega_1) \, A_2^* A_3 e^{-i(k_1+k_2-k_3)z}$$

$$\frac{\partial A_2}{\partial z} = \frac{i}{2k_2} \nabla_T^2 A_2 + \tan \rho_2 \frac{\partial A_2}{\partial x} - \frac{\alpha_2}{2} A_2$$

$$+ \frac{i}{2k_2} \mu_0 \omega_2^2 \, 4\varepsilon_0 d_{\text{eff}}(\omega_2) \, A_1^* A_3 e^{-i(k_1+k_2-k_3)z}$$

$$\frac{\partial A_3}{\partial z} = \frac{i}{2k_3} \nabla_T^2 A_3 + \tan \rho_3 \frac{\partial A_3}{\partial x} - \frac{\alpha_3}{2} A_3$$

$$+ \frac{i}{2k_3} \mu_0 \omega_3^2 \, 4\varepsilon_0 d_{\text{eff}}(\omega_3) \, A_1 A_2 e^{i(k_1+k_2-k_3)z} \tag{4.10}$$

where the subscripts 1, 2 and 3 in the parameters above refer to the frequencies ω_1, ω_2 and ω_3. As before, it is assumed here that ω_3 is the highest and ω_1 is the lowest of the three frequencies and that

$$\omega_3 = \omega_1 + \omega_2. \tag{4.11}$$

For the case of Type 1 *degenerate* three wave mixing process such as second harmonic generation (SHG), the components of nonlinear polarization are obtained from Eqns. 3.20. To avoid confusion, the subscripts p and s are used to denote the characteristics of the lower frequency and the higher frequency waves, respectively, for SHG. Eqn. 4.11 is then replaced by

$$\omega_s = \omega_p + \omega_p = 2\omega_p. \tag{4.12}$$

If A_p and A_s denote the slowly varying amplitudes of the oscillating electric fields at the frequencies ω_p and ω_s, the propagation equations at the two frequencies are

$$\frac{\partial A_p}{\partial z} = \frac{i}{2k_p} \nabla_T^2 A_p + \tan \rho_p \frac{\partial A_p}{\partial x} - \frac{\alpha_p}{2} A_p$$

$$+ \frac{i}{2k_p} \mu_0 \omega_p^2 \, 4\varepsilon_0 d_{\text{eff}} \, A_p^* A_s \, e^{-i(2k_p-k_s)z}$$

$$\frac{\partial A_s}{\partial z} = \frac{i}{2k_s} \nabla_T^2 A_s + \tan \rho_s \frac{\partial A_s}{\partial x} - \frac{\alpha_s}{2} A_s$$

$$+ \frac{i}{2k_s} \mu_0 \omega_s^2 \, 2\varepsilon_0 d_{\text{eff}} \, A_p^2 \, e^{i(2k_p-k_s)z}. \tag{4.13}$$

The nonlinear propagation equations, Eqns. 4.10 and 4.13 contain the effects of diffraction (through the ∇_T^2 terms), beam walk-off, linear absorption and nonlinear frequency mixing. Additional effects arising from other physical processes, such as thermal effects, two-photon absorption or refractive nonlinearities will need to be added to the equations as they become important. Usually such processes are deleterious to efficient frequency mixing. When all the linear and nonlinear effects are included, the propagation equations need to be solved by numerical techniques which will be discussed later in this chapter. In special cases, such as for collimated beams when diffraction terms and beam walk-off can be ignored, the Eqns. 4.10 and 4.13 can be solved analytically. Such analytical expressions can be useful for obtaining insight about the nonlinear processes, as checks on the numerical results and also for quick estimations of efficiencies in practical cases.

The effects of beam focusing on nonlinear frequency conversion is determined primarily by the ∇_T^2 term in the propagation equation. Optimum focusing of light beams is usually necessary for efficient frequency mixing when available power is limited, for example for continuous wave lasers, for which the maximum power available is a few hundred watts or at most in the kilowatt range. For short duration laser pulses, the peak powers available can exceed hundreds of kilowatts. Focusing light too tightly in the nonlinear material would cause damage to the medium and more frequency converted energy can be obtained with collimated beams. In this chapter, both the optimized focusing and collimated cases are discussed, for the processes of sum and difference frequency generations (SFG, DFG) and for second harmonic generation (SHG), which is of course a special case of SFG.

4.1.1   Normalized form of the three wave mixing equations

For simplicity of notations and also to explicitly determine the dependence of frequency conversion efficiencies on the key physical parameters, it is useful to formulate the three wave mixing equations (Eqns. 4.10) in terms of dimensionless parameters. Denoting by A_0 a **real** valued amplitude having the same dimension as the field amplitudes A_1, A_2 and A_3, the normalized amplitude variables u_1, u_2 and u_3 are defined as

$$u_1 \equiv \frac{A_1}{A_0} \qquad u_2 \equiv \frac{A_2}{A_0} \qquad u_3 \equiv \frac{A_3}{A_0}. \qquad (4.14)$$

The actual value of A_0 is unimportant because it is only a normalization parameter which will drop off in the end. The transverse spatial coordinates x and y are normalized here with respect to a distance r_0, which we can leave unspecified at this point.(When the beams are circular Gaussian in transverse shape, r_0 can be assigned to equal the half width at $1/e$ of irradiance/intensity of a Gaussian beam at focus). The spatial coordinate z is normalized with respect to the length of the nonlinear medium, denoted by ℓ. The normalized spatial coordinates are therefore

$$x_1 \equiv \frac{x}{r_0} \qquad y_1 \equiv \frac{y}{r_0} \qquad z_1 \equiv \frac{z}{\ell}. \qquad (4.15)$$

Furthermore, defining the dimensionless parameters

$$a_1 \equiv \alpha_1 \ell, \qquad a_2 \equiv \alpha_2 \ell, \qquad a_3 \equiv \alpha_3 \ell \qquad (4.16)$$

and

$$\sigma \equiv (k_3 - k_1 - k_2)\ell \qquad (4.17)$$

the Eqns. 4.10 are rewritten as

$$\frac{\partial u_1}{\partial z_1} = \frac{i\ell \nabla^2_{T_1} u_1}{2k_1 r_0^2} + \frac{\ell \tan \rho_1}{r_0} \frac{\partial u_1}{\partial x_1} - \frac{a_1}{2} u_1 + \frac{4\pi i d_{\mathrm{eff}} \ell A_0}{\lambda_1 n_1} u_3 u_2^* e^{i\sigma z_1}$$

$$\frac{\partial u_2}{\partial z_1} = \frac{i\ell \nabla^2_{T_1} u_2}{2k_2 r_0^2} + \frac{\ell \tan \rho_2}{r_0} \frac{\partial u_2}{\partial x_1} - \frac{a_2}{2} u_2 + \frac{4\pi i d_{\mathrm{eff}} \ell A_0}{\lambda_2 n_2} u_3 u_1^* e^{i\sigma z_1}$$

$$\frac{\partial u_3}{\partial z_1} = \frac{i\ell \nabla^2_{T_1} u_3}{2k_3 r_0^2} + \frac{\ell \tan \rho_3}{r_0} \frac{\partial u_3}{\partial x_1} - \frac{a_3}{2} u_3 + \frac{4\pi i d_{\mathrm{eff}} \ell A_0}{\lambda_3 n_3} u_1 u_2 e^{-i\sigma z_1}$$

$$(4.18)$$

where $\nabla_{T_1} = r_0 \nabla_T$.

The four terms on the right hand sides of Eqns. 4.18 determine the effects of diffraction, beam walk-off, linear absorption and nonlinear coupling, respectively. The effect of phase mismatch is contained in the nonlinear coupling term. When other nonlinear processes, such as two photon absorption and nonlinear refraction are present, additional terms need to be included. The three coupled equations can be solved numerically using split-step technique. However, when the interacting laser beams are collimated in the region of nonlinear interaction, and when beam walk-offs and linear absorption can be ignored, analytical solutions of the coupled wave equations have been obtained in the ground-breaking paper of nonlinear optics [1]. In the next few sections the results presented in Ref. [1] are rederived for the general case of three wave mixing as well as for the three special cases of sum frequency generation, second harmonic generation and difference frequency generation .

4.2 Solutions To The Three Wave Mixing Equations In The Absence Of Diffraction, Beam Walk-off And Absorption

Diffraction - No	Beam Walk-off - No	Absorption - No
Phase Matching - No	Pump Depletion - Yes	

In Ref. [1] it was shown that analytical solutions to the three wave mixing equations in Eqns. 4.18 can be obtained when diffraction and beam walk-offs can be ignored and when the absorption coefficients at all the three frequencies involved are equal to each other. Since the case of equal values of the absorption coefficients is rare in practice, and also to avoid notational complication, we consider here only the cases of no absorption at any of the wavelengths. Equations 4.18 then reduce to

$$\frac{\partial u_1}{\partial z_1} = \frac{4\pi i d_{\text{eff}} A_0 \ell}{\lambda_1 n_1} u_3 u_2^* e^{i\sigma z_1}$$

$$\frac{\partial u_2}{\partial z_1} = \frac{4\pi i d_{\text{eff}} A_0 \ell}{\lambda_2 n_2} u_3 u_1^* e^{i\sigma z_1} \qquad (4.19)$$

$$\frac{\partial u_3}{\partial z_1} = \frac{4\pi i d_{\text{eff}} A_0 \ell}{\lambda_3 n_3} u_1 u_2 e^{-i\sigma z_1}$$

where u_1, u_2 and u_3 are complex variables which can be written as

$$u_1 \equiv \frac{v_1 e^{i\phi_1}}{\sqrt{n_1 \lambda_1}}, \qquad u_2 \equiv \frac{v_2 e^{i\phi_2}}{\sqrt{n_2 \lambda_2}} \quad \text{and} \quad u_3 \equiv \frac{v_3 e^{i\phi_3}}{\sqrt{n_3 \lambda_3}} \qquad (4.20)$$

and where v_1, v_2, v_3 and ϕ_1, ϕ_2, ϕ_3 are all real. We note here that the variables v_1, v_2 and v_3 are not dimensionless - they have the dimensions of $[\text{length}]^{1/2}$. Moreover, the quantities v_1^2, v_2^2 and v_3^2 in any plane are proportional to the photon flux, i.e., the number of photons crossing through the plane per unit area per unit time (and denoted by N_1, N_2 and N_3, say, at wavelengths λ_1, λ_2 and λ_3 respectively). If I_1, I_2 and I_3 denote the irradiances at the three wavelengths, we have

$$v_1^2 = \frac{I_1 \lambda_1}{2c\varepsilon_0 A_0^2} = \left(\frac{h}{2\varepsilon_0 A_0^2}\right) N_1 \qquad (4.21)$$

$$v_2^2 = \frac{I_2 \lambda_2}{2c\varepsilon_0 A_0^2} = \left(\frac{h}{2\varepsilon_0 A_0^2}\right) N_2 \qquad (4.22)$$

$$v_3^2 = \frac{I_3 \lambda_3}{2c\varepsilon_0 A_0^2} = \left(\frac{h}{2\varepsilon_0 A_0^2}\right) N_3. \qquad (4.23)$$

The three complex (i.e., six real) Eqns. 4.19 can now be rewritten as 4 real equations

$$\frac{\partial v_1}{\partial z_1} = -\beta_1 v_2 v_3 \sin\theta \qquad (4.24)$$

$$\frac{\partial v_2}{\partial z_1} = -\beta_1 v_3 v_1 \sin\theta \qquad (4.25)$$

$$\frac{\partial v_3}{\partial z_1} = \beta_1 v_1 v_2 \sin\theta \qquad (4.26)$$

$$\frac{\partial \theta}{\partial z_1} = \sigma + \beta_1 \cos\theta \left(\frac{v_1 v_2}{v_3} - \frac{v_2 v_3}{v_1} - \frac{v_3 v_1}{v_2}\right) \qquad (4.27)$$

where

$$\theta = \sigma z_1 + \phi_3 - \phi_1 - \phi_2 \qquad (4.28)$$

and

$$\beta_1 = \frac{4\pi d_{\text{eff}} \ell A_0}{\sqrt{n_1 n_2 n_3 \lambda_1 \lambda_2 \lambda_3}}. \qquad (4.29)$$

Equations 4.24 - 4.28 show that

$$v_1 \frac{\partial v_1}{\partial z_1} = v_2 \frac{\partial v_2}{\partial z_1} = -v_3 \frac{\partial v_3}{\partial z_1} = -\beta_1 v_1 v_2 v_3 \sin\theta \qquad (4.30)$$

i.e.,

$$\frac{\partial}{\partial z_1}(v_1^2 - v_2^2) = \frac{\partial}{\partial z_1}(v_2^2 + v_3^2) = \frac{\partial}{\partial z_1}(v_3^2 + v_1^2) = 0. \qquad (4.31)$$

Thus

$$v_2^2 + v_3^2 = m_1, \quad v_3^2 + v_1^2 = m_2 \quad \text{and} \quad v_1^2 - v_2^2 = m_3 \qquad (4.32)$$

where m_1, m_2 and m_3 are constants.

4.2.1 An interlude - the Manley-Rowe relations

The Eqns. 4.31, expressed in terms of the photon fluxes N_1, N_2 and N_3, constitute the *Manley-Rowe* relations. If ΔN_1, ΔN_2 and ΔN_3 denote the changes in photon fluxes at the three frequencies, we obtain from Eqn. 4.31

$$\Delta N_1 = \Delta N_2 = -\Delta N_3. \qquad (4.33)$$

Equations 4.33 show that in a three wave mixing process, changes in the photon numbers of the two lower frequency waves are equal to each other in sign and magnitude, and are equal and opposite of the change in the photon number at the highest frequency. Thus, for SFG, the reduction of photon flux at frequency ω_1 by an amount ΔN_1 corresponds to a reduction of photon flux at frequency ω_2 by an equal amount and an *increase* of photon flux at frequency ω_3 by the same amount. In the process of DFG, with incident beams at frequencies ω_3 and ω_2, the reduction of the photon flux at ω_3 causes a gain in photon flux not only at the generated frequency ω_1 but also at the incident frequency ω_2. The Manley-Rowe relations also let us introduce the inverse process of SFG, known as Optical Parametric Generation (OPG) in which a reduction in the photon flux at one incident beam at frequency ω_3 can lead to the generation of photon fluxes at two frequencies ω_1 and ω_2. For a given ω_3, there are of course an infinite number of values of ω_1 and ω_2 that satisfy Eqn. 4.11. The desired frequencies (ω_1 and ω_2) are usually obtained in the OPG process by providing the proper phase matching conditions for the beams at those frequencies.

4.2.2   Back to solutions of the three wave mixing equations

Using Eqns. 4.24 it can be shown (in a few steps) that

$$\frac{\partial}{\partial z_1}(v_1 v_2 v_3 \cos\theta) = -a \frac{\partial}{\partial z_1}(v_3^2) \tag{4.34}$$

where

$$a \equiv \frac{\sigma}{2\beta_1}. \tag{4.35}$$

Thus,

$$v_1 v_2 v_3 \cos\theta + a\, v_3^2 = \Gamma \tag{4.36}$$

where Γ is a constant. Equation 4.26 can be rewritten as

$$
\begin{aligned}
v_3 \frac{\partial v_3}{\partial z_1} &= \beta_1 v_1 v_2 v_3 \sin\theta \\
&\equiv \beta_1 \sqrt{\mathcal{D}_1(v_3)}
\end{aligned}
\tag{4.37}
$$

where

$$
\begin{aligned}
\mathcal{D}_1(v_3) &= (v_1 v_2 v_3)^2 - (v_1 v_2 v_3 \cos\theta)^2 \\
&= (m_2 - v_3^2)(m_1 - v_3^2)v_3^2 - (\Gamma - a v_3^2)^2
\end{aligned}
\tag{4.38}
$$

using Eqns. 4.32 and Eqn. 4.36. Defining a variable s with

$$s \equiv v_3^2 \tag{4.39}$$

the expression for $\mathcal{D}_1$ is seen to be a cubic polynomial in s, having three roots, which are designated s_a, s_b and s_c, with the stipulation that

$$s_c > s_b > s_a \geq 0. \tag{4.40}$$

Thus

$$
\begin{aligned}
\mathcal{D}_1(s) &= (m_1 - s)(m_2 - s)s - (\Gamma - as)^2 \\
&= (s - s_a)(s - s_b)(s - s_c).
\end{aligned}
\tag{4.41}
$$

Further defining a variable y and a constant γ through the relations

$$y^2 \equiv \frac{s - s_a}{s_b - s_a} \qquad \gamma^2 \equiv \frac{s_b - s_a}{s_c - s_a} \tag{4.42}$$

the Eqn.4.37 can be rewritten as

$$\frac{\partial y}{\partial z_1} = \beta_1 \sqrt{s_c - s_a}\sqrt{(1 - y^2)(1 - \gamma^2 y^2)}. \tag{4.43}$$

Since z_1 goes from 0 to 1, Eqn. 4.43 can be integrated as

$$\beta_2 \equiv \beta_1 \sqrt{s_c - s_a} \tag{4.44}$$

$$= \int_{y_0}^{y_\ell} \frac{dy}{\sqrt{(1 - y^2)(1 - \gamma^2 y^2)}} \tag{4.45}$$

where y_0 and y_ℓ correspond to the values of y at the entrance and the exit of the nonlinear medium, i.e.,

$$y_0 \equiv y(z_1 = 0) \tag{4.46}$$

$$y_\ell \equiv y(z_1 = 1). \tag{4.47}$$

Denoting by $v_1(0)$, $v_2(0)$, $v_3(0)$ and $\theta(0)$ the values of v_1, v_2, v_3 and θ at the entrance to the nonlinear crystal, i.e., at $z_1 = 0$, we have

$$y(0) = \frac{v_3(0)^2 - s_a}{s_b - s_a} \quad \text{and} \quad \Gamma = v_1(0)v_2(0)v_3(0)\cos\theta(0) + av_3(0)^2. \tag{4.48}$$

Equation 4.45 can be rewritten as

$$\beta_2 = u_\ell - u_0 \tag{4.49}$$

$$\tag{4.50}$$

where

$$u_\ell \equiv \int_0^{y_\ell} \frac{dy}{\sqrt{(1 - y^2)(1 - \gamma^2 y^2)}}; \quad u_0 \equiv \int_0^{y_0} \frac{dy}{\sqrt{(1 - y^2)(1 - \gamma^2 y^2)}}. \tag{4.51}$$

From the definition of Jacobian elliptic functions given by Abramowitz and Stegun [3] (Eqn. 16.1.3 and 16.1.5) and repeated below, we have

$$y_\ell = \text{sn}(u_\ell, \gamma^2) \quad \text{and} \quad y_0 = \text{sn}(u_0, \gamma^2). \tag{4.52}$$

4.2.3 Another interlude - Jacobian elliptic functions

In Ref. [1] and also in later books such as by Shen [3] and by Sutherland [5] the second argument of the elliptic function is written as γ instead of γ^2 as it is done here in Eqns. 4.52. Such use of γ as the second argument is inconsistent with the definition of the parameter m $(= \gamma^2)$ in Eqn. 16.1.3 of Ref. [3], and probably more importantly at present, with the definition of the Jacobian elliptic functions with double arguments, ellipj(U,M), used in the commercial computational software MATLAB. For the sake of completeness and consistency, the definitions of the Jacobian elliptic functions used here are explicitly defined below, along with the approximate values for them given in Ref. [3].

With u defined as a function of two variables θ_a and γ through the relation

$$
\begin{aligned}
u &= \int_0^{\theta_a} \frac{d\theta}{\sqrt{1 - \gamma^2 \sin^2 \theta}} \\
&= \int_0^{\sin \theta_a} \frac{dy}{\sqrt{(1 - \gamma^2 y^2)(1 - y^2)}}
\end{aligned}
\tag{4.53}
$$

three Jacobian elliptic functions are defined as

$$
\begin{aligned}
\mathrm{sn}\left(u, \gamma^2\right) &\equiv \sin \theta_a \\
\mathrm{cn}\left(u, \gamma^2\right) &\equiv \cos \theta_a \\
\mathrm{dn}\left(u, \gamma^2\right) &\equiv \sqrt{1 - \gamma^2 \sin^2 \theta_a}.
\end{aligned}
\tag{4.54}
$$

In Ref. [3], the functions sn, cn and dn are also expressed as sn(u|m), cn(u|m) and dn(u|m), with m $\equiv \gamma^2$ *defined* as a positive real number with values between 0 and 1. The expressions for DFG efficiency given in Ref. [3] [Chapter 8, Eqn. 8.3] contain sn functions with imaginary arguments, which are thus inconsistent with the Jacobian elliptic functions defined in Ref. [3] (and also in MATLAB, where the parameter M of the function (ellipj(U,M) is restricted to real numbers between 0 and 1).

The approximate expressions for the three Jacobian elliptic functions for γ^2 close to 0 and for γ^2 close to 1 are given in Ref. [3] as

$$
\underline{\gamma^2 \approx 0}
\tag{4.55}
$$

$$
\begin{aligned}
\mathrm{sn}(u, \gamma^2) &\approx \sin u - \frac{1}{4}\gamma^2 (u - \sin u \cos u) \cos u \\
\mathrm{cn}(u, \gamma^2) &\approx \cos u + \frac{1}{4}\gamma^2 (u - \sin u \cos u) \sin u \\
\mathrm{dn}(u, \gamma^2) &\approx 1 - \frac{1}{4}\gamma^2 \sin^2 u
\end{aligned}
\tag{4.56}
$$

and

$$
\underline{\gamma^2 \approx 1}
\tag{4.57}
$$

$$
\begin{aligned}
\mathrm{sn}(u, \gamma^2) &\approx \tanh u + \frac{1}{4}(1 - \gamma^2)(\sinh u \cosh u - u)\mathrm{sech}^2 u \\
\mathrm{cn}(u, \gamma^2) &\approx \mathrm{sech}\, u - \frac{1}{4}(1 - \gamma^2)(\sinh u \cosh u - u)\tanh u \, \mathrm{sech}\, u \\
\mathrm{dn}(u, \gamma^2) &\approx \mathrm{sech}\, u + \frac{1}{4}(1 - \gamma^2)(\sinh u \cosh u + u)\tanh u \, \mathrm{sech}\, u.
\end{aligned}
\tag{4.58}
$$

The addition theorem for the sn function is given in Ref. [3] as

$$
\mathrm{sn}(u + v, \gamma^2) = \frac{\mathrm{sn}\, u \cdot \mathrm{cn}\, v \cdot \mathrm{dn}\, v + \mathrm{sn}\, v \cdot \mathrm{cn}\, u \cdot \mathrm{dn}\, u}{1 - \gamma^2 \mathrm{sn}^2 u \, \mathrm{sn}^2 v}.
\tag{4.59}
$$

4.2.4 Return to the solution of the coupled three wave mixing equations

Using Eqn. 4.49 and the addition theorem in Eqn. 4.59 we obtain

$$
\begin{aligned}
y_\ell &= \operatorname{sn}(u_\ell, \gamma^2) \\
&= \operatorname{sn}(u_0 + \beta_2, \gamma^2) \\
&= \frac{\operatorname{sn}(u_0, \gamma^2)\,\operatorname{cn}(\beta_2, \gamma^2)\,\operatorname{dn}(\beta_2, \gamma^2) + \operatorname{sn}(\beta_2, \gamma^2)\,\operatorname{cn}(u_0, \gamma^2)\operatorname{dn}(u_0, \gamma^2)}{1 - \gamma^2\operatorname{sn}^2(u_0, \gamma^2)\,\operatorname{sn}^2(\beta_2, \gamma^2)} \\
&= \frac{y_0\,\operatorname{cn}(\beta_2, \gamma^2)\,\operatorname{dn}(\beta_2, \gamma^2) + \operatorname{sn}(\beta_2, \gamma^2)\sqrt{1 - y_0^2}\sqrt{1 - \gamma^2 y_0^2}}{1 - \gamma^2 y_0^2\,\operatorname{sn}^2(\beta_2, \gamma^2)}.
\end{aligned} \tag{4.60}
$$

The values of v_1, v_2, v_3 at the exit of the nonlinear crystal can be obtained from the value of y_ℓ, using Eqns. 4.39, 4.42 and 4.32

$$
\begin{aligned}
v_3(\ell)^2 &= (s_b - s_a)y_\ell^2 + s_a \\
v_2(\ell)^2 &= v_2(0)^2 + v_3(0)^2 - (s_b - s_a)y_\ell^2 - s_a \\
v_1(\ell)^2 &= v_1(0)^2 + v_3(0)^2 - (s_b - s_a)y_\ell^2 - s_a.
\end{aligned} \tag{4.61}
$$

From the first equations in Eqns. 4.21:

$$
\begin{aligned}
I_1 &= \frac{2c\varepsilon_0 A_0^2}{\lambda_1} v_1^2 \\[4pt]
I_2 &= \frac{2c\varepsilon_0 A_0^2}{\lambda_2} v_2^2 \\[4pt]
I_3 &= \frac{2c\varepsilon_0 A_0^2}{\lambda_3} v_3^2.
\end{aligned} \tag{4.62}
$$

Defining a constant I_a which has the units of irradiance, through the relationship

$$
I_a \equiv \frac{c\varepsilon_0 n_1 n_2 n_3 \lambda_1 \lambda_2}{8\pi^2 d_{\text{eff}}^2 \ell^2} \tag{4.63}
$$

and using the definition of β_1 (from Eqn. 4.29) we have

$$
I_a \beta_1^2 = \frac{2c\varepsilon_0 A_0^2}{\lambda_3}. \tag{4.64}
$$

Thus we have

$$
\begin{aligned}
I_1 &= \frac{\lambda_3}{\lambda_1} I_a\, \beta_1^2\, v_1^2 \\[4pt]
I_2 &= \frac{\lambda_3}{\lambda_2} I_a\, \beta_1^2\, v_2^2 \\[4pt]
I_3 &= I_a\, \beta_1^2\, v_3^2.
\end{aligned} \tag{4.65}
$$

The values of the irradiances $I_1(\ell)$, $I_2(\ell)$ and $I_3(\ell)$ are obtained from the values of $v_1(\ell)$, $v_2(\ell)$ and $v_3(\ell)$ in Eqns. 4.61.

In a general three wave mixing process, with light amplitudes at all three frequencies present at the entrance to the nonlinear crystal, the irradiances at the exit of the crystal can be obtained from the value of y_ℓ. When all three incident irradiances are non-zero, the exit irradiances also depend on the phase difference $\phi_3 - \phi_1 - \phi_2$ at the entrance face. However, if any of the three incident irradiances is zero, for example in the processes of *unseeded* sum or difference frequency generation (including unseeded second harmonic generation), the irradiances of the beams at the sample exit are independent of the phase values of the beams at the incident plane. Next we derive the irradiance expressions for the beams at the exit of a nonlinear medium resulting from nonlinear mixing in the cases of unseeded SFG, DFG and SHG. The expressions for the seeded cases, with nonzero values of the incident irradiances at all three frequencies, are of course more complicated, but they can be derived from Eqns. 4.61 using the same procedures.

4.3　Unseeded Sum Frequency Generation ($\omega_1 + \omega_2 = \omega_3$)

If only the two lower frequency beams of light are incident on a nonlinear crystal, i.e., if $v_1(0)$ and $v_2(0)$ are non-zero, with $v_3(0) = 0$, Eqns. 4.48 shows that $\Gamma = 0$. In this case we have from Eqn. 4.32

$$m_1 = v_2(0)^2, \quad m_2 = v_1(0)^2 \quad \text{and} \quad m_3 = v_1(0)^2 - v_2(0)^2 \qquad (4.66)$$

where m_1, m_2 and m_3 are constants.

$$
\begin{aligned}
\mathcal{D}_1(s) &= (m_1 - s)(m_2 - s)s - (as)^2 \\
&= s\{s^2 - s(m_1 + m_2 + a^2) + m_1 m_2\} \\
&\equiv s\,(s^2 - t_1 s + t_2)
\end{aligned}
\qquad (4.67)
$$

where

$$t_1 = m_1 + m_2 + a^2; \quad \text{and} \quad t_2 = m_1 m_2. \qquad (4.68)$$

The three roots of the polynomial $\mathcal{D}_1(s)$ in Eqn.4.67 are given by

$$s_a = 0; \quad s_b = \frac{t_1 - t_3}{2}; \quad s_c = \frac{t_1 + t_3}{2} \qquad (4.69)$$

where

$$t_3 \equiv \sqrt{t_1^2 - 4t_2}. \qquad (4.70)$$

With $s_a = 0$, Eqns. 4.39 and 4.42 show that $y^2 = v_3^2/s_b$ so that with $v_3(0) = 0$ for the unseeded case, we have $y_0 = 0$. Thus from Eqn. 4.51, we obtain $u_2 = 0$ and $u_1 = \beta_2 = \sqrt{s_c \beta_1^2}$. From Eqns. 4.52

$$y_\ell = \mathrm{sn}(\beta_2, \gamma^2) = \mathrm{sn}\left(\sqrt{s_c \beta_1^2}, \gamma^2\right) \tag{4.71}$$

with

$$\gamma^2 = \frac{s_b}{s_c} \tag{4.72}$$

so that from Eqns. 4.39 and 4.42

$$v_3(\ell)^2 = s(\ell) = s_a + (s_b - s_a)y_\ell^2 = s_b \, \mathrm{sn}^2\left(\sqrt{s_c \beta_1^2}, \gamma^2\right) \tag{4.73}$$

from which we find, using the constants in Eqn. 4.32

$$v_1(\ell)^2 = v_1(0)^2 - v_3(\ell)^2, \qquad v_2(\ell)^2 = v_2(0)^2 - v_3(\ell)^2. \tag{4.74}$$

From Eqn. 4.65, the irradiance $I_3(\ell)$ at the exit face of the nonlinear material is then

$$
\begin{aligned}
I_3(\ell) &= I_a \beta_1^2 v_3(\ell)^2 \\
&= I_a \, \beta_1^2 s_b \, \mathrm{sn}^2\left(\sqrt{s_c \beta_1^2}, \gamma^2\right) \\
&= I_a \, \beta_1^2 \left(\frac{t_1 - t_3}{2}\right) \mathrm{sn}^2\left(\sqrt{\beta_1^2\left(\frac{t_1 + t_3}{2}\right)}, \gamma^2\right) \\
&= I_a \, b_1 \, \mathrm{sn}^2(\sqrt{b_2}, \gamma^2)
\end{aligned}
\tag{4.75}
$$

where

$$b_1 \equiv \frac{a_1 - a_3}{2}, \qquad b_2 \equiv \frac{a_1 + a_3}{2}, \qquad \gamma^2 = \frac{b_1}{b_2} \tag{4.76}$$

and

$$
\begin{aligned}
a_1 &\equiv \beta_1^2 t_1 = \left(\frac{\lambda_2}{\lambda_3}\frac{I_2(0)}{I_a} + \frac{\lambda_1}{\lambda_3}\frac{I_1(0)}{I_a} + \frac{\sigma^2}{4}\right) \\
a_2 &\equiv \beta_1^4 t_2 = \beta_1^4 m_1 m_2 = \frac{\lambda_1 \lambda_2}{\lambda_3^2}\frac{I_1(0)I_2(0)}{I_a^2} \\
a_3 &\equiv \beta_1^2 t_3 = \sqrt{a_1^2 - 4a_2}.
\end{aligned}
\tag{4.77}
$$

4.3.1 SFG irradiance for collimated beams with no phase matching ($\sigma \neq 0$) and with no pump depletion

Diffraction - No	Beam Walk-off - No	Absorption - No
Pump Depletion - No	Phase Mismatch - Yes	

When the peak irradiances of the incident beams are small compared to the material parameter I_a the peak irradiance of the generated beam is much smaller than those of the incident beams, and the incident beams are not changed much due to the nonlinear interaction. This is the 'no pump depletion' case, for which we have $a_2 << a_1$, so that

$$a_3 = a_1 \left(1 - \frac{4a_2}{a_1^2} \right)^{1/2}$$

$$\approx a_1 - \frac{2a_2}{a_1}. \tag{4.78}$$

Thus we get

$$b_1 = \frac{a_2}{a_1}, \qquad b_2 = a_1 \tag{4.79}$$

with

$$a_1 = \frac{\sigma^2}{4}, \qquad a_2 = \frac{\lambda_1 \lambda_2}{\lambda_3^2} \frac{I_1(0) I_2(0)}{I_a^2}. \tag{4.80}$$

Since

$$\gamma^2 = \frac{b_1}{b_2} = \frac{a_2}{a_1^2} << 1 \tag{4.81}$$

the Jacobian elliptic function $\mathrm{sn}(\sqrt{b_2}, \gamma^2)$ can be replaced by $\sin \sigma/2$ and we obtain

$$I_3(\ell) = \frac{\lambda_1 \lambda_2}{\lambda_3^2} \frac{I_1(0) I_2(0)}{I_a} \left(\frac{\sin(\sigma/2)}{\sigma/2} \right)^2. \tag{4.82}$$

4.3.2 SFG irradiance for collimated beams with phase matching ($\sigma = 0$) and with pump depletion

Diffraction - No	Beam Walk-off - No	Absorption - No
Phase Mismatch - No	Pump Depletion - Yes	

When $\sigma = 0$, the term a_3 in Eqn. 4.77 takes the simpler form

$$a_3 = \pm \left(\frac{\lambda_2}{\lambda_3} \frac{I_2(0)}{I_a} - \frac{\lambda_1}{\lambda_3} \frac{I_1(0)}{I_a} \right) \tag{4.83}$$

and Eqn. 4.75 can be simplified. Since by definition (from Eqns. 4.40 and 4.277) b_1 is smaller than or equal to b_2, we have the following three cases:

Case 1. $\lambda_1 I_1(0) < \lambda_2 I_2(0)$

In this case,

$$b_1 = \frac{\lambda_1}{\lambda_3}\frac{I_1(0)}{I_a}, \qquad b_2 = \frac{\lambda_2}{\lambda_3}\frac{I_2(0)}{I_a} \qquad (4.84)$$

and

$$I_3(\ell) = \frac{\lambda_1}{\lambda_3}I_1(0)\mathrm{sn}^2\left(\sqrt{\frac{\lambda_2}{\lambda_3}\frac{I_2(0)}{I_a}}, \left(\frac{\lambda_1 I_1(0)}{\lambda_2 I_2(0)}\right)^2\right). \qquad (4.85)$$

Case 2. $\lambda_1 I_1(0) > \lambda_2 I_2(0)$

In this case,

$$b_1 = \frac{\lambda_2}{\lambda_3}\frac{I_2(0)}{I_a}, \qquad b_2 = \frac{\lambda_1}{\lambda_3}\frac{I_1(0)}{I_a} \qquad (4.86)$$

and

$$I_3(\ell) = \frac{\lambda_2}{\lambda_3}I_2(0)\mathrm{sn}^2\left(\sqrt{\frac{\lambda_1}{\lambda_3}\frac{I_1(0)}{I_a}}, \left(\frac{\lambda_2 I_2(0)}{\lambda_1 I_1(0)}\right)^2\right). \qquad (4.87)$$

Case 3. $\lambda_1 I_1(0) = \lambda_2 I_2(0)$

In this case the expression for $I_3(\ell)$ is further simplified and we obtain

$$\begin{aligned}
I_3(\ell) &= \frac{\lambda_1}{\lambda_3}I_1(0)\mathrm{sn}^2\left(\sqrt{\frac{\lambda_1}{\lambda_3}\frac{I_1(0)}{I_a}}, 1\right) \\
&= \frac{\lambda_1}{\lambda_3}I_1(0)\tanh^2\left(\sqrt{\frac{\lambda_1}{\lambda_3}\frac{I_1(0)}{I_a}}\right). \qquad (4.88)
\end{aligned}$$

4.3.3 SFG power and energy conversion efficiency for collimated beams with arbitrary spatial and temporal shapes

Diffraction - No	Beam Walk-off - No	Absorption - No
	Phase Mismatch - Yes	Pump Depletion - Yes

When both the lower frequency beams are collimated and have known transverse distributions, the total *power* at the sum frequency wavelength can be calculated from Eqn. 4.75 by integrating over the generated irradiance distribution. The spatial and temporal distributions of collimated laser beams are usually Gaussian, but in general they are not necessarily so. The actual beam distributions and temporal pulse shapes of the incident beams in nonlinear interactions can of course be experimentally determined and may or may not fit well with analytic expressions. We assume the normalized spatial and temporal distributions of the irradiance of the two beams in the incident plane (at $z = 0$) to be given by the functions $f(x, y)$ and $g(t)$, with appropriate subscripts (1 or 2) indicating the corresponding frequencies (ω_1 or ω_2):

$$I_1(0, x, y, t) = I_{10} f_1(x, y) g_1(t) \qquad I_2(0, r) = I_{20} f_2(x, y) g_2(t). \qquad (4.89)$$

Defining parameters δ, p_1 and K_1 as

$$\delta \equiv \frac{\lambda_2}{\lambda_3} \qquad p_1 \equiv \frac{I_{20}}{I_{10}} \qquad K_1 \equiv \frac{I_{10}}{I_a} \qquad (4.90)$$

and the normalized coordinate $x_1 = x/r_0$, $y_1 = y/r_0$ and $t_1 = t/t_0$, where r_0 and t_0 will be related to laser beam parameters later in specific cases, we find that

$$
\begin{aligned}
a_1(x_1, y_1, t_1) &= \frac{\delta}{\delta - 1} K_1 f_1(x_1, y_1) g_1(t_1) + \delta K_1 p_1 f_2(x_1, y_1) g_2(t_1) + \frac{\sigma^2}{4} \\
a_2(x_1, y_1, t_1) &= \frac{\delta^2}{\delta - 1} p_1 K_1^2 f_1(x_1, y_1) f_2(x_1, y_1) g_1(t_1) g_2(t_1) \qquad (4.91)
\end{aligned}
$$

where

$$\frac{\lambda_1}{\lambda_3} = \frac{\delta}{\delta - 1}, \qquad \text{since} \qquad \frac{1}{\lambda_3} = \frac{1}{\lambda_1} + \frac{1}{\lambda_2}. \qquad (4.92)$$

Defining

$$a_3(x_1, y_1, t_1) \equiv \sqrt{a_1(x_1, y_1, t_1)^2 - 4a_2(x_1, y_1, t_1)} \qquad (4.93)$$

we have

$$
\begin{aligned}
b_1(x_1, y_1, t_1) &= \frac{a_1(x_1, y_1, t_1) - a_3(x_1, y_1, t_1)}{2}, \\
b_2(x_1, y_1, t_1) &= \frac{a_1(x_1, y_1, t_1) + a_3(x_1, y_1, t_1)}{2} \qquad (4.94)
\end{aligned}
$$

and

$$\gamma^2(x_1, y_1, t_1) = \frac{b_1(x_1, y_1, t_1)}{b_2(x_1, y_1, t_1)}. \qquad (4.95)$$

The sum frequency power generated at the exit of the nonlinear material (assuming no reflection loss at the surface) is obtained by integrating over the beam cross-section

$$P_3(\ell, t_1) = \int\limits_{-\infty}^{\infty} \int\limits_{-\infty}^{\infty} dx\,dy\; I_3(\ell)$$

$$= r_0^2 I_a \int\limits_{-\infty}^{\infty} \int\limits_{-\infty}^{\infty} dx_1\,dy_1\; b_1(x_1, y_1, t_1)$$

$$\text{sn}^2(\sqrt{b_2(x_1, y_1, t_1)}, \gamma^2(x_1, y_1, t_1)). \qquad (4.96)$$

Similarly, the incident power values at the two lower frequencies are given by

$$P_1(0, t_1) = I_{10} r_0^2 g_1(t_1) \int\limits_{-\infty}^{\infty} \int\limits_{-\infty}^{\infty} f_1(x_1, y_1) dx_1\,dy_1$$

$$P_2(0, t_1) = I_{20} r_0^2 g_2(t_1) \int\limits_{-\infty}^{\infty} \int\limits_{-\infty}^{\infty} f_2(x_1, y_1) dx_1\,dy_1 \qquad (4.97)$$

so that the power conversion efficiency for SFG

$$\eta_{\text{SFG}}^P \equiv \frac{P_3(\ell, t_1)}{P_1(0, t_1) + P_2(0, t_1)} \qquad (4.98)$$

can be obtained from Eqns. 4.96 and 4.97.

For continuous wave (CW) lasers, the functions $g_1(t_1)$ and $g_2(t_1)$ can be set equal to 1 and the time independent power conversion efficiency can be obtained from Eqn. 4.113.

The sum frequency energy generated at the exit of the nonlinear material is obtained by integrating the generated power over time

$$E_3(\ell) = \int\limits_{-\infty}^{\infty} dt\, P_3(\ell, t)$$

$$= r_0^2 t_0 I_a \int\limits_{-\infty}^{\infty} dt_1 \int\limits_{-\infty}^{\infty} \int\limits_{-\infty}^{\infty} dx_1\,dy_1\; b_1(x_1, y_1, t_1)$$

$$\text{sn}^2(\sqrt{b_2(x_1, y_1, t_1)}, \gamma^2(x_1, y_1, t_1)). \qquad (4.99)$$

The energy values at the two lower frequencies at $z = 0$ are given by

$$E_1(0) = I_{10} r_0^2 t_0 \int\limits_{-\infty}^{\infty} dt_1 g_1(t_1) \int\limits_{-\infty}^{\infty} f_1(x_1, y_1) dx_1\,dy_1$$

$$E_2(0) = I_{20} r_0^2 t_0 \int\limits_{-\infty}^{\infty} dt_1 g_2(t_1) \int\limits_{-\infty}^{\infty} f_2(x_1, y_1) dx_1\,dy_1 \qquad (4.100)$$

so that the energy conversion efficiency for SFG

$$\eta_{SFG}^{E} \equiv \frac{E_3(\ell)}{E_1(0) + E_2(0)} \qquad (4.101)$$

can be obtained from Eqns. 4.99 and 4.100.

4.3.4 SFG power and energy conversion efficiency for collimated Gaussian beams

Diffraction - No	Beam Walk-off - No	Absorption - No
Phase Mismatch - Yes	Pump Depletion - Yes	

We assume that the two incident beams are of circular Gaussian shape in cross-section and have Gaussian temporal profiles with pulse widths t_{01} and t_{02}, defined as the half widths at e^{-1} of maximum irradiance (HWe^{-1}M) for frequencies ω_1 and ω_2, respectively:

$$I_1(0, r, t) = I_{10}e^{-(r/r_{01})^2 - (t/t_{01})^2}$$
$$I_2(0, r, t) = I_{20}e^{-(r/r_{02})^2 - (t/t_{02})^2}. \qquad (4.102)$$

Defining parameters ϱ and τ as

$$\varrho \equiv \frac{r_{01}}{r_{02}}, \qquad \tau \equiv \frac{t_{01}}{t_{02}} \qquad (4.103)$$

and setting r_0 in the last section equal to r_{01} we have

$$f_1(x_1, y_1) = f_1(r_1) = e^{-r_1^2}, \qquad f_2(x_1, y_1) = f_2(r_1) = e^{-\varrho^2 r_1^2} \qquad (4.104)$$

and

$$g_1(t_1) = e^{-t_1^2} \qquad g_2(t_1) = e^{-\tau^2 t_1^2}. \qquad (4.105)$$

Then

$$\int_{-\infty}^{\infty} \int_{-\infty}^{\infty} f_1(x_1, y_1)dx_1 dy_1 = \pi \qquad (4.106)$$

$$\int_{-\infty}^{\infty} g_1(t_1)dt_1 = \sqrt{\pi}. \qquad (4.107)$$

Thus, for laser beams with circular Gaussian spatial distribution,

$$P_1(0, t_1) = I_{10}\pi r_{01}^2 g_1(t_1), \qquad P_2(0, t_1) = I_{20}\pi r_{02}^2 g_2(t_1) \qquad (4.108)$$

and for beams which in addition have Gaussian temporal shapes,

$$E_1(0) = I_{10}\pi\sqrt{\pi}r_{01}^2 t_{01}, \qquad E_2(0) = I_{20}\pi\sqrt{\pi}r_{02}t_{02}. \tag{4.109}$$

For CW lasers (i.e., setting $g_1(t_1) = g_2(t_1) = 1$) with circular Gaussian spatial profiles the total incident power values at the two frequencies of the incident beams are then

$$P_1(0) = I_{10}\pi r_{01}^2, \qquad P_2(0) = I_{20}\pi r_{02}^2 \tag{4.110}$$

so that

$$P_1(0) + P_2(0) \quad = \quad I_{10}\pi r_{01}^2 \left(1 + \frac{p_1}{\varrho^2}\right) \tag{4.111}$$

and

$$E_1(0) + E_2(0) \quad = \quad I_{10}\pi\sqrt{\pi}r_{01}^2 t_{01} \left(1 + \frac{p_1}{\varrho^2\tau}\right). \tag{4.112}$$

The expression for SFG power conversion efficiency (using Eqn. 4.96) is

$$
\begin{aligned}
\eta_{\text{SFG}}^P \quad &\equiv \quad \frac{P_3(\ell, t_1)}{P_1(0) + P_2(0)} \\
&= \quad \frac{2}{K_1\left(1 + \dfrac{p_1}{\varrho^2}\right)} \int_0^\infty r_1\, dr_1\, b_1(r_1) \\
&\qquad \times \mathrm{sn}^2(\sqrt{b_2(r_1)}, \gamma(r_1)^2)
\end{aligned}
\tag{4.113}
$$

where

$$b_1(r_1) = \frac{(a_1(r_1) - a_3(r_1))}{2}, \qquad b_2(r_1) = \frac{(a_1(r_1) + a_3(r_1))}{2} \tag{4.114}$$

and

$$\gamma^2(r_1) = \frac{b_1(r_1)}{b_2(r_1)} \tag{4.115}$$

with

$$
\begin{aligned}
a_1(r_1) \quad &= \quad \frac{\delta}{\delta - 1} K_1 e^{-r_1^2} + \delta K_1 p_1 e^{-\varrho^2 r_1^2} + \frac{\sigma^2}{4} \\
a_2(r_1) \quad &= \quad \frac{\delta^2}{\delta - 1} p_1 K_1^2 e^{-r_1^2}(1 + \varrho^2).
\end{aligned}
\tag{4.116}
$$

For pulsed incident beams with Gaussian temporal shapes along with circular Gaussian spatial shape

$$E_3(\ell) = 2\pi \int_{-\infty}^{\infty} dt \int_0^{\infty} rdr I_3(\ell)$$

$$= 2\pi r_{10}^2 t_{01} I_a \int_{-\infty}^{\infty} dt_1 \int_0^{\infty} r_1 \, dr_1 \, b_1(r_1, t_1)$$

$$\times \, \text{sn}^2(\sqrt{b_2(r_1, t_1)}, \gamma^2(r_1, t_1)) \qquad (4.117)$$

so that the *energy* conversion efficiency for SFG can be written as

$$\eta_{\text{SFG}}^E \equiv \frac{E_3(\ell)}{E_1(0) + E_2(0)}$$

$$= \frac{2}{\sqrt{\pi} K_1 (1 + \frac{p_1}{\varrho^2 \tau})} \int_{-\infty}^{\infty} dt_1 \int_0^{\infty} r_1 \, dr_1 \, b_1(r_1, t_1)$$

$$\times \, \text{sn}^2(\sqrt{b_2(r_1, t_1)}, \gamma^2(r_1, t_1)) \qquad (4.118)$$

where

$$b_1(r_1, t_1) = \frac{a_1(r_1, t_1) - a_3(r_1, t_1)}{2}$$

$$b_2(r_1, t_1) = \frac{a_1(r_1, t_1) + a_3(r_1, t_1)}{2} \qquad (4.119)$$

with

$$\gamma^2(r_1, t_1) = \frac{b_1(r_1, t_1)}{b_2(r_1, t_1)} \qquad (4.120)$$

and

$$a_1(r_1, t_1) = \frac{\delta}{\delta - 1} K_1 e^{-r_1^2} e^{-t_1^2} + \delta K_1 p_1 e^{-\varrho^2 r_1^2} e^{-\tau^2 t_1^2} + \frac{\sigma^2}{4}$$

$$a_2(r_1, t_1) = \frac{\delta^2}{\delta - 1} p_1 K_1^2 e^{-r_1^2(1 + \varrho^2)} e^{-\tau^2 t_1^2}$$

$$a_3(r_1, t_1) = \sqrt{a_1(r_1, t_1)^2 - 4a_2(r_1, t_1)}. \qquad (4.121)$$

4.3.5 SFG power and energy conversion efficiency for collimated Gaussian beams with phase mismatch ($\sigma \neq 0$) and no pump depletion

Diffraction - No	Beam Walk-off - No	Absorption - No
Pump Depletion - No	Phase Mismatch - Yes	

In the nondepleted pump case, the irradiance at the sum frequency given in Eqn. 4.82 can be integrated over the beam area and time to provide the power and energy outputs. Assuming that the incident beams are both circular Gaussian in cross-section and are constant in time, we obtain from Eqns. 4.82 and 4.102

$$P_3(\ell) = 2\pi r_{01}^2 \int_0^\infty I_3(\ell) r_1 dr_1$$

$$= \frac{\pi r_{01}^2}{1+\varrho^2} \frac{\lambda_1 \lambda_2}{\lambda_3^2} \frac{I_{10} I_{20}}{I_a} \left(\frac{\sin(\sigma/2)}{\sigma/2} \right)^2 \tag{4.122}$$

so that

$$\eta_{\text{SFG}}^P = \frac{P_3(\ell)}{P_1(0) + P_2(0)} \tag{4.123}$$

$$= \frac{\delta^2}{\delta - 1} \frac{1}{p_1 K_1 (1+\varrho^2) \left(1 + \frac{p_1}{\varrho^2} \right)} \left(\frac{\sin(\sigma/2)}{\sigma/2} \right)^2. \tag{4.124}$$

Equation 4.122 can be written in terms of the incident powers $P_1(0)$ and $P_2(0)$ as

$$P_3(\ell) = \frac{\lambda_1 \lambda_2}{\lambda_3^2} \frac{P_1(0) P_2(0)}{\pi (r_{01}^2 + r_{02}^2) I_a} \left(\frac{\sin(\sigma/2)}{\sigma/2} \right)^2 \tag{4.125}$$

Assuming that the incident beams are both circular Gaussian in cross-section and are constant in time, i.e., with the incident irradiances given by Eqns. 4.102, we obtain from Eqn. 4.82

$$E_3(\ell) = 2\pi r_{01}^2 t_{01} \int_{-\infty}^{\infty} \int_0^\infty I_3(\ell, t_1) r_1 dr_1 dt_1$$

$$= \frac{\pi \sqrt{\pi} r_{01}^2 t_{01}}{(1+\varrho^2)\sqrt{1+\tau^2}} \frac{\lambda_1 \lambda_2}{\lambda_3^2} \frac{I_{10} I_{20}}{I_a} \left(\frac{\sin(\sigma/2)}{\sigma/2} \right)^2 \tag{4.126}$$

and

$$E_1(0) + E_2(0) = \pi \sqrt{\pi} r_{01}^2 t_{01} I_{10} \left(1 + \frac{p_1}{\varrho^2 \tau} \right) \tag{4.127}$$

so that using Eqn. 4.112

$$\eta_{\text{SFG}}^E = \frac{E_3(\ell)}{E_1(0) + E_2(0)} \tag{4.128}$$

$$= \frac{\delta^2}{\delta - 1} \frac{p_1 K_1}{(1+\varrho^2)(1+\frac{p_1}{\varrho^2 \tau})\sqrt{1+\tau^2}} \left(\frac{\sin(\sigma/2)}{\sigma/2} \right)^2. \tag{4.129}$$

Equation 4.126 can be written in terms of the incident energy values $E_1(0)$ and $E_2(0)$ as

$$E_3(\ell) = \frac{\lambda_1 \lambda_2}{\lambda_3^2} \frac{E_1(0) E_2(0)}{\pi \sqrt{\pi} (r_{01}^2 + r_{02}^2) \sqrt{t_{01}^2 + t_{02}^2}} \frac{1}{I_a} \left(\frac{\sin(\sigma/2)}{\sigma/2} \right)^2. \qquad (4.130)$$

4.3.6 Some results of SFG power and energy conversion efficiency for collimated Gaussian beams

Diffraction - No   Beam Walk-off - No   Absorption - No,
Pump Depletion - Yes   Phase Mismatch - Yes

The power conversion efficiency given in Eqn. 4.113 for the sum frequency generation of two collimated incident beams having circular Gaussian cross-section depends on 5 factors. They are the phase mismatch parameter σ, the wavelength ratio δ, the irradiance ratio p_1, the ratio of the beam radii ϱ and the parameter K_1 which is the ratio of I_{10} to I_a, where I_{10} is the peak value of the irradiance of one of the incident beams and I_a is a constant depending on the material and the light wavelengths. Since the Jacobian elliptic functions are readily available in commercial programming platforms (such as MAT-LAB), it is convenient to write a computer program in MATLAB to evaluate η_3. Such a code is provided in Appendix 1 from which results can be obtained for any particular case of interest. Here we provide some results for the special case of ϱ equal to 1 and the phase mismatch σ equal to zero.

Figure 4.1 shows the SFG power conversion efficiency given in Eqn. 4.113 plotted against K_1 for various values of p_1, with δ (the ratio of middle wavelength to the shortest wavelength) equal to 1.1 and with the phase mismatch parameter σ equal to 0. The curve marked g, with p_1 equal to 10 corresponds to the largest conversion efficiency (exceeding 95%). Figures 4.2 and 4.3 show the results for δ equal to 2.5 and 5, respectively. Maximum values of the SFG conversion efficiency obtained are found to decrease with increasing δ. Values of p_1 at which maximum efficiency is obtained for a given value of K_1 is lower for δ equal to 2.5 and 5, than for δ equal to 1.1.

In Fig. 4.4 the SFG energy conversion efficiency values obtained from Eqn. 4.209 are plotted against K_1 for various values of p_1, with δ equal to 1.1 and with the phase mismatch parameter σ equal to 0. Values of energy conversion efficiency are similar to (while somewhat smaller than) those obtained for the power conversion efficiency, and the curve marked g, with $p_1 = 10$ corresponding to the largest energy conversion efficiency (exceeding 95%).

4.3.7 SFG conversion efficiency for focused Gaussian beams

Diffraction - Yes   Beam Walk-off - Yes   Absorption - Yes
Pump Depletion - Yes   Phase Mismatch - Yes

The efficiency of three wave mixing processes depends on the irradiance(s)

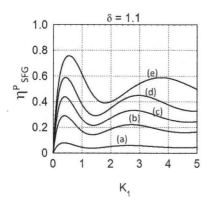
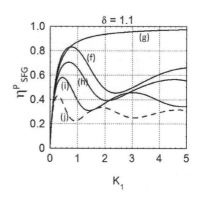

FIGURE 4.1: Power conversion efficiency for sum frequency generation of collimated beams with $\delta = 1.1$, $\sigma = 0$ and $\varrho = 1$ for different values of p_1. (a) $p_1 = 0.1$, (b) $p_1 = 0.5$, (c) $p_1 = 1.0$, (d) $p_1 = 2.0$, (e) $p_1 = 5.0$, (f) $p_1 = 7.5$, (g) $p_1 = 10.0$, (h) $p_1 = 12.5$, (i) $p_1 = 15.0$, (j) $p_1 = 20.0$.

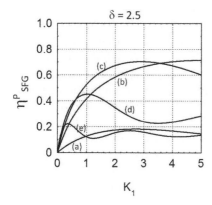
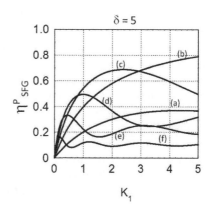

FIGURE 4.2: Power conversion efficiency for sum frequency generation of collimated beams with $\delta = 2.5$, $\sigma = 0$ and $\varrho = 1$ for different values of p_1. (a) $p_1 = 0.1$, (b) $p_1 = 0.5$, (c) $p_1 = 1.0$, (d) $p_1 = 2.0$, (e) $p_1 = 5.0$.

FIGURE 4.3: Power conversion efficiency for sum frequency generation of collimated beams with $\delta = 5.0$, $\sigma = 0$ and $\varrho = 1$ for different values of p_1. (a) $p_1 = 0.1$, (b) $p_1 = 0.25$, (c) $p_1 = 0.5$, (d) $p_1 = 1.0$, (e) $p_1 = 2.0$, (f) $p_1 = 5.0$.

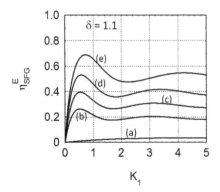
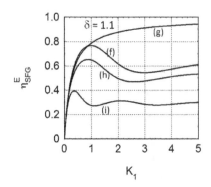

FIGURE 4.4: Energy conversion efficiency for sum frequency generation of collimated beams with $\delta = 1.1$, $\sigma = 0$ and $\varrho = 1$ for different values of p_1. (a) $p_1 = 0.1$, (b) $p_1 = 0.5$, (c) $p_1 = 1.0$, (d) $p_1 = 2.0$, (e) $p_1 = 5.0$, (f) $p_1 = 7.5$, (g) $p_1 = 10.0$, (h) $p_1 = 12.5$, (i) $p_1 = 15.0$, (j) $p_1 = 20.0$.

of the incident beam(s), so it would seem to be desirable to focus the beam tighter in the nonlinear optical medium to increase conversion efficiency. However, the damage threshold of the material imposes an upper limit on the amount of irradiance the material can be exposed to. The irradiance available from high peak power short pulse lasers (typically of nanoseconds or shorter duration) can reach the material damage threshold limit even with beams well collimated through the nonlinear material. For nonlinear interactions of pulsed laser beams, the theory of using collimated beams is therefore adequate in most cases. On the other hand, continuous wave or long pulse lasers (having pulse duration in the tens of microseconds or longer) typically have powers in the tens of Watts (or at most a few hundreds of Watts) and the material damage threshold is often not reached with tightest possible focusing. But for a given amount of incident laser power and a given crystal length, the case of smallest possible spot size will not necessarily lead to the maximum conversion efficiency because beam diffraction will reduce the effective 'interaction' region over which the irradiance is high. In such cases it is important to know the optimum focusing conditions that would maximize the frequency conversion efficiency.

Boyd and Kleinman [2] pioneered the theoretical treatment of nonlinear three wave mixing of focused Gaussian beams. However, their treatment deals only with the small conversion efficiency case, i.e., the case of no pump depletion. For practical applications it is of course desirable to have as high a conversion efficiency as possible, so determination of optimum focusing in the cases of pump depletion is of interest. For that purpose, the normalized three wave mixing equations (Eqns. 4.18) are re-written with the 'focusing parameters' of the incident beams explicitly expressed. The equations are numerically solved, including the effects of diffraction, beam walk-offs, linear absorption, phase mismatch and nonlinear coupling, and the dependence of the generated (sum frequency) power on the focusing parameters of the incident beams for different values of the focusing parameters are studied.

To define the parameters used here, we assume circular Gaussian beams are incident normally on a nonlinear crystal of length ℓ at $z = 0$, focused at a distance f from the incident surface, as shown in Fig. 4.5. The slowly varying amplitude A of the electric field at a point x, y, z is given by

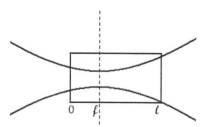

FIGURE 4.5: A Gaussian beam, focused at a distance f from the incident face of a material of length ℓ.

$$A(x,y,z) = \frac{A(0,0,f)}{1 + i(z - f)/z_0} e^{-\dfrac{x^2 + y^2}{2r_0^2(1 + i(z - f)/z_0)}} \qquad (4.131)$$

where r_0 is the 'beam radius' and

$$z_0 = \frac{2\pi r_0^2 n}{\lambda} \qquad (4.132)$$

with λ the wavelength of the beam and n denoting the value of the refractive index of the medium at the wavelength λ. z_0 is also known as the Rayleigh range of the beam.

With the focusing parameter ξ and the relative focusing position μ defined as

$$\xi \equiv \frac{\ell}{2z_0} = \frac{\ell\lambda}{4\pi r_0^2 n}$$

$$\mu \equiv 1 - 2\frac{f}{\ell} \qquad (4.133)$$

the field amplitude is given in terms of the normalized coordinates (defined in Eqns. 4.15) as

$$A(x_1, y_1, z_1) = \frac{A(0,0,f)}{1 + i\xi\{2z_1 - (1 - \mu)\}} e^{-\dfrac{1}{2}\dfrac{x_1^2 + y_1^2}{1 + i\xi\{2z_1 - (1 - \mu)\}}}. \qquad (4.134)$$

If the beam is focused at the center of the crystal, i.e., if $f = \ell/2$, then $\mu = 0$.

Denoting the parameters associated with the two incident beams at frequencies ω_1 and ω_2 by the subscripts 1 and 2 respectively, and assigning

$$r_0 = r_{01}, \qquad A_0 = A_1(0,0,f) \tag{4.135}$$

in Eqns. 4.18, we obtain

$$
\begin{aligned}
\frac{\partial u_1}{\partial z_1} &= i\xi_1\nabla_{T_1}^2 u_1 + \sqrt{2k_1\ell\xi_1}\tan\rho_1\frac{\partial u_1}{\partial x_1} - \frac{a_1}{2}u_1 + i\frac{\lambda_3 n_3}{\lambda_1 n_1}\kappa u_3 u_2^* e^{i\sigma z_1} \\
\frac{\partial u_2}{\partial z_1} &= i\frac{k_1}{k_2}\xi_1\nabla_{T_1}^2 u_2 + \sqrt{2k_1\ell\xi_1}\tan\rho_2\frac{\partial u_2}{\partial x_1} - \frac{a_2}{2}u_2 + i\frac{\lambda_3 n_3}{\lambda_2 n_2}\kappa u_3 u_1^* e^{i\sigma z_1} \\
\frac{\partial u_3}{\partial z_1} &= i\frac{k_1}{k_3}\xi_1\nabla_{T_1}^2 u_3 + \sqrt{2k_1\ell\xi_1}\tan\rho_3\frac{\partial u_3}{\partial x_1} - \frac{a_3}{2}u_1 + i\kappa u_1 u_2 e^{-i\sigma z_1}
\end{aligned}
$$

$$\tag{4.136}$$

where

$$\kappa = \frac{4\pi A_0 d_{\text{eff}}\ell}{\lambda_3 n_3} \tag{4.137}$$

and where we have used

$$\frac{\ell}{r_{01}} = (2k_1\ell\xi_1)^{1/2}. \tag{4.138}$$

κ is a coupling parameter for the three wave mixing equations, similar to β_1 defined in Eqn. 4.29, but unlike β_1, κ is dimensionless.

If I_{10} denotes the peak irradiance of the incident beam at frequency ω_1, we find from Eqn. 4.135

$$I_{10} = 2n_1 c\varepsilon_0|A_1(0,0,f)|^2 = 2n_1 c\varepsilon_0|A_0|^2. \tag{4.139}$$

Using Eqns. 4.137, 4.139 and 4.63 we find that the parameter κ introduced here is related to K_1 defined in Eqn. 4.90:

$$
\begin{aligned}
\kappa^2 &= \frac{16\pi^2 d^2\ell^2}{\lambda_3^2 n_3^2}\frac{I_{10}}{2n_1 c\varepsilon_0} \\
&= \frac{n_2}{n_3}\frac{\delta^2}{\delta - 1}K_1.
\end{aligned}
\tag{4.140}
$$

The three wave mixing Eqns. 4.136 can be solved using the split step technique described in Chapter 7, with the linear propagation part (including diffraction, linear absorption and beam walk-off but ignoring nonlinear coupling) solved by the Fourier transform method and the nonlinear propagation part (ignoring diffraction, linear absorption and beam walk-off) solved using the finite difference technique. For both incident beams circular Gaussian in cross-section, the variables u_1, u_2 and u_3 at the entrance face of the nonlinear medium ($z_1 = 0$) are given by

$$u_1(x_1, y_1, 0) = \frac{1}{1 - i\xi_1\{(1 - \mu_1)\}} e^{-\frac{1}{2}\left(\frac{x_1^2 + y_1^2}{1 - i\xi_1\{(1 - \mu_1)\}}\right)}$$

$$u_2(x_1, y_1, 0) = \sqrt{p_1} \frac{1}{1 - i\xi_2\{(1 - \mu_2)\}} e^{-\frac{\varrho^2}{2}\left(\frac{x_1^2 + y_1^2}{1 - i\xi_2\{(1 - \mu_2)\}}\right)}$$

$$u_3(x_1, y_1, 0) = 0 \qquad (4.141)$$

with p_1 and ϱ defined earlier in Eqns. 4.90 and 4.103. In addition to the five factors listed in Sec. 4.3.6, (σ, δ, p_1, ϱ and K_1) for collimated beams, the generated power in the focused case also depends on the focusing parameters of the two incident beams (ξ_1 and ξ_2), positions of their foci (μ_1 and μ_2) and the refractive index ratios n_1/n_3 and n_2/n_3. When beam walk-offs are included, the walk-off angles at the three frequencies ρ_1, ρ_2 and ρ_3 need to be specified along with the parameter $k_1\ell$. The linear absorption parameters a_1, a_2 and a_3, if non-zero, should also be included in the numerical computation.

The value of the dimensionless variable $u_3(x_1, y_1, 1)$ at the exit surface of the nonlinear material ($z_1 = 1$) can be calculated by numerically solving Eqns. 4.136. The irradiance distribution at the generated frequency ω_3 is then given by

$$\begin{aligned} I_3(x_1, y_1, 1) &= 2n_3 c\varepsilon_0 |A_3(x_1, y_1, 1)|^2 \\ &= 2n_3 c\varepsilon_0 A_0^2 |u_3(x_1, y_1, 1)|^2 \\ &= \frac{n_3}{n_1} I_{10} |u_3(x_1, y_1, 1)|^2. \qquad (4.142) \end{aligned}$$

The SFG power $P_3(\ell)$ at the exit plane is obtained by integrating over the irradiance distribution, i.e.,

$$
\begin{aligned}
P_3(\ell) &= \int\!\!\!\int_{-\infty}^{\infty} I_3(x,y,\ell)\,dx\,dy \\[2mm]
&= \frac{n_3}{n_1}I_{10}r_{01}^2 \int_{-\infty}^{\infty}\!\!\int_{-\infty}^{\infty} |u_3(x_1,y_1,1)|^2\,dx_1\,dy_1 \\[2mm]
&= \frac{n_3}{n_1}\frac{P_1(0)}{\pi} \int_{-\infty}^{\infty}\!\!\int_{-\infty}^{\infty} |u_3(x_1,y_1,1)|^2\,dx_1\,dy_1 \\[2mm]
&= \frac{n_3}{n_1}\frac{P_1(0)P_2(0)}{\pi}\frac{1}{P_2(0)} \int_{-\infty}^{\infty}\!\!\int_{-\infty}^{\infty} |u_3(x_1,y_1,1)|^2\,dx_1\,dy_1 \\[2mm]
&= \frac{P_1(0)P_2(0)}{P_{SF}}h_3 \tag{4.143}
\end{aligned}
$$

where

$$
h_3 \equiv \frac{\xi_2}{\pi K_1 p_1}\int\!\!\!\int_{-\infty}^{\infty} |u_3(x_1,y_1,1)|^2\,dx_1\,dy_1,
$$

$$
P_{SF} \equiv \frac{c\varepsilon_0 n_1^2\lambda_1\lambda_2^2}{32\pi^2 d_{\text{eff}}^2\ell} \tag{4.144}
$$

and where we have used the expression for I_a from Eqn. 4.63 and

$$
\begin{aligned}
P_2(0) &= \pi I_{20}r_{02}^2 \\[2mm]
&= \pi K_1 p_1 I_a \frac{\ell\lambda_2}{4\pi n_2\xi_2}
\end{aligned}
\tag{4.145}
$$

from Eqns. 4.90 and 4.133.

For collimated incident beams, i.e., when ξ_1 and ξ_2 are small, h_3 simplifies to

$$
\begin{aligned}
h_{3(\text{coll})} &= \frac{n_1\lambda_1\lambda_2}{n_3\lambda_3^2}\frac{\xi_2}{1+\varrho^2}\left(\frac{\sin(\sigma/2)}{\sigma/2}\right)^2 \\[2mm]
&= \frac{n_1^2}{n_3}\frac{\lambda_1\lambda_2^2}{\lambda_3^2}\left(\frac{\xi_1\xi_2}{\lambda_2 n_1\xi_1+\lambda_1 n_2\xi_2}\right)\left(\frac{\sin(\sigma/2)}{\sigma/2}\right)^2 \tag{4.146}
\end{aligned}
$$

since

$$
\varrho \equiv \frac{r_{01}}{r_{02}} = \sqrt{\frac{\lambda_1 n_2\xi_2}{\lambda_2 n_1\xi_1}} \tag{4.147}
$$

Using Eqn.4.143 and the expression for P_{SF} (Eqn.4.144) we thus obtain the sum frequency power for the case of collimated incident Gaussian beams to be

$$P_{3(\text{coll})}(\ell) = \left(\frac{32\pi^2 d_{\text{eff}}^2 \ell}{c\varepsilon_0 n_3 \lambda_3^2}\right)\left(\frac{\xi_1 \xi_2}{\lambda_2 n_1 \xi_1 + \lambda_1 n_2 \xi_2}\right)\left(\frac{\sin(\sigma/2)}{\sigma/2}\right)^2 P_1(0)P_2(0).$$

(4.148)

4.3.8 Optimization of focusing parameters for SFG

The effects of focusing on the power conversion efficiency of SFG can be determined by calculating h_3 for different values of the focusing parameters ξ_1 and ξ_2. For given values of ξ_1 and ξ_2, the value of σ at which h_3 maximizes is in general non-zero. We denote by $h_{3m}(\xi_1, \xi_2)$ the value of h_3 maximized with respect to the parameter σ, and by $\sigma_m(\xi_1, \xi_2)$ the value of σ at that maximum. For a given ξ_2, the function $h_{3m}(\xi_1, \xi_2)$ has a maximum with respect to the variable ξ_1, and say we denote this maximum of $h_{3m}(\xi_1, \xi_2)$ by $h_{3mm}(\xi_2)$, and by $\xi_{1m}(\xi_2)$ the value of ξ_1 at that maximum. The corresponding value of σ that maximizes $h_{3m}(\xi_1, \xi_2)$ with respect to ξ_1 is denoted by $\sigma_{mm}(\xi_2)$. Going one step further in the optimization process, the symbol h_{3mmm} is used to denote the maximum value of $h_{3mm}(\xi_2)$ with respect to ξ_2, with ξ_{2m} denoting the value of ξ_2 at which the maximum occurs.

A computer program that calculates h_3 for a set of ξ_1, ξ_2 and σ (for given values of the other parameters) is provided in the appendix. The program also determines the values of $h_{3mm}(\xi_2)$, $\xi_{1m}(\xi_2)$ and $\sigma_{mm}(\xi_2)$.

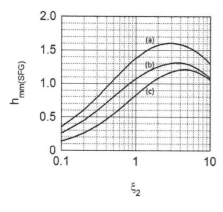

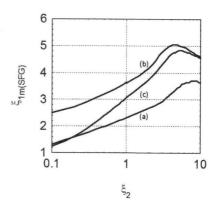

FIGURE 4.6: h_{3mm} as a function of ξ_2 for $\delta = 1.5$, (a) $K_1 = 0.0001$, $p_1 = 0.1$, (b) $K_1 = 1.0$, $p_1 = 0.1$ and (c) $K_1 = 1.0$, $p_1 = 2.0$

FIGURE 4.7: ξ_{1m} as a function of ξ_2 for $\delta = 1.5$, (a) $K_1 = 0.0001$, $p_1 = 0.1$, (b) $K_1 = 1.0$, $p_1 = 0.1$ and (c) $K_1 = 1.0$, $p_1 = 2.0$

Figures 4.6, 4.7 and 4.8 show the dependence of $h_{3mm}(\xi_2)$, $\xi_{1m}(\xi_2)$ and

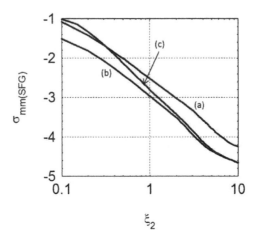

FIGURE 4.8: σ_{mm} as a function of ξ_2 for $\delta = 1.5$, (a) $K_1 = 0.0001$, $p_1 = 0.1$, (b) $K_1 = 1.0$, $p_1 = 0.1$ and (c) $K_1 = 1.0$, $p_1 = 2.0$

$\sigma_{mm}(\xi_2)$ on ξ_2 for the case of no pump depletion ($K_1 = 0.0001$, $p_1 = 0.1$) and for the case when pump depletion is present ($K_1 = 1$, $p_1 = 2$), for $\delta = 1.5$. We note that for the case of no pump depletion, ($K_1 = 0.0001$, $p_1 = 0.1$), the maximum of h_3 is obtained at $\xi_{1m} = 3.7$, $\xi_{2m} = 1.8$, with the value of σ at this maximum being -3.24 and h_{3mmm}, the maximum value of h_3, equal to 1.522.

When pump depletion from only one of the incident beams is present, ($K_1 = 1.0$, $p_1 = 0.1$), with $\delta = 1.5$, the maximum of h_3 is obtained at $\xi_1 = 6.2$, $\xi_2 = 2.9$, with the value of σ at this maximum being -4.09 and h_{3mmm} equal to 1.303.

When pump depletion takes place from both the incident beams, (with $K_1 = 1.0$, $p_1 = 2.0$), with $\delta = 1.5$, the maximum of h_3 is obtained at $\xi_1 = 6.6$, $\xi_2 = 3.3$, with the value of σ at this maximum being -4.49 and h_{3mmm} equal to 1.148.

These results are somewhat different from the prediction of Boyd and Kleinman theory [2](where ξ_1 and ξ_2 values are assumed to be equal to each other and the optimum value of ξ_1, ξ_2 is found to be 2.84, and the value of h optimized with respect to ξ_1 and σ is 1.06). Optimization for other cases of interest can be obtained by running the computer program for those specific cases.

4.4 Unseeded Second Harmonic Generation ($2\omega_p = \omega_s$)

Second harmonic generation (SHG) is of course a special case of SFG and the solutions already obtained for SFG reduce to the results for SHG with the two incident beams made identical in frequency and their irradiance properly normalized. However, in keeping with tradition and to avoid possible confusion arising from the complexity of notations used, the SHG case will be treated here by itself. This also allows a reader interested only in the SHG (and not the SFG) process to skip many of the details of the SFG section presented above and get the results directly in this section.

With the normalized amplitudes of the pump and the second harmonic fields defined as

$$u_p \equiv \frac{A_p}{A_0} \qquad \text{and} \qquad u_s \equiv \frac{A_s}{A_0} \qquad (4.149)$$

where again A_0 is a real number to be defined later, the SHG Eqns. 4.13 can be re-written as

$$\frac{\partial u_p}{\partial z_1} = \frac{i\ell}{2k_p r_0^2}\nabla_{T_1}^2 u_p + \frac{\ell \tan\rho_p}{r_0}\frac{\partial u_p}{\partial x_1} - \frac{a_p}{2}u_p + \frac{4\pi i d_{\text{eff}}\ell A_0}{\lambda_p n_p}u_s u_p^* e^{i\sigma z_1}$$

$$\frac{\partial u_s}{\partial z_1} = \frac{i\ell}{2k_s r_0^2}\nabla_{T_1}^2 u_s + \frac{\ell \tan\rho_s}{r_0}\frac{\partial u_s}{\partial x_1} - \frac{a_s}{2}u_s + \frac{2\pi i d_{\text{eff}}\ell A_0}{\lambda_s n_s}u_p^2 e^{-i\sigma z_1}$$

$$(4.150)$$

with

$$\sigma = (k_s - 2k_p)\ell, \qquad a_p \equiv \alpha_p \ell \qquad \text{and} \qquad a_s \equiv \alpha_s \ell. \qquad (4.151)$$

The complex variables u_p and u_s are expressed in terms of the real variables v_p, v_s, ϕ_p and ϕ_s through the relations

$$u_p \equiv \frac{v_p e^{i\phi_p}}{\sqrt{n_p \lambda_p}} \qquad \text{and} \qquad u_s \equiv \frac{v_s e^{i\phi_s}}{\sqrt{n_s \lambda_s}} \qquad (4.152)$$

and the variables v_p and v_s are related to the irradiances I_p, I_s and the photon fluxes N_p, N_s of the pump and the second harmonic beams through the relations

$$v_p^2 = \frac{I_p \lambda_p}{2c\varepsilon_0 A_0^2} = \left(\frac{h}{2\varepsilon_0 A_0^2}\right)N_p \qquad (4.153)$$

$$v_s^2 = \frac{I_s \lambda_s}{2c\varepsilon_0 A_0^2} = \left(\frac{h}{2\varepsilon_0 A_0^2}\right)N_s. \qquad (4.154)$$

4.4.1 Solution of SHG equations in the absence of diffraction, beam walk-off and absorption

Diffraction - No	Beam Walk-off - No	Absorption - No
Pump Depletion - Yes	Phase Mismatch - Yes	

Ignoring diffraction, beam walk-off and linear absorption in the Eqns.4.150, we obtain the four real equations

$$\frac{\partial v_p}{\partial z_1} = -\beta_1' v_p v_s \sin\theta \tag{4.155}$$

$$\frac{\partial v_s}{\partial z_1} = \frac{1}{2}\beta_1' v_p^2 \sin\theta \tag{4.156}$$

$$\frac{\partial \phi_p}{\partial z_1} = \beta_1' v_s \cos\theta \tag{4.157}$$

$$\frac{\partial \phi_s}{\partial z_1} = \frac{\beta_1'}{2}\frac{v_p^2}{v_s}\cos\theta \tag{4.158}$$

that is

$$\frac{\partial \theta}{\partial z_1} = \sigma + \beta_1' \cos\theta\left(\frac{v_p^2}{2\,v_s} - 2\,v_s\right) \tag{4.159}$$

where

$$\theta = \sigma\,z_1 + \phi_s - 2\,\phi_p \tag{4.160}$$

and

$$\beta_1' = \frac{4\pi d_{\text{eff}} l A_0}{\sqrt{n_p^2 n_s \lambda_p^2 \lambda_s}}. \tag{4.161}$$

Defining a constant I_a' which has the units of irradiance, through the relationship

$$I_a' \equiv \frac{c\varepsilon_0 n_p^2 n_s \lambda_p^2}{8\pi^2 d_{\text{eff}}^2 l^2} \tag{4.162}$$

and using the definition of β_1' (from Eqn. 4.161) we have

$$I_a' {\beta_1'}^2 = \frac{2c\varepsilon_0 A_0^2}{\lambda_s}. \tag{4.163}$$

The expressions for β_1 and I_a used in the SFG (Eqn. 4.29 and 4.63) case reduce to β_1' and I_a' in the SHG case (Eqns. 4.161 and 4.163) if the variables associated with the beams at frequencies ω_1 and ω_2 are replaced by those associated with the pump beam, and the variables associated with the sum frequency beam (at frequency ω_3) are replaced with those associated with the second harmonic beam.

Thus we have from Eqns. 4.153 and 4.163

$$I_p(\ell) = I_a' {\beta_1'}^2 v_p(\ell)^2, \qquad I_s(\ell) = I_a' {\beta_1'}^2 v_s(\ell)^2. \tag{4.164}$$

Equations 4.155 - 4.160 show that

$$v_p \frac{\partial v_p}{\partial z_1} = -2 \, v_s \frac{\partial v_s}{\partial z_1} = -\beta_1' v_p^2 v_s \sin \theta \tag{4.165}$$

i.e.,

$$\frac{\partial}{\partial z_1} (v_p^2 + 2 \, v_s^2) = 0. \tag{4.166}$$

Thus

$$v_p^2 + 2 \, v_s^2 = m_1 \tag{4.167}$$

where m_1 is a constant.

4.4.2 Another interlude - the Manley-Rowe relations for SHG

Equations 4.166, expressed in terms of the photon fluxes N_p and N_s, constitute the *Manley-Rowe* relations for SHG. If ΔN_p and ΔN_s denote the changes in photon fluxes at the two frequencies, we obtain from Eqn.4.166

$$\Delta N_p = -2 \, \Delta N_s \tag{4.168}$$

which shows that for SHG, changes in the photon flux of the lower frequency wave are equal to each other in sign and magnitude, and they are equal and opposite of the change in the photon number at the highest frequency. Thus, for SHG, if $\Delta N_p = -2$, then $\Delta N_s = +1$, i.e., each photon generated at the second harmonic frequency ω_s corresponds to the destruction of two photons at the fundamental (or pump) frequency ω_p.

4.4.3 Back to the solutions of SHG equations

Defining a variable v_p', with

$$v_p' = \frac{v_p}{\sqrt{2}}. \tag{4.169}$$

Equations 4.155, 4.156 and 4.159 can be rewritten as

$$\frac{\partial v_p'}{\partial z_1} = -\beta_1' v_p' v_s \sin \theta \tag{4.170}$$

$$\frac{\partial v_s}{\partial z_1} = \beta_1' v_p'^2 \sin \theta \tag{4.171}$$

$$\frac{\partial \phi_p}{\partial z_1} = \beta_1' v_s \cos \theta \tag{4.172}$$

$$\frac{\partial \phi_s}{\partial z_1} = \frac{v_p'^2}{v_s} \cos \theta \tag{4.173}$$

that is

$$\frac{\partial \theta}{\partial z_1} = \sigma + \beta_1' \cos\theta \left(\frac{v_p'^2}{v_s} - 2\, v_s \right). \tag{4.174}$$

The coupled amplitude equations (Eqns. 4.170, 4.171) and the phase equation (Eqn. 4.174) for SHG are the same ones as those for SFG (Eqns. 4.155, 4.156 and 4.28), with the replacement of symbols 1 and 2 by p and the symbol 3 by s. Since the solutions for the equations in the SFG case have already been obtained (Eqns. 4.61) we can get the solutions for the SHG case by appropriate substitution of the symbols in the solutions. For the case of *unseeded* SHG, i.e., with $v_s(0) = 0$, we obtain from Eqn. 4.32 (and Eqn. 4.167)

$$m_1 = m_2 = v_p'(0)^2 = \frac{v_p(0)^2}{2} \tag{4.175}$$

so that Eqn. 4.68 becomes

$$t_1' = 2m_1 + \left(\frac{\sigma}{2\beta_1'} \right)^2 \quad \text{and} \quad t_2' = m_1^2 \tag{4.176}$$

and using Eqn. 4.77

$$
\begin{aligned}
a_1' &\equiv \beta_1'^2 t_1' = 2m_1 \beta_1'^2 + \left(\frac{\sigma}{2} \right)^2 = 2\mathrm{w} + \left(\frac{\sigma}{2} \right)^2 \\
a_2' &\equiv \beta_1'^4 t_2' = \beta_1'^4 m_1^2 = \mathrm{w}^2 \\
a_3' &\equiv \sqrt{a_1'^2 - 4a_2'} = \sigma \sqrt{\mathrm{w} + \left(\frac{\sigma}{4} \right)^2}
\end{aligned} \tag{4.177}
$$

where we have defined

$$\mathrm{w} \equiv m_1 \beta_1'^2 = \frac{I_p(0)}{I_a'}. \tag{4.178}$$

The irradiance at frequency ω_s at the crystal exit face is then

$$I_s(\ell) = I_a'\, b_1'\, \mathrm{sn}^2(\sqrt{b_2'}, \gamma'^2) \tag{4.179}$$

where, from Eqn. 4.76

$$b_1' \equiv \frac{a_1' - a_3'}{2}, \qquad b_2' \equiv \frac{a_1' + a_3'}{2}, \qquad \gamma'^2 = \frac{b_1'}{b_2'}. \tag{4.180}$$

As shown in Ref. [1], the expression for $Is(\ell)$ for this case can be expressed in an alternate form, since

$$b_1'\, b_2' = \frac{a_1'^2 - a_3'^2}{4} = a_2' = \mathrm{w}^2. \tag{4.181}$$

Thus,

$$\gamma'^2 = \frac{b_1'}{b_2'} = \frac{b_1'^2}{w^2} \tag{4.182}$$

that is

$$b_1' = w \, \gamma'. \tag{4.183}$$

Using Eqn. 4.177 we find that

$$\frac{a_1' + a_3'}{2} = \left(\sqrt{w + \left(\frac{\sigma}{4} \right)^2} + \left(\frac{\sigma}{4} \right)^2 \right)^2 \tag{4.184}$$

so that

$$\sqrt{b_2'} = \sqrt{\frac{a_1' + a_3'}{2}} = \sqrt{w + \left(\frac{\sigma}{4} \right)^2} + \left(\frac{\sigma}{4} \right)^2 \tag{4.185}$$

and using Eqn. 4.179

$$I_s(\ell) = I_a' \, w \, \gamma' \, \mathrm{sn}^2 \left(\sqrt{w + \left(\frac{\sigma}{4} \right)^2} + \left(\frac{\sigma}{4} \right)^2, \gamma'^2 \right) \tag{4.186}$$

i.e.,

$$I_s(\ell) = I_p(0) \, \gamma' \, \mathrm{sn}^2 \left(\sqrt{\frac{I_p(0)}{I_a'} + \left(\frac{\sigma}{4} \right)^2} + \left(\frac{\sigma}{4} \right)^2, \gamma'^2 \right). \tag{4.187}$$

From the photon flux conservation relation Eqn. 4.167 we have

$$v_p(\ell)^2 + 2v_s(\ell)^2 = v_p(0)^2 + 2v_s(0)^2. \tag{4.188}$$

Since for the unseeded SHG case, $v_s(0) = 0$, we have

$$I_p(\ell) = \frac{2c\varepsilon_0 A_0^2}{\lambda_p} v_p(\ell)^2 = \frac{2c\varepsilon_0 A_0^2}{\lambda_p} (v_p(0)^2 - 2v_s(\ell)^2) = I_p(0) - I_s(\ell) \tag{4.189}$$

using the relation $\lambda_p = 2\lambda_s$ from Eqn. 4.12.

Some special cases (such as no pump depletion or no phase mismatch) for SHG will be considered next following the same treatment used for SFG earlier.

4.4.4 SHG irradiance for collimated beams with no phase matching ($\sigma \neq 0$) and with no pump depletion

Diffraction - No	Beam Walk-off - No	Absorption - No
Pump Depletion - No	Phase Mismatch - Yes	

When the peak irradiance of the incident beam is small compared to the material parameter I'_a, i.e., for $w \ll 1$ in Eqn. 4.178, we have $a'_2 \ll a'_1$ and as in Eqn. 4.78

$$
\begin{aligned}
a'_3 &= a'_1 \left(1 - \frac{4a'_2}{a'^2_1} \right)^{1/2} \\
&\approx a'_1 - \frac{2a'_2}{a'_1}.
\end{aligned}
$$ (4.190)

Thus we get from Eqn. 4.180

$$
b'_1 = \frac{a'_2}{a'_1}, \qquad b'_2 = a'_1
$$ (4.191)

with

$$
a'_1 = \frac{\sigma^2}{4}, \qquad a'_2 = \left(\frac{I_p(0)}{I_a} \right)^2.
$$ (4.192)

Since

$$
\gamma'^2 = \frac{b'_1}{b'_2} = \frac{a'_2}{a'^2_1} \ll 1
$$ (4.193)

the Jacobian elliptic function $\mathrm{sn}(\sqrt{b'_2}, \gamma'^2)$ can be replaced by $\sin \sigma/2$ and we obtain

$$
I_s(\ell) = \frac{I_p(0)^2}{I'_a} \left(\frac{\sin(\sigma/2)}{\sigma/2} \right)^2.
$$ (4.194)

4.4.5 SHG irradiance for collimated beams with phase matching ($\sigma = 0$) and with pump depletion

Diffraction - No	Beam Walk-off - No	Absorption - No
Phase Mismatch - No	Pump Depletion - Yes	

When $\sigma = 0$, Eqns. 4.177 show that

$$
a'_1 = 2w, \qquad a'_2 = w^2 \quad \text{and} \quad a'_3 = 0
$$ (4.195)

so that from Eqn. 4.180

$$
b'_1 = b'_2 = w \quad \text{and} \quad \gamma'^2 = \frac{b'_1}{b'_2} = 1
$$ (4.196)

and we obtain

$$
\begin{aligned}
I_s(\ell) &= I'_a \, w \, \mathrm{sn}^2(\sqrt{w}, 1) \\
&= I'_a \, w \, \tanh^2(\sqrt{w}) \\
&= I_p(0) \, \tanh^2 \left(\sqrt{\frac{I_p(0)}{I'_a}} \right).
\end{aligned}
$$ (4.197)

In the case of no pump depletion (and no phase mismatch), i.e., for w << 1 and $\sigma = 0$, Eqn. 4.197 reduces to the relation

$$I_s(\ell) = \frac{I_p(0)^2}{I_a'}. \tag{4.198}$$

4.4.6 SHG power and energy conversion efficiency for collimated beams

Diffraction - No	Beam Walk-off - No	Absorption - No
Phase Mismatch - Yes	Pump Depletion - Yes	

We assume the normalized spatial and temporal distributions of the irradiance of the pump beam in the incident plane (at $z = 0$) to be given by the functions $f_p(x, y)$ and $g_p(t)$, i.e.,

$$I_p(0, x, y, t) = I_{p0} f_p(x, y) g_p(t) \tag{4.199}$$

Defining the parameter K_1' as

$$K_1' \equiv \frac{I_{p0}}{I_a'} \tag{4.200}$$

and using normalized spatial and temporal coordinates, $x_1 = x/r_0$, $y_1 = y/r_0$ and $t_1 = t/t_0$, (with r_0 and t_0 to be defined later) we find from Eqn. 4.177 that

$$
\begin{aligned}
a_1'(x_1, y_1, t_1) &= 2K_1' f_p(x_1, y_1) g_p(t_1) + \left(\frac{\sigma}{2}\right)^2 \\
a_2'(x_1, y_1, t_1) &= \{K_1' f_p(x_1, y_1) g_p(t_1)\}^2 \\
a_3'(x_1, y_1, t_1) &\equiv \sqrt{a_1'(x_1, y_1, t_1)^2 - 4a_2'(x_1, y_1, t_1)} \\
&= \sigma \sqrt{K_1' f_p(x_1, y_1) g_p(t_1) + \left(\frac{\sigma}{4}\right)^2}.
\end{aligned}
\tag{4.201}
$$

With

$$
\begin{aligned}
b_1'(x_1, y_1, t_1) &= \frac{a_1'(x_1, y_1, t_1) - a_3'(x_1, y_1, t_1)}{2}, \\
b_2'(x_1, y_1, t_1) &= \frac{a_1'(x_1, y_1, t_1) + a_3'(x_1, y_1, t_1)}{2}
\end{aligned}
\tag{4.202}
$$

and

$$\gamma'^2(x_1, y_1, t_1) = \frac{b_1'(x_1, y_1, t_1)}{b_2'(x_1, y_1, t_1)} \tag{4.203}$$

we obtain from Eqn. 4.187

$$
\begin{aligned}
I_s(\ell, x_1, y_1, t_1) &= I_{p0} f_p(x_1, y_1) g_p(t_1) \, \gamma'(x_1, y_1, t_1)^2 \\
&\times \mathrm{sn}^2 \left(\sqrt{K_1' f_p(x_1, y_1) g_p(t_1) + \left(\frac{\sigma}{4}\right)^2} + \left(\frac{\sigma}{4}\right)^2, \gamma'^2(x_1, y_1, t_1) \right).
\end{aligned}
$$

$$(4.204)$$

The second harmonic power generated at the exit of the nonlinear material, assuming no reflection loss at the surface, is obtained by integrating over the beam area

$$
\begin{aligned}
P_s(\ell, t_1) &= \int_{-\infty}^{\infty} \int_{-\infty}^{\infty} dx\, dy\, I_s(\ell, x, y, t) \\
&= r_0^2 I_{p0} g_p(t_1) \int_{-\infty}^{\infty} \int_{-\infty}^{\infty} dx_1\, dy_1 \; f_p(x_1, y_1) \, \gamma'(x_1, y_1, t_1) \\
&\quad \mathrm{sn}^2 \left(\sqrt{b_2'(x_1, y_1, t_1)}, \gamma'^2(x_1, y_1, t_1) \right).
\end{aligned}
$$

$$(4.205)$$

The incident pump power is given by

$$
P_p(0, t_1) = I_{p0} r_0^2 g_p(t_1) \int_{-\infty}^{\infty} \int_{-\infty}^{\infty} f_p(x_1, y_1) dx_1\, dy_1 \qquad (4.206)
$$

so that the power conversion efficiency for SHG can be obtained from Eqns. 4.205 and 4.206.

For continuous wave (CW) lasers, the functions $g_p(t_1)$ in Eqns. 4.201 can be set equal to 1 and the time independent power conversion efficiency can also be obtained from Eqn. 4.205.

The energy of the generated second harmonic beam at the exit of the nonlinear material is obtained by integrating the generated power over time

$$
\begin{aligned}
E_s(\ell) &= \int_{-\infty}^{\infty} dt\, P_s(\ell, t_1) \\
&= r_0^2 t_0 I_{p0} \int_{-\infty}^{\infty} dt_1 \int_{-\infty}^{\infty} \int_{-\infty}^{\infty} dx_1\, dy_1 \; f_p(x_1, y_1) g_p(t_1) \, \gamma'(x_1, y_1, t_1) \\
&\times \mathrm{sn}^2 \left(\sqrt{K_1' f_p(x_1, y_1) g_p(t_1) + \left(\frac{\sigma}{4}\right)^2} + \left(\frac{\sigma}{4}\right)^2, \gamma'^2(x_1, y_1, t_1) \right).
\end{aligned}
$$

$$(4.207)$$

The energy of the pump beam at $z = 0$ is given by

$$E_p(0) = I_{p0}r_0^2 t_0 \int_{-\infty}^{\infty} dt_1 g_p(t_1) \int_{-\infty}^{\infty} f_p(x_1, y_1) dx_1 dy_1 \qquad (4.208)$$

so that the energy conversion efficiency for SHG is

$$\eta_{SHG}^E \equiv \frac{E_s(\ell)}{E_p(0)} \qquad (4.209)$$

can be obtained from Eqns. 4.207 and 4.208.

4.4.7 SHG power and energy conversion efficiency for collimated Gaussian beams

Diffraction - No	Beam Walk-off - No	Absorption - No
Phase Mismatch - Yes	Pump Depletion - Yes	

For an incident beam that is spatially and temporally Gaussian

$$I_p(0, r, t) = I_{p0} e^{-(r/r_{0p})^2 - (t/t_{0p})^2} \qquad (4.210)$$

setting r_0 in the last section equal to r_{0p} and t_0 equal to t_{0p} we obtain

$$f_1(x_1, y_1) = e^{-r_1^2} \qquad g_1(t_1) = e^{-t_1^2} \qquad (4.211)$$

$$P_p(0, t_1) = I_{p0}\pi r_{0p}^2 e^{-(t^2/t_{0p}^2)} \qquad (4.212)$$

and

$$E_p(0) = I_{p0}\pi\sqrt{\pi} r_{0p}^2 t_{0p}. \qquad (4.213)$$

For CW beams with circular Gaussian spatial distribution, Eqns. 4.201, 4.202 and 4.203 are rewritten as

$$a_1'(r_1) = 2K_1' e^{-r_1^2} + \left(\frac{\sigma}{2}\right)^2$$

$$a_2'(r_1) = \left(K_1' e^{-r_1^2}\right)^2$$

$$a_3'(r_1) \equiv \sqrt{a_1'(r_1)^2 - 4a_2'(r_1)} = \sigma\sqrt{K_1' e^{-r_1^2} + \left(\frac{\sigma}{4}\right)^2} \qquad (4.214)$$

we have

$$b_1'(r_1) = \frac{a_1'(r_1) - a_3'(r_1)}{2},$$

$$b_2'(r_1) = \frac{a_1'(r_1) + a_3'(r_1)}{2} \qquad (4.215)$$

and

$$\gamma'(r_1) = \frac{b_1'(r_1)}{b_2'(r_1)}. \tag{4.216}$$

The expression for SHG power conversion efficiency is then obtained from Eqn. 4.205 as

$$
\begin{aligned}
\eta_{SHG}^P &= \frac{P_s(\ell, t_1)}{P_p(0, t_1)} \\
&= 2 \int_0^\infty r_1 \, dr_1 \, e^{-r_1^2} \, \gamma'(r_1) \\
&\times \mathrm{sn}^2\left(\sqrt{K_1' e^{-r_1^2} + \left(\frac{\sigma}{4}\right)^2} + \left(\frac{\sigma}{4}\right)^2 , \gamma'^2(r_1) \right).
\end{aligned} \tag{4.217}
$$

Similarly, for pulsed incident beams with Gaussian spatial and temporal distributions, the Eqns. 4.201, 4.202 and 4.203 are rewritten as

$$
\begin{aligned}
a_1'(r_1, t_1) &= 2K_1' e^{-r_1^2} e^{-t_1^2} + \left(\frac{\sigma}{2}\right)^2 \\
a_2'(r_1, t_1) &= \left(K_1' e^{-r_1^2} e^{-t_1^2} \right)^2 \\
a_3'(r_1, t_1) &\equiv \sqrt{a_1'(r_1, t_1)^2 - 4a_2'(r_1, t_1)} \\
&= \sigma \sqrt{K_1' e^{-r_1^2} e^{-t_1^2} + \left(\frac{\sigma}{4}\right)^2}
\end{aligned} \tag{4.218}
$$

we have

$$
\begin{aligned}
b_1'(r_1, t_1) &= \frac{a_1'(r_1, t_1) - a_3'(r_1, t_1)}{2}, \\
b_2'(r_1, t_1) &= \frac{a_1'(r_1, t_1) + a_3'(r_1, t_1)}{2}
\end{aligned} \tag{4.219}
$$

and

$$\gamma'^2(r_1, t_1) = \frac{b_1'(r_1, t_1)}{b_2'(r_1, t_1)} \tag{4.220}$$

using which we obtain

$$
\begin{aligned}
E_s(\ell) &= 2\pi \int_{-\infty}^\infty dt \int_0^\infty r dr I_s(\ell) \\
&= 2\pi r_{0p}^2 t_{0p} I_{p0} \int_{-\infty}^\infty dt_1 \int_0^\infty r_1 \, dr_1 \, e^{-r_1^2} e^{-t_1^2} \, \gamma'(r_1, t1) \\
&\times \mathrm{sn}^2\left(\sqrt{K_1' e^{-r_1^2} e^{-t_1^2} + \left(\frac{\sigma}{4}\right)^2} + \left(\frac{\sigma}{4}\right)^2 , \gamma'^2(r_1, t_1) \right)
\end{aligned} \tag{4.221}
$$

so that the energy conversion efficiency for SHG of collimated pulsed Gaussian beams can be written as

$$
\begin{aligned}
\eta_{\text{SHG}}^{E} &\equiv \frac{E_s(\ell)}{E_p(0)} \\
&= \frac{2}{\sqrt{\pi}} \int_{-\infty}^{\infty} dt_1 \int_0^{\infty} r_1 \, dr_1 \; e^{-r_1^2} e^{-t_1^2} \, \gamma'(r_1, t1) \\
&\quad \times \; \text{sn}^2 \left(\sqrt{K_1' e^{-r_1^2} e^{-t_1^2} + \left(\frac{\sigma}{4}\right)^2} + \left(\frac{\sigma}{4}\right)^2, \gamma'^2(r_1, t_1) \right).
\end{aligned}
\tag{4.222}
$$

Equations 4.217 and 4.222 show that the power and energy conversion efficiencies η_{SHG}^{P} and η_{SHG}^{E} for the SHG of collimated Gaussian beams depend only on the parameter K_1' and the phase mismatch σ. The values of η_{SHG}^{P} and η_{SHG}^{E} for K_1' ranging from 0 to 50 are plotted in Fig. 4.10 along with the values of peak irradiance efficiency η_{SHG}^{I}, defined as

$$
\begin{aligned}
\eta_{\text{SHG}}^{I} &\equiv \frac{I_s(\ell, 0, 0)}{I_p(0)} \\
&= \tanh^2 \sqrt{K_1'}.
\end{aligned}
\tag{4.223}
$$

The values of $K_1'/\eta_{\text{SHG}}^{P}$ and $K_1'/\eta_{\text{SHG}}^{E}$ are found to increase with K_1' for K_1' in the range of 0 to 50, as shown in Fig. 4.10 for $\sigma = 0$.

From quadratic fits to the plots of $K_1'/\eta_{\text{SHG}}^{P}$ and $K_1'/\eta_{\text{SHG}}^{E}$ against K_1' shown in Fig. 4.10, we find that instead of using the previously given integral expressions, the values of η_{SHG}^{P} and η_{SHG}^{E} can also be obtained from the simple formulas:

$$
\eta_{\text{SHG}}^{P} = \frac{K_1'}{2 + 0.9627 \, K_1' + 0.00053 \, K_1'^{2}}
\tag{4.224}
$$

$$
\eta_{\text{SHG}}^{E} = \frac{K_1'}{2.828 + 1.0157 \, K_1' - 1.48 \times 10^{-4} K_1'^{2}}.
\tag{4.225}
$$

4.4.8 SHG power and energy conversion efficiency for collimated Gaussian beams with phase matching ($\sigma = 0$) in presence of pump depletion

Diffraction - No	Beam Walk-off - No	Absorption - No
	Phase Mismatch - No	Pump Depletion - Yes

When the phase mismatch σ is equal to 0, the Jacobian Elliptic function sn reduces to the hyperbolic tangent function (tanh) and the expressions for

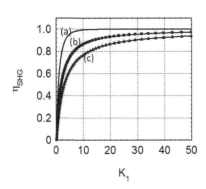

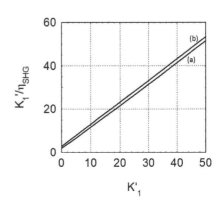

FIGURE 4.9: The values of SHG conversion efficiencies for collimated Gaussian beams plotted against K_1' for $\sigma = 0$. (a) Peak irradiance conversion efficiency $\tanh^2 \sqrt{K_1'}$, (b) power conversion efficiency values η_{SHG}^P calculated from the fitted equation are shown by square dots, and the results from the integral expression are shown by the solid line (c) energy conversion efficiency η_{SHG}^E values calculated from the fitted equation are shown by square dots, with the results from the integral expression shown by the solid line.

FIGURE 4.10: (a) K_1'/η_{SHG}^P and (b) K_1'/η_{SHG}^E plotted as functions of K_1'

power and energy conversion efficiencies are simplified:

$$\eta_{SHG}^P = \frac{P_s(\ell, t_1)}{P_p(0, t_1)}$$

$$= 2 \int_0^\infty r_1 \, dr_1 \, e^{-r_1^2} \tanh^2\left(\sqrt{K_1'} \, e^{-r_1^2}\right) \qquad (4.226)$$

using Eqn. 4.197 for the case of no phase mismatch.

For pulsed incident beams with Gaussian temporal and spatial shapes

$$E_s(\ell) = 2\pi \int_{-\infty}^{\infty} dt \int_0^\infty r \, dr \, I_s(\ell)$$

$$= 2\pi r_{0p}^2 t_{0p} I_{p0} \int_{-\infty}^{\infty} dt_1 \int_0^\infty r_1 \, dr_1 \, e^{-r_1^2} \, e^{-t_1^2}$$

$$\times \quad \tanh^2\left(\sqrt{K_1'} \, e^{-r_1^2} \, e^{-t_1^2}\right) \qquad (4.227)$$

so that the energy conversion efficiency for SHG can be written as

$$\eta_{SHG}^E \equiv \frac{E_s(\ell)}{E_p(0)}$$

$$= \frac{2}{\sqrt{\pi}} \int_{-\infty}^{\infty} dt_1 \int_0^\infty r_1 \, dr_1 \, e^{-r_1^2} \, e^{-t_1^2}$$

$$\times \quad \tanh^2\left(\sqrt{K_1'} \, e^{-r_1^2} \, e^{-t_1^2}\right). \qquad (4.228)$$

When $K_1' \ll 1$, i.e., for the case of no pump depletion, using the relation $\tanh x \approx x$ for small x, we have

$$\eta_{SHG}^P = 2K_1' \int_0^\infty r_1 \, dr_1 \, e^{-2r_1^2}$$

$$= \frac{K_1'}{2}$$

$$= \frac{I_{p0}}{2I_a'} \qquad (4.229)$$

and

$$\eta_{SHG}^E = \frac{2K_1'}{\sqrt{\pi}} \int\limits_{-\infty}^{\infty} dt_1 \int\limits_{0}^{\infty} r_1 \, dr_1 \, e^{-2r_1^2} e^{-2t_1^2}$$

$$= \frac{K_1'}{2\sqrt{2}}$$

$$= \frac{I_{p0}}{2\sqrt{2}I_a'} \tag{4.230}$$

where we have used

$$\int\limits_{0}^{\infty} r_1 dr_1 e^{-2r_1^2} = \frac{1}{4} \quad \text{and} \quad \int\limits_{-\infty}^{\infty} dt_1 e^{-2t_1^2} = \sqrt{\frac{\pi}{2}}. \tag{4.231}$$

4.4.9 SHG power and energy conversion efficiency for collimated Gaussian beams with no pump depletion

Diffraction - No	Beam Walk-off - No	Absorption - No
Pump Depletion - No	Phase Mismatch - Yes	

In the nondepleted pump case, the irradiance at the second harmonic frequency given in Eqn. 4.194 can be integrated over the beam cross-section and over time to provide the power and energy outputs. Assuming that the incident beam is circular Gaussian in cross-section and constant in time,

$$P_s(\ell) = 2\pi r_{0p}^2 \int\limits_{0}^{\infty} I_s(\ell) r_1 dr_1$$

$$= 2\pi r_{0p}^2 \frac{I_{p0}^2}{I_a'} \left(\frac{\sin(\sigma/2)}{\sigma/2} \right)^2 \int\limits_{0}^{\infty} r_1 dr_1 e^{-2(r^2/r_{0p}^2)}$$

$$= \frac{\pi}{2} r_{0p}^2 \frac{I_{p0}^2}{I_a'} \left(\frac{\sin(\sigma/2)}{\sigma/2} \right)^2 \tag{4.232}$$

using Eqn. 4.231.

The incident pump power is

$$P_p(0) = \pi I_{p0} r_{0p}^2 \tag{4.233}$$

so the power conversion efficiency is

$$
\begin{aligned}
\eta_{\text{SHG}}^{P} &= \frac{P_s(\ell)}{P_p(0)} \\
&= \frac{I_{p0}}{2I_a'} \left(\frac{\sin(\sigma/2)}{\sigma/2} \right)^2 \\
&= \frac{4\pi^2 d_{\text{eff}}^2 \ell^2 I_{p0}}{c\varepsilon_0 n_p^2 n_s \lambda_p^2} \left(\frac{\sin(\sigma/2)}{\sigma/2} \right)^2 .
\end{aligned}
\tag{4.234}
$$

Assuming that the temporal profile of the incident beam is Gaussian, i.e., with the incident irradiances given by Eqns. 4.102, we obtain

$$
\begin{aligned}
E_s(\ell) &= 2\pi r_{0p}^2 t_{0p} \int_{-\infty}^{\infty} dt_1 \int_{0}^{\infty} I_s(\ell) r_1 dr_1 \\
&= \frac{\pi\sqrt{\pi}}{2\sqrt{2}} \frac{I_{p0}^2}{I_a'} \left(\frac{\sin(\sigma/2)}{\sigma/2} \right)^2
\end{aligned}
\tag{4.235}
$$

using Eqn. 4.231.

The incident pump energy is

$$
E_p(0) = \pi\sqrt{\pi} r_{0p}^2 t_{0p} I_{p0}
\tag{4.236}
$$

so that the energy conversion efficiency is

$$
\eta_{\text{SHG}}^{E} = \frac{E_s(\ell)}{E_p(0)}
\tag{4.237}
$$

$$
\begin{aligned}
&= \frac{I_{p0}}{2\sqrt{2} I_a'} \left(\frac{\sin(\sigma/2)}{\sigma/2} \right)^2 \\
&= \frac{2\sqrt{2}\pi^2 d_{\text{eff}}^2 \ell^2 I_{p0}}{c\varepsilon_0 n_p^2 n_s \lambda_p^2} \left(\frac{\sin(\sigma/2)}{\sigma/2} \right)^2 .
\end{aligned}
\tag{4.238}
$$

4.4.10 SHG conversion efficiency for focused Gaussian beams

Diffraction - Yes	Beam Walk-off - Yes	Absorption - yes
Pump Depletion - Yes	Phase Mismatch - Yes	

As for the case of SFG, to optimize the SHG power conversion efficiency of CW beams, focusing effects need to be considered. The procedure for SHG focusing optimization is similar to the procedure described for SFG, and is simpler in the unseeded case because there is only one incident beam. Assuming that the amplitude of the pump beam is

$$A_p(x_1, y_1, z_1) = \frac{A_p(0,0,f_p)}{1 + i\xi_p\{2z_1 - (1-\mu_p)\}} e^{-\frac{1}{2}\frac{x_1^2 + y_1^2}{1 + i\xi_p\{2z_1 - (1-\mu_p)\}}}$$

(4.239)

where the 'focusing parameter' (denoted by ξ_p) and the relative focusing position (denoted by μ_p) are defined as

$$\xi_p \equiv \frac{\ell}{2k_p r_{0p}^2} = \frac{\ell\lambda_p}{4\pi r_{0p}^2 n_p}$$

$$\mu_p \equiv 1 - 2\frac{f_p}{\ell}$$

(4.240)

and where f_p denotes the distance from the incident surface at which the incident pump beam is focused in the absence of nonlinearity. Choosing

$$r_0 = r_{0p}, \qquad \text{and} \qquad A_0 = A_p(0,0,f_p)$$

(4.241)

the normalized amplitudes of the incident beams at the incident face of the nonlinear medium are given by

$$u_p(x_1, y_1, 0) = \frac{1}{1 - i\xi_p\{(1-\mu_p)\}} e^{-\frac{1}{2}\left(\frac{x_1^2 + y_1^2}{1 - i\xi_p\{(1-\mu_p)\}}\right)}$$

$$u_s(x_1, y_1, 0) = 0.$$

(4.242)

Equations 4.150 for the normalized amplitudes of the pump and the SHG beams can then be written as

$$\frac{\partial u_p}{\partial z_1} = i\xi_p \nabla_{T_1}^2 u_p + B_p \sqrt{8\xi_p} \frac{\partial u_p}{\partial x_1} - \frac{a_p}{2} u_1 + i\frac{n_s}{n_p}\kappa_s u_p^* u_s e^{i\sigma z_1}$$

$$\frac{\partial u_s}{\partial z_1} = i\frac{k_p}{k_s}\xi_p \nabla_{T_1}^2 u_s + B_s \sqrt{8\xi_p} \frac{\partial u_s}{\partial x_1} - \frac{a_s}{2} u_s + i\kappa_s u_p^2 e^{-i\sigma z_1}$$

(4.243)

where

$$\kappa_s = \frac{4\pi d_{\text{eff}}\ell A_0}{n_s\lambda_p}, \quad B_p = \frac{\sqrt{\ell k_p}}{2}\tan\rho_p, \quad B_s = \frac{\sqrt{\ell k_p}}{2}\tan\rho_s$$

(4.244)

and where we have used

$$\frac{\ell}{r_{0p}} = \sqrt{2k_p\ell\xi_p}.$$

(4.245)

The value of the dimensionless variable $u_s(x_1, y_1, 1)$ at the exit of the nonlinear material ($z_1 = 1$) can be calculated by numerically solving Eqns. 4.243. The irradiance distribution at the frequency ω_s is then given by

$$
\begin{aligned}
I_s(x_1, y_1, 1) &= 2n_s c\varepsilon_0 |A_s(x_1, y_1, 1)|^2 \\
&= 2n_s c\varepsilon_0 A_0^2 |u_s(x_1, y_1, 1)|^2 \\
&= \frac{n_s}{n_p} I_{p0} |u_s(x_1, y_1, 1)|^2.
\end{aligned}
\tag{4.246}
$$

The SHG power $P_s(\ell)$ at the exit plane is obtained by integrating over the irradiance distribution, i.e.,

$$
\begin{aligned}
P_s(\ell) &= \int_{-\infty}^{\infty} \int_{-\infty}^{\infty} I_s(x, y, l) \, dxdy \\
&= \frac{n_s}{n_p} I_{p0} r_{0p}^2 \int_{-\infty}^{\infty} \int_{-\infty}^{\infty} |u_s(x_1, y_1, 1)|^2 \, dx_1 dy_1 \\
&= \frac{n_s}{n_p} \frac{P_p(0)}{\pi} \int_{-\infty}^{\infty} \int_{-\infty}^{\infty} |u_s(x_1, y_1, 1)|^2 \, dx_1 dy_1 \\
&= \frac{n_s P_p(0)}{\pi n_p} \frac{\pi \kappa_s^2}{2\xi_p} h_s
\end{aligned}
\tag{4.247}
$$

where we define

$$
h_s \equiv \frac{2\xi_p}{\pi \kappa_s^2} \int_{-\infty}^{\infty} \int_{-\infty}^{\infty} |u_s(x_1, y_1, 1)|^2 \, dx_1 dy_1.
\tag{4.248}
$$

Defining h_s this way makes it identical with the well established h function given in Ref. [2] for the case of no pump depletion.

Using Eqns. 4.244 and 4.240 $\kappa_s^2 / 2\xi_p$ can be written out as

$$
\begin{aligned}
\frac{\kappa_s^2}{2\xi_p} &= \frac{16\pi^2 d_{\text{eff}}^2 \ell^2 A_0^2}{n_s^2 \lambda_p^2} \frac{4\pi r_{0p}^2 n_p}{2\ell \lambda_p} \\
&= \frac{P_p(0)}{P_a} \equiv K_2
\end{aligned}
\tag{4.249}
$$

where

$$
P_a \equiv \frac{c\varepsilon_0 n_s^2 \lambda_p^3}{16\pi^2 d_{\text{eff}}^2 \ell}
\tag{4.250}
$$

and using

$$
P_p(0) = \pi r_{0p}^2 I_{p0} = 2\pi r_{0p}^2 n_p c\varepsilon_0 \, A_{p0}^2.
\tag{4.251}
$$

The parameter K_2 defined above is the incident power normalized by the constant P_a, just as K_1' is the incident peak irradiance normalized by the constant I_a'. The relation between K_2 and K_1' can be derived to be

$$K_2 = \frac{1}{2}\frac{n_p}{n_s}\frac{K_1'}{\xi_p}. \tag{4.252}$$

Equations 4.247 and 4.248 can be re-written as

$$P_s(\ell) = \frac{n_s}{n_p}\frac{P_p(0)^2}{P_a}h_s \tag{4.253}$$

and

$$h_s \equiv \frac{1}{\pi K_2}\int\limits_{-\infty}^{\infty}\int\limits_{-\infty}^{\infty}|u_s(x_1,y_1,1)|^2 dx_1 dy_1 \tag{4.254}$$

with the power conversion efficiency in the focused case given by

$$\eta_s \equiv \frac{P_s(\ell)}{P_p(0)} = \frac{n_s}{n_p}K_2 h_s. \tag{4.255}$$

h_s depends on ξ_p, k_p/k_s, $k_p l$, ρ_p, ρ_s, $\alpha_p l$, σ and $\alpha_s l$, as does the function h in Ref. [2]. In addition, h_s depends on the parameter K_2 since the values of $u_s(x_1, y_1, 1)$ are determined by solving Eqn. 4.243 which depend on κ_s and using Eqn. 4.249, κ_s can be replaced by $\sqrt{\xi_p K_2}$.

For the case of small κ_s (no pump depletion), no phase mismatch ($\sigma = 0$), no beam walk-offs, no absorption and no diffraction ($\xi_p \ll 1$) it can be easily shown by integrating Eqn. 4.243 that

$$u_s(x_1,y_1,1) = i\kappa_s u_p^2(x_1,y_1,0) \tag{4.256}$$

from which we find (using Eqn. 4.242 and integrating over x_1 and y_1) that h_s (in Eqn. 4.254) becomes equal to ξ_p.

Thus, Eqn. 4.253 gives the second harmonic power for the collimated Gaussian pump beam case to be

$$\begin{aligned}
P_{s(\text{coll})}(\ell) &= \frac{n_s}{n_p}\frac{P_p(0)^2}{P_a}\xi_p\left(\frac{\sin(\sigma/2)}{\sigma/2}\right)^2 \\
&= P_p(0)^2\left(\frac{16\pi^2 d_{\text{eff}}^2 \ell}{c\varepsilon_0 n_s n_p \lambda_p^3}\right)\xi_p\left(\frac{\sin(\sigma/2)}{\sigma/2}\right)^2. \tag{4.257}
\end{aligned}$$

4.4.11   An interlude - Boyd and Kleinman theory for SHG

For the case of small K_2, i.e., for the 'no pump depletion' case of low conversion efficiency, Boyd and Kleinman [2] presented the theory describing the dependence of the SHG power on the beam focusing in terms of a double integral expression, which we denote here by h_{BK}. Including the effects of beam walk-off, phase mismatch and focusing, the integral given by Boyd and Kleinman can be written as

$$h_{BK}(\sigma_{BK}, B, \xi_p) = \frac{1}{4\xi_p} \int\limits_{-\xi_p}^{\xi_p} \int\limits_{-\xi_p}^{\xi_p} \frac{e^{i\sigma_{BK}(\tau - \tau')}e^{-B^2(\tau - \tau')^2/\xi_p}}{(1 + i\tau)(1 - i\tau')} d\tau d\tau'$$

(4.258)

where

$$B = (\rho\sqrt{lk_p})/2, \quad \text{and} \quad \sigma_{BK} = \frac{\sigma}{2\xi_p}.$$

(4.259)

The phase mismatch parameter σ used here differs from the parameter σ_{BK} used by Boyd and Kleinman by the factor of $2\xi_p$.

It has been verified that the function h_s defined in Eqn. 4.254 is numerically the same as the Boyd and Kleinman function h_{BK} when K_2 is small. In Ref. [2] the function h_{BK} is maximized with respect to σ_{BK} for a wide range of values of ξ_p and B and it is shown that at small values of ξ_p, i.e., for the collimated beam case, h_{BK} is maximum at σ_{BK} close to 1. But as ξ_p increases, approaching and exceeding 1, the value of σ_{BK} at which h_{BK} is maximized falls, approximately as $\tan^{-1}(\xi_p/2)/(\xi_p/2)$.

4.4.12   Return to the case of SHG for focused Gaussian beams including pump depletion effects

We have followed a treatment similar to that in Ref. [2], and have determined the value of h_s as a function of σ for a range of values of K_2 and ξ_p. The value of σ at which h_s is a maximum denoted by σ_m and that maximum value of h_s is denoted by h_{sm}. For each value of K_2, the values of h_{sm} and σ_m are then found for a range of ξ_p, and the value of ξ_p at which h_{sm} is maximum is denoted by ξ_{pm}, that maximum value of h_{sm} is denoted h_{smm} and the corresponding value of σ_m is denoted σ_{mm}.

Figure 4.11 shows the plots of h_{sm} against ξ_p for values of the parameter K_2 ranging from 0.001 to 20, for the case of no walk-off. At small values of K_2, the results are identical with the values given in Ref. [2], i.e., ξ_{pm} is equal to 2.84. The values of σ_m plotted against ξ_p for a few different values of K_2 are shown in Fig. 4.12.

Following Eqn. 4.255 we define

$$\eta_{smm} \equiv \frac{n_s}{n_p} K_2 h_{smm}$$

(4.260)

and plot the values of h_{smm}, η_{smm}, ξ_{pm} and σ_{mm} in Figs. 4.13, 4.14, 4.15 and 4.16 as functions of the parameter K_2 for several values of the parameter B_s of the SHG beam walk-off parameter. In all of these cases, linear absorption values a_1, a_2 and the pump walk-off B_p are set equal to zero and the beam is assumed to be focused at the center of the crystal. At low values of K_2, i.e.,

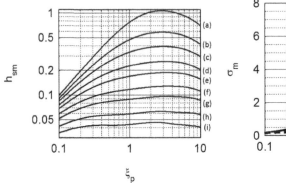

FIGURE 4.11: The values of h_{sm} plotted against ξ_p for $B_p = B_s = 0$ and different values of K_2 are shown in the curves (a) through (i). The values of K_2 are (a) 0.001 (b) 1, (c) 2, (d) 3.5, (e) 5, (f) 7.5, (g) 10, (h) 15 and (i) 20.

FIGURE 4.12: The values of σ_m plotted against ξ_p for $B_p = B_s = 0$ and different values of K_2 are shown in the curves (a) through (i). The values of K_2 are (a) 0.001 (b) 1, (c) 2, (d) 3.5, (e) 5, (f) 7.5, (g) 10, (h) 15 and (i) 20, . The curves (a) - (d) are shown by solid lines, and (e) - (i) by dashed lines.

for the case of no pump depletion, and for no walk-off and absorption, h_{smm} is 1.066, ξ_{pm} is 2.84 and σ_{mm} is 3.26, which are same as the results obtained in [2].

4.4.13   Optimum value of the focusing parameter

Since

$$\xi_p = \frac{\ell}{2z_{0p}} = \frac{\ell\lambda_p}{4\pi n_p r_{0p}^2} \tag{4.261}$$

we have

$$\pi r_{0p}^2 = \frac{\ell\lambda_p}{4\pi\xi_p}. \tag{4.262}$$

For a fundamental beam with power $P_p(0)$, the irradiance at focus in the absence of pump depletion is

$$I_p = \frac{P_p(0)}{\pi r_{0p}^2} = \frac{4n_p\xi_p P_p(0)}{\ell\lambda_p}. \tag{4.263}$$

The maximum value of the irradiance to which a material can be subjected without damage is called the damage threshold, which is an intrinsic material property like the refractive index or the nonlinear susceptibility. Denoting by I_D the damage threshold irradiance for a material, it is desirable to have

$$I_p \leq I_D. \tag{4.264}$$

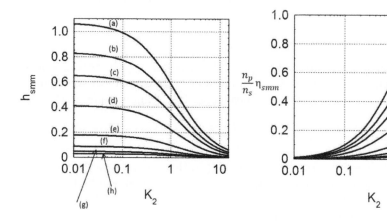

FIGURE 4.13: The values of h_{smm} plotted against K_2 for $B_p = 0$ and different values of B_s are shown in the curves (a) through (f). The values of B_s are (a) 0, (b) 0.5, (c) 1.0, (d) 2.0, (e) 4.0, (f) 6.0, (g) 8.0 and (h) 10.0

FIGURE 4.14: The values of η_{smm} plotted against K_2 for $B_p = 0$ and different values of B_s are shown in the curves (a) through (h). The values of B_s are (a) 0, (b) 0.5, (c) 1.0, (d) 2.0, (e) 4.0, (f) 6.0, (g) 8.0, (h) 10.

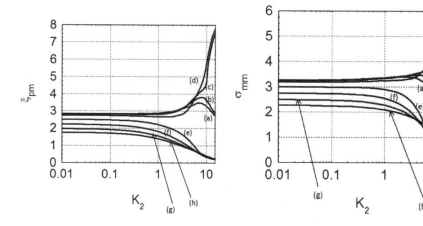

FIGURE 4.15: The values of ξ_{pm} plotted against K_2 for $B_p = 0$ and different values of B_s are shown in the curves (a) through (h). The values of B_s are (a) 0, (b) 0.5, (c) 1.0, (d) 2.0, (e) 4.0, (f) 6.0, (g) 8.0, (h) 10.

FIGURE 4.16: The values of σ_{mm} plotted against K_2 for $B_p = 0$ and different values of B_s are shown in the curves (a) through (h). The values of B_s are (a) 0, (b) 0.5, (c) 1.0, (d) 2.0, (e) 4.0, (f) 6.0, (g) 8.0, (h) 10.

Therefore $\xi_{p(\text{max})}$, the maximum value of ξ_p before the material undergoes laser damage is

$$\xi_{p(\text{max})} \equiv \frac{\ell \lambda_p I_D}{4 n_p P_p(0)} \qquad (4.265)$$

from Eqn. 4.263. If the value of $\xi_{p(\text{max})}$ calculated for a given crystal is less than 2.84, then at $\xi_p = 2.84$, the maximum value of the available power cannot be used without damage to the nonlinear material. Defining a power P_{opt} by the relation

$$P_{\text{opt}} = \frac{1}{2.84} \frac{\ell \lambda_p I_D}{4 n_p} = \frac{\ell \lambda_p I_D}{11.36 n_p} \qquad (4.266)$$

if the available power $P_p(0)$ is smaller than P_{opt}, then the value of the focusing parameter should be chosen to be approximately 2.84, as shown by Fig. 4.11. If $P_p(0)$ is larger than P_{opt} then the value of the focusing parameter that gives maximum SHG output is $\xi_{p(\text{max})}$, given in Eqn. 4.265.

For typical nonlinear optical materials, I_D is of the order of 10 MW/cm². Say for example for a given crystal, ℓ is 3 cm and n_p is 2.5. Then the value of P_{opt} at pump wavelength of 1 μm is 105 W and at pump wavelength of 10 μm it is about 1 KW. To obtain the maximum SHG output, continuous wave lasers having power less than 100 W at wavelength of 1 μm or less than 1 KW at wavelength of 10 μm need to be optimally focused, with ξ_p close to 2.84. For pulsed or high power CW lasers having powers exceeding these values, the focusing parameter needs to be chosen based on Eqn. 4.265, which corresponds to a minimum beam radius $r_{0p(\text{min})}$ given by

$$r_{0p(\text{min})} = \sqrt{\frac{P_p(0)}{\pi I_D}}. \qquad (4.267)$$

4.4.14 Analytical (fitted) expressions for SHG conversion efficiency h_{sm}, optimized with respect to σ

Figures 4.11 and 4.12 show the dependence of h_{sm} and σ_m on the focusing parameter ξ_p for different values of the parameter K_2. From numerical fits, Chen and Chen [7] have derived simple analytical expressions for the Boyd and Kleinman's h_m (the function h optimized with respect to the parameter σ_{BK}) as functions of the focusing parameter ξ and the walk-off parameter B. Such expressions are helpful in providing rapid estimates of optimum conversion efficiency values without the hassle of finding and optimizing the double integral. Some analytical expressions are presented here for the cases of non-negligible K_2, that is for cases of conversion efficiencies exceeding 20%. In these calculations, the walk-off parameters B_p and B_s are assumed to be zero. For ξ_p below 0.1, the analytical results presented for collimated Gaussian beams apply.

Results shown in Fig. 4.11 were fitted to polynomials in ξ_p to a high degree of accuracy, for the two regions of ξ_p from 0.1 to 3 and from 3 to 10 and for

K_2 ranging from 0.001 to 10 and 10 to 15 in each of the two regions of ξ_p:

$$h_{sm} = \sum_{i=0}^{i_{\max}} a_i(K_2)(\xi_p)^i \tag{4.268}$$

where the coefficients $a_i(K_2)$ are themselves expressed as polynomials in K_2:

$$a_i(K_2) = \sum_{j=0}^{j_{\max}} a_{ij}(K_2)^j. \tag{4.269}$$

The values of $i_{\max}$, $j_{\max}$ and the coefficients a_{ij} obtained by fitting the results to polynomials are given in Tables 4.1 to 4.6. In Figs. 4.17 and 4.18 the solid

For $\xi_p = 0.1$ to 3 and K_2 between 0.001 and 10, best fits are obtained with $i_{\max} = 7$, $j_{\max} = 5$.

i,j	0	1	2	3	4	5
0	6.470580E−04	−4.957141E−03	6.403381E−03	−1.554784E−03	1.566007E−04	−5.735528E−06
1	9.876282E−01	8.636837E−02	−1.023427E−01	2.287404E−02	−2.208452E−03	7.898531E−05
2	8.283880E−02	−1.427297E+00	5.578214E−01	−1.010859E−01	8.921360E−03	−3.040955E−04
3	−5.951945E−01	1.564190E+00	−4.915830E−01	7.425448E−02	−5.634121E−03	1.703238E−04
4	4.239164E−01	−8.520984E−01	2.068540E−01	−2.105366E−02	8.273428E−04	−3.757703E−06
5	−1.496197E−01	2.643728E−01	−4.526465E−02	4.719247E−04	4.449459E−04	−2.670666E−05
6	2.740197E−02	−4.452655E−02	4.505111E−03	9.194969E−04	−1.904648E−04	9.082265E−06
7	−2.070942E−03	3.161322E−03	−1.072878E−04	−1.332789E−04	2.137464E−05	−9.560528E−07

In this and the next Table, the index i denotes the row numbers from 0 to 7 and j denotes the column numbers from 0 to 5. Thus $a_{00} = 6.470580E − 04$, $a_{01} = −4.957141E − 03$, $a_{02} = 6.403381EE − 03$, $a_{10} = 9.876282E − 01$ etc.

TABLE 4.1. The coefficients a_{ij}.

For $\xi_p = 3$ to 10 and K_2 between 0.001 and 10, best fits are obtained with $i_{\max} = 7$, $j_{\max} = 5$.

i,j	0	1	2	3	4	5
0	5.144203E−02	3.650789E−01	−2.177787E−01	5.166190E−02	−5.453847E−03	2.109647E−04
1	1.125839E+00	−1.152364E+00	5.137998E−01	−1.104180E−01	1.110062E−02	−4.163983E−04
2	−4.938994E−01	4.955360E−01	−2.187276E−01	4.687785E−02	−4.715486E−03	1.772918E−04
3	1.140237E−01	−1.098777E−01	4.761560E−02	−1.015114E−02	1.021542E−03	−3.853803E−05
4	−1.564199E−02	1.416587E−02	−5.953983E−03	1.256640E−03	−1.263598E−04	4.785994E−06
5	1.270687E−03	−1.068503E−03	4.304175E−04	−8.943101E−05	8.971941E−06	−3.414691E−07
6	−5.646857E−05	4.378460E−05	−1.673497E−05	3.410418E−06	−3.416142E−07	1.310447E−08
7	1.059853E−06	−7.573143E−07	2.710439E−07	−5.367457E−08	5.352811E−09	−2.072869E−10

TABLE 4.2. The coefficients a_{ij}.

lines show the results of calculation of h_{sm} through evaluation of h_s using Eqn. 4.254 and numerical solution of the coupled wave equations, integration over the beam cross-section and optimization with respect to σ. The filled circles show the results of the polynomial fits in Eqn. 4.268. The values of h_{sm} evaluated at K_2 equal to 0.001, 1, 2.5, 5, 7.5, 10, 11, 12, 13, 14 and 15 were used to obtain the polynomial fits. Results at K_2 equal to 0.5 and 14.5 were not used in the fitting process, but comparison of the results at these two values of K_2 with the polynomial fits show that the deviation is less than 1 %.

For $\xi_p = 0.1$ to 2 and K_2 between 10 and 15, best fits are obtained with $i_{max} = 7$, $j_{max} = 3$.

i, j	0	1	2	3
0	$-5.72831E-03$	$5.23145E-03$	$-3.23601E-04$	$6.68627E-06$
1	$1.23041E+00$	$-1.32400E-01$	$6.31460E-03$	$-1.17391E-04$
2	$-2.83067E+00$	$2.45512E-01$	$-1.03793E-02$	$1.83473E-04$
3	$4.18112E+00$	$-3.31320E-01$	$1.39327E-02$	$-2.59595E-04$
4	$-3.90885E+00$	$3.19249E-01$	$-1.47759E-02$	$3.04047E-04$
5	$2.21215E+00$	$-1.98075E-01$	$1.02553E-02$	$-2.27269E-04$
6	$-6.85252E-01$	$6.81823E-02$	$-3.85830E-03$	$8.94189E-05$
7	$8.85576E-02$	$-9.70659E-03$	$5.85304E-04$	$-1.39359E-05$

TABLE 4.3. The coefficients a_{ij}.

For $\xi_p = 2$ to 7 and K_2 between 10 and 15, best fits are obtained with $i_{max} = 4$, $j_{max} = 4$

i, j	0	1	2	3	4
0	2.64297E-01	-2.95889E-02	9.60255E-04	2.18835E-05	-1.32628E-06
1	6.32118E-02	-1.63174E-02	2.21851E-03	-1.29020E-04	2.72578E-06
2	-1.10880E-03	-2.28719E-04	-1.02670E-04	1.00891E-05	-2.66634E-07
3	-1.64926E-03	5.12868E-04	-3.58420E-05	9.82352E-07	-5.49791E-09
4	2.78555E-04	-8.80389E-05	9.27285E-06	-4.48004E-07	8.21204E-09

TABLE 4.4. The coefficients a_{ij}.

4.4.15 Analytical expressions for SHG conversion efficiency h_{smm}, optimized with respect to σ and ξ_p

The values of $1/h_{smm}$ plotted against K_2 for different values of B_s are shown in Fig. 4.19. All the curves can be fitted to a cubic polynomial in K_2 with error less than 1 %.

From the cubic fits we find that the power conversion efficiency (η_{smm}) of a focused Gaussian beam (in the absence of linear absorption) optimized with respect to the phase mismatch parameter σ as well as the pump beam focusing parameter ξ_p, can be expressed as a simple function of the normalized pump power K_2. For the case of no walk-off ($B_p = B_s = 0$) and $K_2 = 0 \, to \, 15$, we have

$$\eta_{smm} = \frac{n_s}{n_p} \frac{K_2}{0.9265 + 0.777 \, K_2 + 0.0224 \, K_2^2 - 5.4 \times 10^{-4} K_2^3}. \qquad (4.270)$$

For $B_p = 0$, and B_s ranging from 0 to 10, the following relations hold

$$\eta_{smm}(B_s) = \frac{n_s}{n_p} \frac{K_2}{b_0(B_s) + b_1(B_s) \, K_2 + b_2(B_s) \, K_2^2 + \dots} \qquad (4.271)$$

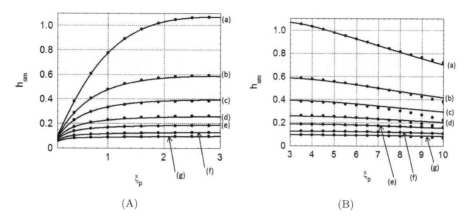

FIGURE 4.17: Figures (A) and (B) show the dependence of h_{sm} on ξ_p for $B_p = B_s = 0$ and different values of K_2: (a) 0.001, (b) 0.5, (c) 1, (d) 2.5, (e) 5, (f) 7.5 and (g) 10. The solid lines are the results of the detailed solution of the coupled wave equations and the filled circles are obtained from the fitted polynomials.

where the coefficients b_0, b_1, b_2 etc. can be expressed as polynomials in B_s

$$\text{for } i = 0 \text{ to } i_{\max} \qquad b_i(B_s) = \sum_{j=0}^{j_{\max}} b_{ij}(B_s)^j. \qquad (4.272)$$

The values of the coefficients b_{ij} are found from numerical fits and are given in the tables below:

$B_s = 0$ to 2, $K_2 = .01$ to 10

i,j	0	1	2
0	9.2107E-01	3.9712E-01	1.5300E-01
1	8.0193E-01	4.0787E-01	1.3994E-01
2	1.5563E-02	1.2779E-02	1.4181E-02

TABLE 4.5. The coefficients b_{ij}.

$B_s = 2$ to 10, $K_2 = .01$ to 10

i,j	0	1	2
0	4.4712E-01	-1.3623E+00	3.5200E+00
1	1.7033E-01	1.0595E-01	9.5081E-01
2	-1.3967E-02	3.2874E-01	-4.3312E-01
3	1.1886E-03	-1.9556E-02	2.9195E-02

TABLE 4.6. The coefficients b_{ij}.

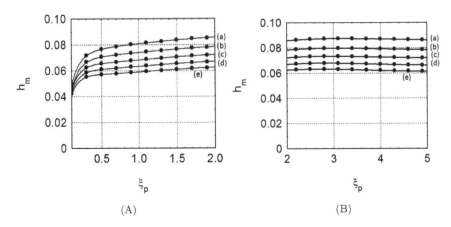

FIGURE 4.18: Figures (A) and (B) show the dependence of h_{sm} on ξ_p for $B_p = B_s = 0$ and different values of K_2: (a) 11, (b) 12, (c) 13, (d) 14 and (e) 15.The solid lines are the results of the detailed solution of the coupled wave equations and the filled circles are obtained from the fitted polynomials.

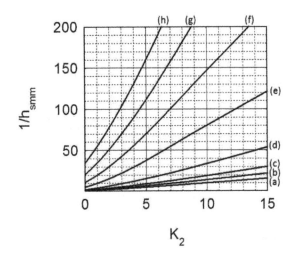

FIGURE 4.19: The values of $1/h_{smm}$ plotted against K_2 for $B_p = 0$ and different values of B_s are shown in the curves (a) through (h). The values of B_s are (a) 0, (b) 0.5, (c) 1.0, (d) 2.0, (e) 4.0, (f) 6.0, (g) 8.0 and (h) 10.0.

To check the validity of the fitting formulas, we determine the value of the optimized conversion efficiency η_{smm} using the formula as well as through the full numerical analysis. We find that for $B_s = 5$ and $K_2 = 7.5$, for example, Eqn. 4.272 gives the value of 0.0122 for h_{smm}, whereas the numerical solution gives a value of 0.0121, showing that the fits are quite accurate.

It may be worth emphasizing here that the formulas for the power conversion efficiency given in Eqns. 4.270 and 4.272, should not be extrapolated beyond the parameter space in which they were derived, i.e., they are valid *only* for K_2 between 0 and 10, and B_s between 0 and 10.

4.5 Unseeded Difference Frequency Generation $(\omega_1 = \omega_3 - \omega_2)$

We assume here that the lowest frequency beam at frequency ω_1 is being generated by the nonlinear mixing of the two higher frequency beams at frequencies ω_3 and ω_1.

For *unseeded* difference frequency generation, i.e., with $v_1(0) = 0$, we have from Eqn. 4.32 the constants m_1, m_2 and m_3 to be given as

$$m_1 = v_2(0)^2 + v_3(0)^2, \quad m_2 = v_3(0)^2 \quad \text{and} \quad m_3 = -v_2(0)^2 \quad (4.273)$$

and from Eqns. 4.48 we have

$$\Gamma = av_3(0)^2 = am_2 \quad (4.274)$$

where, as before, a is $\sigma/(2\beta_1)$, and β_1 is defined in Eqn. 4.29. The expression for $\mathcal{D}_1(s)$ in Eqn. 4.41 is then given by

$$
\begin{aligned}
\mathcal{D}_1(s) &= (m_1 - s)(m_2 - s)s - (am_2 - as)^2 \\
&= (m_2 - s)\{s(m_1 - s) - a^2(m_2 - s)\} \\
&= (s - m_2)\{s^2 - s(m_1 + a^2) + a^2 m_2\} \\
&\equiv (s - m_2)(s^2 - t_1'' s + t_2'') \quad (4.275)
\end{aligned}
$$

where

$$t_1'' = m_1 + a^2; \quad \text{and} \quad t_2'' = a^2 m_2. \quad (4.276)$$

The three roots of the polynomial $\mathcal{D}_1(s)$ in Eqn.4.275, in increasing magnitudes, are given by

$$s_a = \frac{t_1'' - t_3''}{2}, \quad s_b = m_2 \quad s_c = \frac{t_1'' + t_3''}{2} \quad (4.277)$$

where

$$t_3'' \equiv \sqrt{t_1''^2 - 4t_2''}. \quad (4.278)$$

From the expressions for s_a, s_b, s_c and β_1 the values of γ, y_0 and β_2 are found using Eqns. 4.42, 4.46 and 4.44 and then y_ℓ is determined using Eqn. 4.60.

Since $s_b = m_2 = v_3(0)^2$ and $s(0) = v_3(0)^2$ we have $s(0) = s_b$ and

$$y_0 = \frac{s(0) - s_a}{s_b - s_a} = 1. \tag{4.279}$$

The irradiances $I_1(\ell)$, $I_2(\ell)$ and $I_3(\ell)$ at the exit face of the nonlinear medium can then be determined using Eqns. 4.61 and 4.62. Since $v_1(0)^2 = 0$ for the case of unseeded DFG being discussed here, and since $s_b = m_2 = v_3(0)^2$ we have

$$
\begin{aligned}
v_1(\ell)^2 &= v_3(0)^2 - (s_b - s_a)y_\ell^2 - s_a \\
&= (v_3(0)^2 - s_a)(1 - y_\ell^2) \tag{4.280} \\
v_2(\ell)^2 &= v_2(0)^2 + v_3(0)^2 - (s_b - s_a)y_\ell^2 - s_a \\
&= v_2(0)^2 + (v_3(0)^2 - s_a)(1 - y_\ell^2) \tag{4.281} \\
v_3(\ell)^2 &= (s_b - s_a)y_\ell^2 + s_a \tag{4.282}
\end{aligned}
$$

and using Eqns. 4.65 we obtain

$$
\begin{aligned}
I_1(\ell) &= \frac{\lambda_3}{\lambda_1} I_a \beta_1^2 v_1(\ell)^2 \\
&= \frac{\lambda_3}{\lambda_1} I_a (1 - y_\ell^2)(\beta_1^2 v_3(0)^2 - \beta_1^2 s_a) \\
&= \frac{\lambda_3}{\lambda_1} I_a \left(\frac{I_3(0)}{I_a} - b''_1 \right) (1 - y_\ell^2) \tag{4.283}
\end{aligned}
$$

$$
\begin{aligned}
I_2(\ell) &= \frac{\lambda_3}{\lambda_2} I_a \beta_1^2 v_2(\ell)^2 \\
&= \frac{\lambda_3}{\lambda_2} I_a \left(\beta_1^2 v_2(0)^2 + (1 - y_\ell^2)(\beta_1^2 v_3(0)^2 - \beta_1^2 s_a) \right) \\
&= \frac{\lambda_3}{\lambda_2} I_a \left(\frac{\lambda_2}{\lambda_3} \frac{I_2(0)}{I_a} + \left(\frac{I_3(0)}{I_a} - b''_1 \right)(1 - y_\ell^2) \right) \tag{4.284}
\end{aligned}
$$

$$
\begin{aligned}
I_3(\ell) &= I_a \beta_1^2 v_3(\ell)^2 \\
&= I_a \left(\beta_1^2 s_a + y_\ell^2 (\beta_1^2 s_b - \beta_1^2 s_a) \right) \\
&= I_a \left(b''_1 + \left(\frac{I_3(0)}{I_a} - b''_1 \right) y_\ell^2 \right) \tag{4.285}
\end{aligned}
$$

where

$$b''_1 \equiv \beta_1^2 s_a = \frac{a''_1 - a''_3}{2} \tag{4.286}$$

and

$$
\begin{aligned}
a''_1 &\equiv \beta_1^2 t''_1 = \left(\frac{\lambda_2}{\lambda_3} \frac{I_2(0)}{I_a} + \frac{I_3(0)}{I_a} + \frac{\sigma^2}{4} \right) \\
a''_2 &\equiv \beta_1^4 a^2 m_2 = \frac{\sigma^2}{4} \frac{I_3(0) I_2(0)}{I_a} \\
a''_3 &\equiv \sqrt{a''_1{}^2 - 4a''_2} = \beta_1^2 t''_3. \tag{4.287}
\end{aligned}
$$

We then have

$$\beta_2 = \sqrt{\beta_1^2(s_c - s_a)} = \sqrt{\beta_1^2 t''_3} = \sqrt{a''_3} \qquad (4.288)$$

$$
\begin{aligned}
\gamma^2 &= \frac{s_b - s_a}{s_c - s_a} \\
&= \frac{\beta_1^2 m_2 - \beta_1^2 s_a}{\beta_1^2(s_c - s_a)} \\
&= \frac{\dfrac{I_3(0)}{I_a} - b''_1}{a''_3}.
\end{aligned} \qquad (4.289)
$$

Since from Eqn. 4.279 $y_0 = 1$, we have from Eqn. 4.60

$$y_\ell = \frac{\operatorname{cn}(\beta_2, \gamma^2) \, \operatorname{dn}(\beta_2, \gamma^2)}{1 - \gamma^2 \operatorname{sn}^2(\beta_2, \gamma^2)}. \qquad (4.290)$$

From this value of y_ℓ and using b''_1 from Eqn. 4.286, the irradiances of the three beams at the exit plane are calculated using Eqns. 4.283 - 4.285. In the next two subsections it is shown that these irradiance expressions reduce to simpler forms in special cases, thereby bolstering the case for their validity.

4.5.1 DFG irradiance for collimated beams with no phase matching ($\sigma \neq 0$) and with no pump depletion

Diffraction - No	Beam Walk-off - No	Absorption - No
Pump Depletion - No	Phase Mismatch - Yes	

In the case of no pump depletion, the irradiances of the incident beams (at frequencies ω_3 and ω_2) are small compared to the material parameter I_a and so is the the irradiance of the generated beam (at frequency ω_1). Thus in this case the constants m_1 and m_2 may be considered small with respect to the normalized phase mismatch parameter a ($= \sigma/2\beta_1$. We have

$$
\begin{aligned}
t''^2_3 &= t''^2_1 - 4t''_2 \\
&= (m_1 + a^2)^2 - 4m_2 a^2 \\
&= a^4(1 + q)
\end{aligned} \qquad (4.291)
$$

where

$$q \equiv \frac{m_1^2 + 2a^2 m_4}{a^4} \quad \text{and} \quad m_4 \equiv m_1 - 2m_2. \qquad (4.292)$$

Thus we get, keeping terms up to the quadratic in m_4 (and m_1)

$$q^2 \simeq \frac{4m_4^2}{a^4} \qquad (4.293)$$

using which we have

$$
\begin{aligned}
t_3'' &= a^2(1+q)^{1/2} \\
&\simeq a^2\left(1 + \frac{q}{2} - \frac{q^2}{8}\right) \\
&= a^2\left(1 + \frac{m_1^2 + 2a^2 m_4}{2a^4} - \frac{4m_4^2}{8a^4}\right) \\
&= a^2 + m_4 + \frac{m_1^2 - m_4^2}{2a^2}.
\end{aligned}
\tag{4.294}
$$

Thus we get

$$
\begin{aligned}
s_a &= \frac{t_1'' - t_3''}{2} \\
&= \frac{m_1 - m_4}{2} - \frac{m_1^2 - m_4^2}{4a^2} \\
&= m_2 - \frac{m_1^2 - m_4^2}{4a^2} \\
&= m_2 - \frac{(m_1 + m_4)(m_1 - m_4)}{4a^2} \\
&= m_2 - \frac{v_2(0)^2 v_3(0)^2}{a^2}.
\end{aligned}
\tag{4.295}
$$

Since

$$
\begin{aligned}
s_b &= m_2 \\
s_c &= \frac{t_1'' + t_3''}{2} \\
&\simeq a^2
\end{aligned}
\tag{4.296}
$$

we have

$$
\begin{aligned}
\beta_2 &= \sqrt{\beta_1^2(s_c - s_a)} \\
&\approx \sqrt{\beta_1^2 a^2} \\
&= \frac{\sigma}{2}
\end{aligned}
\tag{4.297}
$$

$$
\begin{aligned}
\gamma^2 &= \frac{s_b - s_a}{s_c - s_a} \\
&= \frac{v_2(0)^2 v_3(0)^2}{a^4} \\
&\simeq 0.
\end{aligned}
\tag{4.298}
$$

Since $cn(\beta_2, \gamma^2)$ goes to $\cos\beta_2$ and $dn(\beta_2, \gamma^2)$ goes to 1 as γ goes to 0 (from Eqn. 4.55) we have from Eqn. 4.290

$$y_\ell = \cos\beta_2. \tag{4.299}$$

Thus from Eqn. 4.280 we have

$$
\begin{aligned}
v_1(\ell)^2 &= (m_2 - s_a)(1 - y_\ell^2) \\
&= \frac{v_2(0)^2 v_3(0)^2}{a^2} \sin^2\beta_2 \\
&= (v_2(0)^2\, v_3(0)^2)\beta_1^2 \left(\frac{\sin\sigma/2}{\sigma/2}\right)^2
\end{aligned}
\tag{4.300}
$$

using the definition $a = \sigma/(2\beta_1)$. Thus using Eqns. 4.65 we have

$$
\begin{aligned}
I_1(\ell) &= \frac{\lambda_3}{\lambda_1} I_a \beta_1^2 v_1(\ell)^2 \\
&= \frac{\lambda_3}{\lambda_1} I_a \beta_1^4 v_2(0)^2\, v_3(0)^2 \left(\frac{\sin\sigma/2}{\sigma/2}\right)^2 \\
&= \frac{\lambda_3}{\lambda_1} I_a \frac{\lambda_2 I_2(0)}{\lambda_3 I_a} \frac{I_3(0)}{I_a} \left(\frac{\sin\sigma/2}{\sigma/2}\right)^2 \\
&= \frac{\lambda_2}{\lambda_1} \frac{I_2(0)I_3(0)}{I_a} \left(\frac{\sin\sigma/2}{\sigma/2}\right)^2
\end{aligned}
\tag{4.301}
$$

which is the same result as in Ref. [5].

4.5.2 DFG irradiance for collimated beams with phase matching ($\sigma = 0$) in presence of pump depletion

Diffraction - No	Beam Walk-off - No	Absorption - No
Phase Mismatch - No	Pump Depletion - Yes	

For the phase matched case, i.e., with $\sigma = 0$, the expressions for a''_1, a''_2 and a''_3 in Eqns. 4.287 are simpler, leading to simpler expressions for γ:

$$
\begin{aligned}
a''_1 &= \left(\frac{\lambda_2}{\lambda_3}\frac{I_2(0)}{I_a} + \frac{I_3(0)}{I_a}\right) \\
a''_2 &= 0 \\
a''_3 &= a''_1
\end{aligned}
\tag{4.302}
$$

so that

$$b''_1 = 0, \qquad s_a = 0 \tag{4.303}$$

and

$$\gamma^2 = \frac{\dfrac{I_3(0)}{I_a}}{\left(\dfrac{\lambda_2}{\lambda_3}\dfrac{I_2(0)}{I_a} + \dfrac{I_3(0)}{I_a}\right)} = \frac{\lambda_3 I_3(0)}{\lambda_2 I_2(0) + \lambda_3 I_3(0)}. \tag{4.304}$$

With $b''_1 = 0$, Eqns. 4.283 to 4.285 reduce to

$$I_1(\ell) = \frac{\lambda_3}{\lambda_1} I_3(0)(1 - y_\ell^2) \tag{4.305}$$

$$I_2(\ell) = I_2(0) + \frac{\lambda_3}{\lambda_2} I_3(0)(1 - y_\ell^2) \tag{4.306}$$

$$I_3(\ell) = I_3(0) y_\ell^2. \tag{4.307}$$

When both incident beams are of comparable strength, and pump depletion cannot be ignored, γ obtained from Eqn. 4.304 cannot be further simplified and y_l needs to be evaluated from the general expression given in Eqn. 4.290. However, considerable simplification of the expression for y_l can be obtained in two special cases of $\lambda_3 I_3(0) >> \lambda_2 I_2(0)$ and $\lambda_2 I_2(0) >> \lambda_3 I_3(0)$:

Case 1. $\underline{\lambda_3 I_3(0) >> \lambda_2 I_2(0)}$

Defining a quantity

$$\epsilon_2 \equiv \frac{\lambda_2 I_2(0)}{\lambda_3 I_3(0)}. \tag{4.308}$$

Equation 4.304 is rewritten as

$$\begin{aligned} \gamma^2 &= \frac{1}{1 + \epsilon_2} \\ &\simeq 1 - \epsilon_2 \end{aligned} \tag{4.309}$$

for small ϵ_2. For γ close to 1, the Jacobian elliptic functions cn, dn and sn can be replaced by sech, sech and tanh functions respectively (Eqn. 4.57) and Eqn. 4.290 can be written as

$$\begin{aligned} y_\ell &= \frac{\text{sech}^2 \beta_2}{1 - \gamma^2 \tanh^2 \beta_2} \\ &= \frac{1}{1 + \epsilon_2 \sinh^2 \beta_2} \\ &\simeq 1 - \epsilon_2 \sinh^2 \beta_2 \end{aligned} \tag{4.310}$$

where we have used the fact that ϵ_2 is small. Eqns. 4.302 and 4.288 show that in this case

$$\beta_2 = \sqrt{\frac{I_3(0)}{I_a}}. \tag{4.311}$$

Thus

$$y_\ell^2 \simeq 1 - 2\epsilon_2 \sinh^2 \sqrt{\frac{I_3(0)}{I_a}} \tag{4.312}$$

and we have from Eqns. 4.283, 4.284 and 4.285 (using Eqn. 4.303)

$$
\begin{aligned}
I_1(\ell) &= \frac{\lambda_3}{\lambda_1} I_a \left(\frac{I_3(0)}{I_a} \right) 2\epsilon_2 \sinh^2 \beta_2 \\
&= 2\frac{\lambda_2}{\lambda_1} I_2(0) \sinh^2 \sqrt{\frac{I_3(0)}{I_a}} \tag{4.313}
\end{aligned}
$$

$$
\begin{aligned}
I_2(\ell) &= I_2(0) + \frac{\lambda_3 I_3(0)}{\lambda_2} 2\epsilon_2 \sinh^2 \beta_2 \\
&= I_2(0) + 2I_2(0) \sinh^2 \sqrt{\frac{I_3(0)}{I_a}} \tag{4.314}
\end{aligned}
$$

$$
\begin{aligned}
I_3(\ell) &= I_3(0) y_\ell^2 \\
&= I_3(0)(1 - 2\epsilon_2 \sinh^2 \beta_2) \\
&= I_3(0) - 2\frac{\lambda_2}{\lambda_3} I_2(0) \sinh^2 \sqrt{\frac{I_3(0)}{I_a}}. \tag{4.315}
\end{aligned}
$$

Case 2. $\lambda_2 I_2(0) >> \lambda_3 I_3(0)$

Defining a quantity

$$\epsilon_3 \equiv \frac{\lambda_3 I_3(0)}{\lambda_2 I_2(0)}. \tag{4.316}$$

Equation 4.304 is rewritten as

$$\gamma^2 = \frac{\epsilon_3}{1 + \epsilon_3} \simeq \epsilon_3 \tag{4.317}$$

when ϵ_3 is small. For γ small, the Jacobian elliptic functions cn and sn can be replaced by cos and sin functions respectively and the function dn is ≈ 1 (Eqn. 4.55). Equation 4.290 can then be written as

$$y_\ell \simeq \cos \beta_2. \tag{4.318}$$

Equations 4.302 and 4.288 show that in this case

$$\beta_2 = \sqrt{\frac{\lambda_2 I_2(0)}{\lambda_3 I_a}}. \tag{4.319}$$

Thus we have from Eqns. 4.283, 4.284 and 4.285

$$
\begin{aligned}
I_1(\ell) &= \frac{\lambda_3}{\lambda_1} I_a \left(\frac{I_3(0)}{I_a} \right) \sin^2 \beta_2 \\
&= 2 \frac{\lambda_3}{\lambda_1} I_3(0) \sin^2 \sqrt{\frac{\lambda_2 I_2(0)}{\lambda_3 I_a}} \qquad (4.320)
\end{aligned}
$$

$$
I_2(\ell) = I_2(0) + \frac{\lambda_3}{\lambda_2} I_3(0) \sin^2 \sqrt{\frac{\lambda_2 I_2(0)}{\lambda_3 I_a}} \qquad (4.321)
$$

$$
I_3(\ell) = I_3(0) \cos^2 \sqrt{\frac{\lambda_2 I_2(0)}{\lambda_3 I_a}}. \qquad (4.322)
$$

We find that in the DFG process, if the photon number of the highest frequency beam is much larger than that of the middle frequency beam $(\lambda_3 I_3(0) \gg \lambda_2 I_2(0))$ the irradiance $(I_1(\ell))$ of the generated beam at the lowest frequency (ω_1) varies as a hyperbolic sine (sinh) function of the irradiance $I_3(0)$, whereas if the photon number of the highest frequency beam is much smaller than that of the middle frequency beam $(\lambda_3 I_3(0) \ll \lambda_2 I_2(0))$ the irradiance $(I_1(\ell))$ of the generated beam at the lowest frequency (ω_1) varies as a sinusoidal (sin) function of the irradiance $I_2(0)$.

The power and energy conversion efficiencies for collimated beams of arbitrary (as well as Gaussian) spatial and temporal profiles can be worked out by straightforward extension of the SFG cases. Since the results for the collimated beam cases can be obtained from the general focused case, we present here the formulas for obtaining the power conversion efficiency for the focused Gaussian beam case of DFG. The computer program for the DFG of focused Gaussian beams is provided in the appendix.

4.5.3 DFG conversion efficiency for focused Gaussian beams

Diffraction - Yes	Beam Walk-off - Yes	Absorption - Yes
Pump Depletion - Yes	Phase Mismatch - Yes	

Assuming again that the two incident beams at frequencies ω_2 and ω_3 are interacting in a nonlinear medium to generate a beam at frequency ω_1 and denoting the parameters associated with the incident beams at frequencies ω_2 and ω_3 by the subscripts 2 and 3 respectively, and assigning

$$
r_0 = r_{03}, \qquad A_0 = A_3(0,0,f) \qquad (4.323)
$$

in Eqns. 4.18, we obtain

$$\frac{\partial u_1}{\partial z_1} = i\frac{k_3}{k_1}\xi_3\nabla_{T_1}^2 u_1 + \sqrt{2k_3\ell\xi_3}\tan\rho_1\frac{\partial u_1}{\partial x_1} - \frac{a_1}{2}u_1 + i\frac{\lambda_3 n_3}{\lambda_1 n_1}\kappa u_3 u_2^* e^{i\sigma z_1}$$

$$\frac{\partial u_2}{\partial z_1} = i\frac{k_3}{k_2}\xi_3\nabla_{T_1}^2 u_2 + \sqrt{2k_3\ell\xi_3}\tan\rho_2\frac{\partial u_2}{\partial x_1} - \frac{a_2}{2}u_2 + i\frac{\lambda_3 n_3}{\lambda_2 n_2}\kappa u_3 u_1^* e^{i\sigma z_1}$$

$$\frac{\partial u_3}{\partial z_1} = i\xi_3\nabla_{T_1}^2 u_3 + \sqrt{2k_3\ell\xi_3}\tan\rho_3\frac{\partial u_3}{\partial x_1} - \frac{a_3}{2}u_1 + i\kappa u_1 u_2 e^{-i\sigma z_1}$$

$$(4.324)$$

where

$$\kappa = \frac{4\pi A_0 d_{\text{eff}}\ell}{\lambda_3 n_3} \tag{4.325}$$

is the same dimensionless parameter as defined before in Eqn. 4.137 and where we have used

$$\frac{\ell}{r_{03}} = (2k_3\ell\xi_3)^{1/2}. \tag{4.326}$$

If I_{30} denotes the peak irradiance of the incident beam at frequency ω_3, we find from Eqn. 4.323

$$I_{30} = 2n_3 c\varepsilon_0 |A_3(0,0,f)|^2 = 2n_3 c\varepsilon_0 |A_0|^2. \tag{4.327}$$

Defining the parameters p_3 and K_3 and using the same definition of the parameter δ as in Eqn. 4.90

$$p_3 \equiv \frac{I_{20}}{I_{30}} \qquad K_3 \equiv \frac{I_{30}}{I_a} \qquad \delta \equiv \frac{\lambda_2}{\lambda_3} \tag{4.328}$$

and using Eqns. 4.325, 4.327 and 4.63, we find that the parameter κ is related to K_3 :

$$\kappa^2 = \frac{16\pi^2 d^2 l^2}{\lambda_3^2 n_3^2}\frac{I_{30}}{2n_3 c\varepsilon_0}$$

$$= \frac{n_1 n_2}{n_3^2}\frac{\delta^2}{\delta - 1}K_3. \tag{4.329}$$

The three wave mixing Eqns. 4.136 can be solved using the split step technique described in Chapter 7, with the linear propagation part (including diffraction, linear absorption and beam walk-off but ignoring nonlinear coupling) solved by the Fourier transform method and the nonlinear propagation part (ignoring diffraction, linear absorption and beam walk-off) solved using the finite difference technique. For both incident beams circular Gaussian in cross-section, the variables u_1, u_2 and u_3 at the entrance face of the nonlinear medium ($z_1 = 0$) are given by

$$u_1(x_1, y_1, 0) = 0$$

$$u_2(x_1, y_1, 0) = \sqrt{p_3}\frac{1}{1 - i\xi_2\{(1 - \mu_2)\}}e^{-\frac{\varrho_1^2}{2}\left(\frac{x_1^2 + y_1^2}{1 - i\xi_2\{(1 - \mu_2)\}}\right)}$$

$$u_3(x_1, y_1, 0) = \frac{1}{1 - i\xi_3\{(1 - \mu_3)\}}e^{-\frac{1}{2}\left(\frac{x_1^2 + y_1^2}{1 - i\xi_3\{(1 - \mu_3)\}}\right)} \quad (4.330)$$

where

$$\varrho_1 \equiv \frac{r_{03}}{r_{02}}. \quad (4.331)$$

The generated power in the focused case depends on the factors σ, δ, p_3, ϱ_1, K_3 and on the focusing parameters of the two incident beams (ξ_2 and ξ_3), positions of their foci (μ_2 and μ_3) and the refractive index ratios n_1/n_3 and n_2/n_3. When beam walk-offs are included, the walk-off angles at the three frequencies ρ_1, ρ_2 and ρ_3 need to be specified along with the parameter $k_3\ell$. The linear absorption parameters a_1, a_2 and a_3, if non-zero, should also be included in the numerical computation.

The value of the dimensionless variable $u_1(x_1, y_1, 1)$ at the exit surface of the nonlinear material ($z_1 = 1$) can be calculated by numerically solving Eqns. 4.324. The irradiance distribution at the generated frequency ω_1 is then given by

$$\begin{aligned}I_1(x_1, y_1, 1) &= 2n_1c\varepsilon_0|A_1(x_1, y_1, 1)|^2 \\ &= 2n_1c\varepsilon_0A_0^2|u_1(x_1, y_1, 1)|^2 \\ &= \frac{n_1}{n_3}I_{30}|u_1(x_1, y_1, 1)|^2. \quad (4.332)\end{aligned}$$

The DFG power $P_1(\ell)$ at the exit plane is obtained by integrating over the irradiance distribution, i.e.,

$$
\begin{aligned}
P_1(\ell) &= \int\!\!\!\int_{-\infty}^{\infty} I_1(x,y,l)\,dx\,dy \\[2mm]
&= \frac{n_1}{n_3} I_{10} r_{03}^2 \int_{-\infty}^{\infty}\!\!\int_{-\infty}^{\infty} |u_1(x_1,y_1,1)|^2 dx_1 dy_1 \\[2mm]
&= \frac{n_1}{n_3}\frac{P_3(0)}{\pi} \int_{-\infty}^{\infty}\!\!\int_{-\infty}^{\infty} |u_1(x_1,y_1,1)|^2 dx_1 dy_1 \\[2mm]
&= \frac{n_1}{n_3}\frac{P_2(0)P_3(0)}{\pi}\frac{1}{P_2(0)} \int_{-\infty}^{\infty}\!\!\int_{-\infty}^{\infty} |u_1(x_1,y_1,1)|^2 dx_1 dy_1 \\[2mm]
&= \frac{P_2(0)P_3(0)}{P_{DF}} h_1
\end{aligned}
\tag{4.333}
$$

where

$$
h_1 \equiv \frac{\xi_2}{\pi K_3 p_3} \int\!\!\!\int_{-\infty}^{\infty} |u_1(x_1,y_1,1)|^2 dx_1 dy_1,
\tag{4.334}
$$

$$
P_{DF} \equiv \frac{c\varepsilon_0 n_3^2 \lambda_1 \lambda_2^2}{32\pi^2 d_{\text{eff}}^2 \ell}
\tag{4.335}
$$

and where we have used the expression for I_a from Eqn. 4.63 and

$$
\begin{aligned}
P_2(0) &= \pi I_{20} r_{02}^2 \\[2mm]
&= \pi K_3 p_3 I_a \frac{\ell\lambda_2}{4\pi n_2 \xi_2}
\end{aligned}
\tag{4.336}
$$

from Eqns. 4.328 and 4.133 .

When the incident beams are collimated, i.e., for ξ_2 and $\xi_3 \ll 1$, h_1 goes to

$$
\begin{aligned}
h_{1(\text{coll})} &= \frac{n_3\lambda_2}{n_1\lambda_1}\frac{\xi_2}{1+\varrho_1^2}\left(\frac{\sin(\sigma/2)}{\sigma/2}\right)^2 \\[2mm]
&= \frac{n_3^2\lambda_2^2}{n_1\lambda_1}\left(\frac{\xi_2\xi_3}{n_3\xi_3\lambda_2 + n_2\xi_2\lambda_3}\right)\left(\frac{\sin(\sigma/2)}{\sigma/2}\right)^2
\end{aligned}
\tag{4.337}
$$

since

$$
\varrho_1 \equiv \frac{r_{03}}{r_{02}} = \sqrt{\frac{\lambda_3 n_2 \xi_2}{\lambda_2 n_3 \xi_3}}
\tag{4.338}
$$

Using Eqn.4.333 and the expression for P_{DF} in Eqn.4.335 we thus obtain the difference frequency power for the case of collimated incident beams to be

$$P_{1(\text{coll})}(\ell) = \left(\frac{32\pi^2 d_{\text{eff}}^2 \ell}{c\varepsilon_0 n_1 \lambda_1^2}\right) \left(\frac{\xi_2 \xi_3}{\lambda_2 n_3 \xi_3 + \lambda_3 n_2 \xi_2}\right) \left(\frac{\sin(\sigma/2)}{\sigma/2}\right)^2 P_2(0) P_3(0).$$

$$(4.339)$$

Bibliography

[1] A. Yariv, *Quantum Electronics*, John Wiley and Sons, Inc., 1978.

[2] J. A. Armstrong, N. Bloembergen, J. Ducuing and P. S. Pershan, Interactions between Light Waves in a Nonlinear Dielectric, *Phys. Rev.* **127**, 1918, 1962.

[3] M. Abramowitz and I.A. Stegun *Handbook of Mathematical Functions*, Dover Publications, Inc., 1972.

[4] Y.R. Shen, *The Principles of Nonlinear Optics*, John Wiley and Sons, New York, 1984.

[5] R. L. Sutherland, *Handbook of Nonlinear Optics*, Marcel Dekkar, Inc., 1996.

[6] G. D. Boyd and D. A. Kleinman, Parametric Interaction of Focused Gaussian Light Beams, *J. Appl. Phys.* **39**, 3597, 1968)

[7] Y. F. Chen and Y. C. Chen, Analytical functions for the optimization of second-harmonic generation and parametric generation by focused Gaussian beams, *Appl. Phys. B* **76**, 645–647, 2003.

5

Quasi-Phase Matching

5.1   Quasi Phase Matching, QPM

Nonlinear frequency conversion efficiency depends sensitively on the phase mismatch parameter σ defined in Eqn. 4.17. For collimated light beams, the dependence is $[\sin(\sigma/2)/(\sigma/2)]^2$ which is maximum at $\sigma = 0$ and falls to half its maximum value at σ equal to 2.8. For optimally focused incident beam(s), the dependence of the conversion efficiency on σ is similar, although the maximum is achieved at a nonzero value of σ (as shown in Ref. [2]). Since the ratio of the length of the nonlinear material (typically centimeters) to that of light wavelength (between 0.3 and 10 μm) is large, even small values of dispersion in the material can make σ quite large, and thereby reduce the conversion efficiency.

Choosing an optically anisotropic crystal as the nonlinear medium, the process of birefringent phase matching is utilized to reduce the value of σ to a small number and thereby achieve highly efficient frequency conversion. However, several materials that have relatively large values of the d coefficient compared to those of commonly used anisotropic crystals, along with other desirable properties such as low absorption at the relevant wavelengths and availability in adequate lengths are optically *isotropic*. The key limitation of such materials as efficient nonlinear frequency converter is their inability to allow birefringent phase matching. Even for anisotropic materials, (for example LiNbO$_3$) the crystal structure ($3m$) can be such that the highest values of the d coefficient cannot be utilized for a birefringent phase matching configuration. For LiNbO$_3$ the magnitude of the d_{33} coefficient is more than ten times higher than that of other coefficients [2] but the contribution from this coefficient to d_{eff} is zero, as shown in Table 3.10 in Chapter 3.

The technique of quasi phase-matching (QPM) can be used in such materials to provide efficient frequency conversion. In this technique, the nonlinear medium is configured as a layered structure, with the d_{eff} values of consecutive layers *equal* in magnitude but *opposite* in sign to each other and the thickness of each layer equal (or close) to the coherence length of the medium for the three wave mixing process. Such a layered and periodic structure gives rise to efficient frequency mixing utilizing the entire length of a nonlinear material, as can be shown with a plane wave and nondepleted pump analysis of the three wave mixing equations.

5.1.1　Plane wave analysis of quasi phase matching

Equations 4.19 in Chapter 4 show that (in the absence of pump depletion, diffraction, beam walk-off and absorptions) the field amplitudes in a three wave mixing process can be expressed as

$$\frac{\partial u}{\partial z_1} = iae^{\pm i\sigma z_1} \tag{5.1}$$

where u represents the electric field amplitude of one of the three interacting beams, and the amplitudes of the other two are assumed to be constant, so that the coupling coefficient a in Eqn. 5.1 is a constant. z_1 is the normalized distance, equal to z/ℓ, where z is the distance of propagation and ℓ is length of the nonlinear medium. The negative sign in the exponent of Eqn. 5.1 corresponds to the case when u is the amplitude of the light beam with highest frequency (ω_3) and the positive sign holds if u represents the amplitude of the beam with one of the lower frequency (ω_1 or ω_2).

Suppose the nonlinear medium is divided into N equally spaced layers, as shown in Fig. 5.1, so that the thickness of each layer is ℓ/N. If the value of the normalized distance z_1 at the end of the first layer is denoted by z_{11}, we have

$$z_{11} = \frac{\ell/N}{\ell} = \frac{1}{N}. \tag{5.2}$$

FIGURE 5.1: The nonlinear medium divided into equally spaced layers.

The values of z_1 at the ends of the subsequent layers (denoted by z_{21}, z_{31} etc., as shown in Fig. 5.1) are multiples of z_{11}, i.e.,

$$z_{21} = 2z_{11} = \frac{2}{N}$$
$$z_{31} = 3z_{11} = \frac{3}{N}$$
$$\cdots$$
$$z_{N1} = Nz_{11} = 1. \tag{5.3}$$

For phase matched case, i.e., with σ equal to zero, the solution of Eqn. 5.1 assuming $u(0) = 0$ is

$$u(z_1) = iaz_1 \tag{5.4}$$

and the value of u at the end of N layers is

$$u(z_{N1}) = iaz_{N1} = ia. \tag{5.5}$$

If phase matching cannot be achieved in the medium, for example for propagation in an isotropic crystal, then $\sigma \neq 0$. The solution of Eqn. 5.1 is then given by

$$u(z_1) = \pm \frac{a}{\sigma} \left(e^{\pm i\sigma z_1} - 1 \right) \tag{5.6}$$

so that at the end of the first layer ($z_1 = z_{11}$)

$$|u(z_{11})|^2 = \frac{4a^2}{\sigma^2} \sin^2 \frac{\sigma z_{11}}{2} \tag{5.7}$$

and at the end of N layers (at $z_1 = z_{N1} = 1$)

$$|u(z_{N1})|^2 = \frac{4a^2}{\sigma^2} \sin^2 \frac{\sigma z_{N1}}{2} = a^2 \text{sinc}^2 \left(\frac{\sigma}{2\pi} \right) \tag{5.8}$$

where

$$\text{sinc}(x/\pi) \equiv \sin(x)/x \tag{5.9}$$

in accordance with the definition of the sinc function in Ref. [3] (and in MAT-LAB). Equations 5.7 and 5.8 show that when σ is not equal to zero, irrespective of the number of layers, i.e., of the crystal thickness, the irradiance (proportional to $|u|^2$) achieved by frequency conversion at the end of N layers is at most equal to $4a^2/\sigma^2$, which is also the maximum irradiance at the end of the first layer.

Equation 5.7 also shows that the maximum value of $|u(z_{11})|^2$ is obtained when σz_{11} equals π i.e., with

$$z_{11} = \frac{\pi}{\sigma}. \tag{5.10}$$

The physical length of the first layer that corresponds to this maximum is ℓz_{11}, which satisfies

$$\ell z_{11} = \frac{\ell \pi}{\sigma} = \frac{\pi}{k_3 - k_1 - k_2} = \ell_c \tag{5.11}$$

using the definitions of coherence length ℓ_c and phase mismatch σ in Eqns. 2.44 and 4.17.

The dispersion values for optical materials are such that the coherence lengths in the visible to mid-infrared wavelengths (say up to about 5 μm) are usually in the range of tens of micrometers or smaller. In the longer wavelengths, say for SHG of 9 - 11 μm, the coherence lengths can be longer but usually do not exceed 200 μm. So although nonlinear crystals can be grown to lengths of several centimeters, the conversion efficiency obtained from them in the presence of phase mismatch would only be at most that obtained from a thin layer of thickness of around 200 μm.

The technique of QPM provides a way to alleviate this situation. Suppose the layers of the nonlinear material are configured in a way that the sign of the constant a in Eqn. 5.1 alternates from layer to layer, in addition to the condition that $\sigma z_{11} = \pi$. Then we have at the end of the first layer

$$
\begin{aligned}
u(z_{11}) &= \pm\frac{a}{\sigma}\left(e^{\pm i\sigma z_{11}} - 1\right) \\
&= \mp 2\frac{a}{\sigma} \tag{5.12}
\end{aligned}
$$

and at the end of the second layer (using the solution given in Eqn. 5.6 and $z_{21} = 2z_{11}$)

$$
\begin{aligned}
u(z_{21}) &= u(z_{11}) \pm \frac{-a}{\sigma}\left(e^{\pm i\sigma z_{21}} - e^{\pm i\sigma z_{11}}\right) \\
&= \mp 4\frac{a}{\sigma} \tag{5.13}
\end{aligned}
$$

and continuing, at the end of the third layer (with $z_{31} = 3z_{11}$)

$$
\begin{aligned}
u(z_{31}) &= u(z_{21}) \pm \frac{a}{\sigma}\left(e^{\pm i\sigma z_{31}} - e^{\pm i\sigma z_{21}}\right) \\
&= \mp 6\frac{a}{\sigma}. \tag{5.14}
\end{aligned}
$$

Thus at the end of N layers, the field is

$$
u(z_{N1}) = \mp 2N\frac{a}{\sigma} = -\mp\frac{2Naz_{11}}{\pi} = \mp\frac{2}{\pi}a \tag{5.15}
$$

using the relation $Nz_{11} = 1$ from Eqn. 5.3.

Thus in this case, the field amplitude is only a factor of $2/\pi$ smaller than that obtained with zero phase mismatch (Eqn. 5.5) and efficient frequency conversion can be achieved even though σ is non-zero. Since the generated irradiance is proportional to the square of the amplitude, the factor of $2/\pi$ reduces the frequency conversion efficiency to 40 % of what would be obtained in the phase matched case, if all other factors remained the same. According to Ref. [4], this is why the process is called 'quasi' (meaning almost but not quite full) phase matching.

5.2 Effects Of Focusing And Pump Depletion On Quasi Phase Matched SHG

The QPM process was discussed above through the solution of Eqn. 5.1, which is an approximation of the coupled wave equations with the effects of diffraction, absorption, beam walk-off and pump depletion ignored. Absorption and walk-off are usually small in nonlinear crystals commonly used in QPM, but focusing of light is usually necessary to increase the frequency conversion efficiency, especially for continuous wave (CW) beams of light. For conversion efficiencies exceeding, say 20%, the effects of pump depletion also cannot be ignored. In Chapter 4 it was seen that for tight focusing cases (i.e., for focusing parameter ξ_p of incident Gaussian beams more than 1) the optimum value of the phase mismatch parameter σ is non-zero. This raises the question of what the individual layer thicknesses to achieve optimum conversion efficiency should be for a QPM process.

To answer this question, we consider here the case of SHG and closely follow the treatment in Chapter 4, Section 4.4.10. The coupled wave equations given in Eqns. 4.243 are solved numerically with a circular Gaussian incident field as given in Eqn. 4.242 to obtain h_s (defined in Eqn. 4.248) for arbitrary values of the focusing parameters ξ_p. For simplicity, the beam walk-off and absorption parameters are assumed here to be zero, although results can be easily obtained from the computer programs for the cases when their values are non-zero.

In Section 4.4.10, the coupled wave equations were solved using the finite-difference method, in which the propagation of the beams through the homogeneous nonlinear medium was assumed to occur in a finite number of steps (denoted by m_z, say). The value of m_z needed to achieve convergence of the results was ≈ 50 for most cases and the exact value chosen for m_z was not important, as long as convergence of results was obtained.

The same computer program can be used to describe SHG in a QPM material with m_z set equal to N, the number of layers in a QPM structure, and with the sign of the coupling coefficient made to alternate for consecutive layers. In contrast with the case of bulk phase matching, the exact value of N is a key parameter in QPM.

The plane-wave and approximate analysis in the last subsection showed that the thickness of each layer in the QPM structure needs to be equal to the coherence length ℓ_c (or an odd multiple of it), which is fixed by the material parameters and the wavelengths of light. For a nonlinear medium of length ℓ, the values of N can then be only ℓ/ℓ_c (or $\ell/3\ell_c$, $\ell/5\ell_c$ etc. if odd multiples of ℓ_c are chosen).

Suppose $\mathfrak{z}_1$ denotes the thickness of a layer in a nonlinear medium of length ℓ having N such layers of equal thickness. Then,

$$N = \frac{\ell}{\mathfrak{z}_1} = \frac{1}{z_{11}} \tag{5.16}$$

from Fig. 5.1 and Eqn. 5.3.

With $\mathfrak{z}_1$ not necessarily equal to the coherence length ℓ_c, a parameter δ_1 is defined to denote the fractional difference of $\mathfrak{z}_1$ from ℓ_c, i.e.,

$$\delta_1 \equiv \frac{\mathfrak{z}_1 - l_c}{l_c}. \tag{5.17}$$

In the last subsection it was found that for plane waves in the absence of pump depletion, the frequency conversion efficiency is maximized when the layer thickness $\mathfrak{z}_1$ is equal to the coherence length ℓ_c, i.e., for δ_1 equal to 0. To determine the optimum value of δ_1 that maximizes SHG conversion efficiency in the presence of diffraction and pump depletion, the value of h_s (in 4.248) is calculated as a function of δ_1 for different values of ξ_p. However, for the QPM case, σ is not an independent input parameter in the coupled wave Eqns. 4.243. It is instead calculated from the independent parameters N and δ_1.

Since by definition,

$$\frac{\ell_c}{\ell} = \frac{\pi}{(k_3 - k_1 - k_2)\ell} = \frac{\pi}{\sigma} \tag{5.18}$$

we have from Eqns. 5.16 and 5.17

$$\frac{\mathfrak{z}_1}{\ell} = \frac{(1 + \delta_1)\ell_c}{\ell} = \frac{(1 + \delta_1)\pi}{\sigma} = \frac{1}{N} \tag{5.19}$$

so that

$$\sigma = \pi N (1 + \delta_1). \tag{5.20}$$

5.2.1 Quasi phase matched SHG for collimated beams, with $\delta_1 \neq 0$, and with no pump depletion

From Eqns. 5.16 and 5.20, we find that

$$\sigma z_{11} = \pi + \pi \delta_1. \tag{5.21}$$

In Eqns. 5.12 to 5.15, σz_{11} was assumed to be equal to π, i.e., δ_1 was assumed to be 0. We derive here the expression for the amplitude u at the end of N layers when δ_1 is not equal to 0. For simplicity, we treat only the case with the positive sign in Eqn. 5.1, so that from Eqns. 5.12 and 5.21

$$
\begin{aligned}
u(z_{11}) &= \frac{a}{\sigma}\left(e^{i\sigma z_{11}} - 1\right) \\
&= -\frac{a}{\sigma}(e^{i\pi\delta_1} + 1).
\end{aligned} \tag{5.22}
$$

Similarly, from Eqns. 5.13 and 5.21

$$u(z_{21}) = -\frac{a}{\sigma} - \frac{2a}{\sigma}e^{i\pi\delta_1} - \frac{a}{\sigma}e^{2i\pi\delta_1} \tag{5.23}$$

and continuing, at the end of the third layer (with $z_{31} = 3z_{11}$)

$$u(z_{31}) = -\frac{a}{\sigma} - \frac{2a}{\sigma}e^{i\pi\delta_1} - \frac{2a}{\sigma}e^{2i\pi\delta_1} - \frac{a}{\sigma}e^{3i\pi\delta_1}. \tag{5.24}$$

Thus at the end of N layers, the field is

$$
\begin{aligned}
u(z_{N1}) \;=\; & -\frac{a}{\sigma} - \frac{a}{\sigma} e^{Ni\pi\delta_1} \\
& - \frac{2a}{\sigma}\left(e^{i(N-1)\pi\delta_1} + e^{i(N-2)\pi\delta_1} + \cdots + e^{2i\pi\delta_1} + e^{i\pi\delta_1}\right). \quad (5.25)
\end{aligned}
$$

Defining $w \equiv e^{i\pi\delta_1}$ we have

$$
\begin{aligned}
u(z_{N1}) \;=\; & -\frac{a}{\sigma}\left(1 + w^N\right) - \frac{2a}{\sigma}\left(w + w^2 + w^3 \cdots w^{(N-1)}\right) \\
\;=\; & -\frac{a}{\sigma}\left(1 + w^N\right) - \frac{2a}{\sigma}\left(\frac{1 - w^N}{1 - w} - 1\right) \\
\;=\; & -\frac{a}{\sigma}\frac{\left(1 - w^N\right)\left(1 + w\right)}{1 - w} \\
\;=\; & -\frac{a}{\sigma}\frac{\left(1 - e^{iN\pi\delta_1}\right)\left(1 + e^{i\pi\delta_1}\right)}{1 - e^{i\pi\delta_1}}
\end{aligned} \quad (5.26)
$$

from which we get

$$
|u(z_{N1})|^2 = \frac{4a^2}{\sigma^2}\sin^2\frac{N\delta_1\pi}{2}\cot^2\frac{\delta_1\pi}{2}. \quad (5.27)
$$

Since

$$
a = \kappa_s u_p^2 \quad (5.28)
$$

and for collimated Gaussian beams

$$
u_p = e^{-\dfrac{x_1^2 + y_1^2}{2}} \quad (5.29)
$$

with x_1 and y_1 the Cartesian coordinates x and y normalized with respect to the pump beam radius r_{0p}, we have from Eqn. 4.248 in Chapter 4, substituting $u_s(x_1, y_1, 1)$ by $u(z_{N1})$

$$
\begin{aligned}
h_s \;\equiv\; & \frac{2\xi_p}{\pi\kappa_s^2}\int_{-\infty}^{\infty}\int_{-\infty}^{\infty}|u_s(x_1, y_1, 1)|^2 \; dx_1 dy_1 \\
\;=\; & \frac{2\xi_p}{\pi\kappa_s^2}\frac{4\kappa_s^2}{\sigma^2}\sin^2\frac{N\delta_1\pi}{2}\cot^2\frac{\delta_1\pi}{2} \\
& \times \int_{-\infty}^{\infty}\int_{-\infty}^{\infty}e^{-2(x_1^2 + y_1^2)}dx_1 dy_1 \\
\;=\; & \frac{4}{\pi^2}\xi_p\frac{\sin^2\dfrac{N\delta_1\pi}{2}\cot^2\dfrac{\delta_1\pi}{2}}{N^2(1 + \delta_1)^2} \quad (5.30)
\end{aligned}
$$

using Eqn. 5.20. Thus we find that

$$\text{as} \quad \delta_1 \to 0, \qquad h_s \to \frac{4}{\pi^2}\xi_p. \tag{5.31}$$

5.2.2   Effects of focusing and pump depletion on quasi phase matched SHG

Equation 4.247 in Chapter 4 shows that the dependence of the power output at the converted frequency on focusing, pump depletion, absorption and phase mismatch is contained in the parameter h_s defined in Eqn. 4.248. h_s in turn is determined from $u_s(x_1, y_1, 1)$, which is obtained from the numerical solution of Eqns. 4.150.

Assuming for simplicity that the walk-off angles and absorptions are zero, the values of h_s are calculated here for different values of ξ_p, K_2, N and δ_1, with the change in the computer program that the parameter δ_1 is used as a new input variable instead of σ, with σ in the program calculated from Eqn. 5.20. Another change in the program is to include a multiplicative factor of $(-1)^{i_z}$ in the coupling coefficient κ_s, where i_z denotes the index of the steps taken in the z direction, and ranges from 1 to N. Moreover, for each value of i_z, the nonlinear part of the equation is solved by the finite difference method for a number of steps, say denoted by N_1, where the exact value of N_1 is not as salient as that of N, as long as convergence of the results is obtained. Typically N_1 of 50 is sufficient for the accuracy that is consistent with other calculations.

The dependence of h_s on δ_1 for collimated pump beam with no depletion

The Fig. 5.2 shows the dependence of h_s on the parameter δ_1 for five different values of N (ranging from 100 to 500) for the case of collimated beams (ξ_p equal to 0.1) and for no pump depletion (K_2 equal to 0.01). Under these approximations, dependence of h_s on δ_1 is given by Eqn. 5.30. The width of the h_s versus δ_1 curve decreases with increasing values of N, but the peak value of h_s occurs at δ_1 close to 0, i.e., for collimated beams the optimum layer thickness is very close to the coherence length. The analytical expression for h_s in this collimated beam case is given in Eqn. 5.30 and the results obtained from the numerical solution of the coupled beam equations match the analytical results.

The dependence of h_s on δ_1 for collimated pump beam *with* depletion

The effect of pump depletion on the value of h_s for collimated beams is shown in Fig. 5.3. The dependence is similar to that in the case of no pump depletion and the peak value of h_s still occurs at δ_1 close to 0, although the values of h_s are smaller for K_2 close to or greater than 1.

The dependence of h_s on δ_1 for *focused* pump beam with no pump depletion

For $K_2 = 0.01$ and with $N = 400$ the effect of increasing the pump focusing parameter ξ_p is shown in Fig. 5.4 for ξ_p equal to 0.1, 0.5, 1, 3, 5 and 10. The

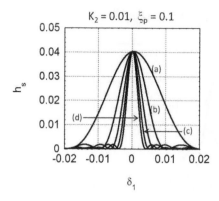

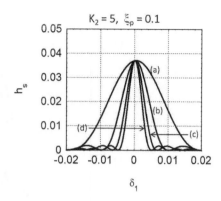

FIGURE 5.2: The parameter h_s plotted against δ_1 for the collimated beam case $\xi_p = 0.1$, for different values of N: (a) $N = 100$, (b) $N = 200$, (c) $N = 300$, (d) $N = 400$ (e) $N = 500$.

FIGURE 5.3: The parameter h_s plotted against δ_1 for the collimated beam case $\xi_p = 0.1$, for $K_2 = 5$ and for different values of N: (a) $N = 100$, (b) $N = 200$, (c) $N = 300$, (d) $N = 400$ (e) $N = 500$

peak value of h_s increases with ξ_p increasing from 0.1 to 3 and then decreases for higher values of ξ_p. As focusing becomes increasingly tighter, the shapes of the h_s versus δ_1 curves change and the value of δ_1 at which h_s is maximized, (denoted by δ_m, say), increases.

Figure 5.5 shows the dependence of h_s on the parameter δ_1 for five different values of N (ranging from 100 to 500) for the focused pump beam ($\xi_p = 3$) with no pump depletion ($K_2 = 0.01$). Similar to the case of collimated beams, the widths of the h_s versus δ_1 curves decrease with increasing N. δ_m is for the focused case is non-zero, and it is found to be dependent on N, decreasing with increasing N.

The dependence of h_s on δ_1 for *focused* pump beam *with* pump depletion

Figure 5.6 shows the dependence of h_s on δ_1 for the focused beam case ($\xi_p = 3$) with pump depletion ($K_2 = 5$) for N ranging from 100 to 400. The results are similar to the case of no pump depletion except that the values of h_s are lower.

Denoting by h_{sm} the values of h_s optimized with respect to the parameter δ_1, the dependence of h_{sm} on ξ_p is shown in Fig. 5.7 for $N = 400$ and for three different values of the parameter K_2. The maximum of h_{sm} occurs at ξ_p approximately equal to 3 for the case of no pump depletion ($K_2 = 0.01$) as well as for the case of strong pump depletion ($K_2 = 5$), and the peak value of h_{sm} for the case of no pump depletion is 0.433, which is $4/\pi^2$ of 1.066, the maximum value of h_{sm} predicted by Boyd and Kleinman.

The dependence of δ_{1m} on N at ξ_p equal to 3 is shown in Fig. 5.8 and the

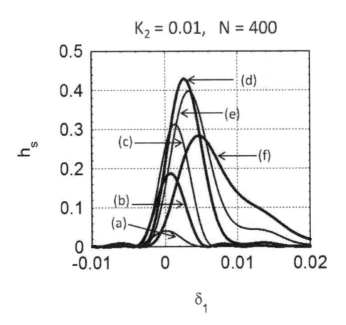

FIGURE 5.4: The parameter h_s plotted against δ_1 for different values of ξ_p, for $N = 400$ and $K_2 = 0.01$: (a) $\xi_p = 0.1$, (b) $\xi_p = 0.5$, (c) $\xi_p = 1$, (d) $\xi_p = 3$ (e) $\xi_p = 5$ and (f) $\xi_p = 10$.

dependence of δ_{1m} on ξ_p at N equal to 400 is shown in Fig. 5.9. For a given N and ξ_p, δ_{1m} was found to be approximately independent of K_2 (at least for K_2 up to 5).

For the optimally focused case ($\xi_p \approx 3$) and for N equal to 400, δ_{1m} is equal to 0.0027, implying that the thickness of the layers need to be increased by 0.27% from the value of the coherence length to obtain maximum conversion efficiency. As an example, the coherence length for frequency doubling of a 9.6μm wavelength laser in GaAs is 109.8μm. For the optimally focused pump case (i.e., with ξ_p equal to 3) in a quasi-phase matched structure with 400 layers, (i.e. for a structure of length about 4cm), the thickness of each layer needs to be 0.3μm (0.27%) larger than the coherence length, i.e., the layer thickness needs to be 110.1μm (instead of 109.8μm). Figure 5.5 shows that due to this small difference in the layer thickness, the conversion efficiency can be about 50% lower than the optimum value achievable.

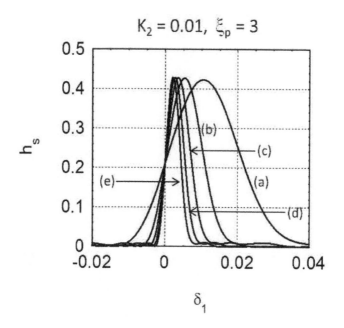

FIGURE 5.5: The parameter h_s plotted against δ_1 for the focused beam case $\xi_p = 3.0$, for $K_2 = 0.01$ and for different values of N: (a) $N = 100$, (b) $N = 200$, (c) $N = 300$, (d) $N = 400$ (e) $N = 500$

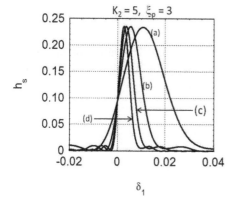

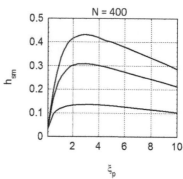

FIGURE 5.6: The parameter h_s plotted against δ_1 for the focused beam case with $\xi_p = 3.0$, for $K_2 = 5.0$, for different values of N: (a) $N = 100$, (b) $N = 200$, (c) $N = 300$, (d) $N = 400$.

FIGURE 5.7: h_{sm} plotted against the pump focusing parameter ξ_p, for $N = 400$ and for three values of K_2, (a) $K_2 = 0.01$, (b) $K_2 = 1$ and (c) $K_2 = 5$.

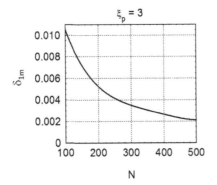

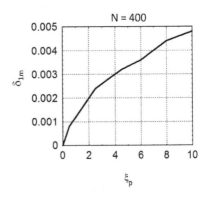

FIGURE 5.8: The values of δ_m plotted against N for $\xi_p = 3.0$ for $K_2 = 0.01$ and 5.0

FIGURE 5.9: The values of δ_m plotted against ξ_p for $N = 400$, and $K_2 = 0.01$ and 5.0

Bibliography

[1] G. D. Boyd and D. A. Kleinman, Parametric Interaction of Focused Gaussian Light Beams, *J. Appl. Phys.* **39**, 3597, 1968.

[2] V.G. Dmitriev, G.G. Gurzadyan, D.N. Nikogosyan, *Handbook of Nonlinear Optical Crystals*, Springer, Berlin, 1999.

[3] J. W. Goodman, *Introduction to Fourier Optics*, Third Edition, Roberts & Company, Greenwood Village, CO, 2005

[4] F. Zernike and J. E. Midwinter, *Applied Nonlinear Optics*, John Wiley and Sons, New York, 1973.

Bibliography

6

Optical Parametric Oscillation

6.1 Optical Parametric Oscillation

An Optical Parametric Oscillator (OPO) is a device in which one or more of the nonlinearly interacting light beams in an optical parametric generation (OPG) process are resonated, usually by placing the nonlinear medium between mirrors with high reflectivity coatings for the resonated wavelength or wavelengths. This process, called Optical Parametric Oscillation (and also designated OPO) was first demonstrated in 1965 by Giordmaine & Miller [1] and it is an important method for generating fixed frequency or tunable coherent radiation in parts of the spectrum where directly generated laser beams are not available.

An OPO is called a Singly Resonant Oscillator (SRO), Doubly Resonant Oscillator (DRO) or a Triply Resonant Oscillator (TRO) depending on the number of resonated beams at different frequencies. In an OPO, the highest frequency beam (which also has the highest power) is called the 'pump', the middle frequency beam is called the 'signal' and the lowest frequency beam is called the 'idler'. Inherent fluctuations in the cavity length (due to temperature or acoustic vibrations) can cause shifts in the output frequencies and fluctuations of output power of a DRO or a TRO, whereas the SRO is relatively stable against such cavity fluctuations [2]. In this book, we will consider only the case of the SRO and will determine the pump power threshold, conversion efficiency and the output powers for pump and signal beams which are plane waves, collimated circular Gaussian beams and focused circular Gaussian beams.

The expressions for the OPO thresholds and efficiencies are available in many books (such as Refs. [2], [3], [4] and [5]). They were first calculated by Siegman [6] assuming all three interacting beams to be plane waves with uniform transverse distribution. Siegman's analysis was extended to the case of collimated Gaussian beams in the absence of diffraction, Poynting vector walk-off or linear absorption but including pump depletion by Bjorkholm [7]. Boyd and Ashkin [8] provided a method by which the effects of focusing, walk-off and absorption can be included in the analysis, but instead of analytic expressions, the results were to be obtained in terms of (easily calculable) double integrals. The work of Boyd and Ashkin was later elaborated by Boyd and Kleinman [9], but in these semi-analytical treatments the effect of pump depletion was ignored, and only the threshold of the OPO process (and not the efficiencies) could be obtained.

Harris [10] showed that for plane waves, the threshold for oscillation of an OPO can be calculated by solving the coupled amplitude equations either by assuming that there is no pump depletion (A_3 independent of A_1 and A_2) which is valid at the initiation of oscillation, or by assuming that once oscillation is established, the slowly varying amplitude of the oscillating beam (say, the signal beam for an SRO) is independent of the pump irradiance value (A_2 independent of A_1 and A_3). Using the second method, not only the threshold, but also the efficiency of the OPO process can be calculated. In both approaches, one of the three amplitudes is independent of the other two, so only two coupled equations need to be solved. Analytical solutions can then be obtained for plane waves and collimated beams even in the presence of pump depletion, phase mismatch and linear absorption.

Here we will first present the analysis using the second approach of Harris [10] (also presented by Kreuzer [11] at about the same time) to determine the oscillation threshold and efficiency of a plane-wave SRO. The results will then be extended to the case where the pump and the signal beams have collimated Gaussian distribution of irradiance. Finally, the case of focused circular Gaussian beams will be considered, with the Boyd and Kleinman analysis extended to the case of large pump powers so that optimum focusing conditions that maximize the OPO efficiency can be obtained.

6.1.1 Plane wave analysis of OPO (SRO) including phase mismatch and pump depletion

If the signal beam (with frequency ω_2) is resonating, the amplitude u_2 is independent of u_1 and u_3 in the three-wave-mixing equations in their normalized form (given by Eqns. 4.19). Without the diffraction, walk-off and the absorption terms, these equations reduce to

$$\frac{\partial u_1}{\partial z_1} = i\frac{4\pi d_{\text{eff}} A_0 \ell}{\lambda_1 n_1} u_3 u_2^* e^{i\sigma z_1}$$

$$\frac{\partial u_3}{\partial z_1} = i\frac{4\pi d_{\text{eff}} A_0 \ell}{\lambda_3 n_3} u_1 u_2 e^{-i\sigma z_1} \tag{6.1}$$

where σ (equal to $(k_3 - k_2 - k_1)\ell$) is the phase mismatch parameter. Writing u_2 as

$$u_2 = w_2 e^{i\phi_2} \tag{6.2}$$

where w_2 and ϕ_2 are real, and assuming that the idler amplitude is zero at the entrance face of the nonlinear crystal (i.e., $u_1(0) = 0$), the solutions of

Eqns. 6.1 can be shown to be

$$u_1(z_1) = i\kappa_1 u_3(0)\sqrt{\frac{n_3\lambda_3}{n_1\lambda_1}}\frac{\sin(\beta z_1)}{\beta}e^{i\sigma z_1/2}e^{-i\phi_2}$$

$$u_3(z_1) = u_3(0)\left(\cos(\beta z_1) + \frac{i\sigma}{2}\frac{\sin(\beta z_1)}{\beta}\right)e^{-i\sigma z_1/2} \tag{6.3}$$

with

$$\beta \equiv \sqrt{\kappa_1^2 + \frac{\sigma^2}{4}} \quad \text{and} \quad \kappa_1 \equiv \frac{4\pi d_{\text{eff}}\ell A_0 w_2}{\sqrt{n_1 n_3 \lambda_1 \lambda_3}}. \tag{6.4}$$

There are a couple of ways to obtain these solutions. One way is to use the ABDP method given in Ref. [1] and described in detail in Chapter 4 for three coupled beams, which can be adapted in a straightforward manner to provide the solution for the case of two coupled beams here. The other way, described in Ref. [3] is to assume exponential solutions to the Eqns. 6.1, plugging them in these equations and solving the resulting algebraic equations for the coefficients of the exponentials.

Harris [10] and Kreuzer [11] find the oscillation condition by equating the fractional single pass signal gain (G) to the roundtrip fractional signal loss (ϵ_2). They define the fractional gain G as

$$G = \frac{\Delta I_2}{I_2(0)} \approx \frac{\Delta I_2}{I_2} \tag{6.5}$$

where ΔI_2, the change in the signal irradiance I_2 in a single pass through the nonlinear material is related through the Manley-Rowe relationship to ΔI_1, the change in the idler irradiance in a single pass,

$$\Delta I_2 = \frac{\lambda_1}{\lambda_2}\Delta I_1. \tag{6.6}$$

When there is no incident idler beam, the change in the idler irradiance in a single pass is equal to the idler irradiance at the exit, which is obtained from Eqn. 6.3 by setting $z_1 = 1$

$$\begin{aligned} \Delta I_1 &= 2n_1 c\varepsilon_0 A_0^2 |u_1(1)|^2 \\ &= 2n_1 c\varepsilon_0 A_0^2 |u_3(0)|^2 \frac{n_3\lambda_3}{n_1\lambda_1}\kappa_1^2\left(\frac{\sin\beta}{\beta}\right)^2 \\ &= \frac{\lambda_3}{\lambda_1}I_3(0)\frac{I_2}{I_b}\text{sinc}^2(\beta/\pi) \end{aligned} \tag{6.7}$$

where using the definition of κ_1 given in Eqn. 6.4 we have written

$$\kappa_1^2 = \frac{I_2}{I_b} \tag{6.8}$$

with

$$I_b = \frac{n_1 n_2 n_3 c \varepsilon_0 \lambda_1 \lambda_3}{8\pi^2 d_{\text{eff}}^2 \ell^2} \tag{6.9}$$

and used the relation $I_3(0) = 2n_3 c \varepsilon_0 |u_3(0)|^2$. The sinc function used here is again as defined in Eqn. 5.9.

The parameter β defined in Eqn. 6.4 can be written as

$$\beta = \sqrt{\frac{I_2}{I_b} + \frac{\sigma^2}{4}}. \tag{6.10}$$

Using Eqns. 6.5 - 6.7 we have

$$\begin{aligned} G = \frac{\lambda_1}{\lambda_2} \frac{\Delta I_1}{I_2} &= \frac{\lambda_1}{\lambda_2} \frac{\lambda_3}{\lambda_1} \frac{I_3(0)}{I_b} \text{sinc}^2(\beta/\pi) \\ &= \frac{\lambda_3}{\lambda_2} \frac{I_3(0)}{I_b} \text{sinc}^2(\beta/\pi). \end{aligned} \tag{6.11}$$

When oscillation of the signal is established, the fractional gain G is equal to the fractional loss ϵ_2, i.e.,

$$\frac{\lambda_3}{\lambda_2} \frac{I_3(0)}{I_b} \text{sinc}^2(\beta/\pi) = \epsilon_2 \tag{6.12}$$

so that

$$I_3(0) = \frac{\lambda_2}{\lambda_3} \frac{\epsilon_2 I_b}{\text{sinc}^2(\beta/\pi)}. \tag{6.13}$$

As pointed out by Harris [10], at the threshold of oscillation, $I_2 \approx 0$, which (using Eqn. 6.10) leads to β equal to $\sigma/2$, and the threshold value of pump irradiance to be

$$I_{3t} = \frac{\lambda_2}{\lambda_3} \frac{\epsilon_2 I_b}{\text{sinc}^2\left(\frac{\sigma}{2\pi}\right)} = \frac{\epsilon_2}{\text{sinc}^2\left(\frac{\sigma}{2\pi}\right)} \frac{n_1 n_2 n_3 c \varepsilon_0 \lambda_1 \lambda_2}{8\pi^2 d_{\text{eff}}^2 \ell^2}. \tag{6.14}$$

If N_I denotes the ratio of $I_3(0)$ to I_{3t}, i.e., the number of times the incident pump irradiance is above threshold value of I_{3t}, we have

$$N_I \equiv \frac{I_3(0)}{I_{3t}} = \frac{\text{sinc}^2\left(\frac{\sigma}{2\pi}\right)}{\text{sinc}^2(\beta/\pi)}. \tag{6.15}$$

When the phase mismatch σ is equal to zero, the expression for the threshold value of pump irradiance (Eqn. 6.14) matches the expression obtained assuming no pump depletion, i.e., by solving the coupled three wave mixing equations (Eqn. 6.1) with u_3 constant ([3], [4]).

Also, from the definition of β in Eqn. 6.4, $\beta = \kappa_1$ when σ goes to zero, so that from Eqn. 6.15,

$$
\begin{aligned}
N_I(\sigma \to 0) &= \frac{1}{\text{sinc}^2(\kappa_1/\pi)} \\
&= \frac{\kappa_1^2}{\sin^2 \kappa_1} = \frac{I_2/I_b}{\sin^2\left(\sqrt{I_2/I_b}\right)}.
\end{aligned}
\tag{6.16}
$$

6.1.2 SRO efficiency and threshold for collimated Gaussian beams

To extend the plane wave analysis presented above to the case of collimated pump and signal beams, we replace Eqn. 6.2 by

$$
u_2(r) = w_{20} e^{i\phi_2} e^{-\frac{r^2}{2r_{02}^2}} = w_{20} e^{i\phi_2} e^{-mS}
\tag{6.17}
$$

and replace the incident pump amplitude $u_3(0)$ by

$$
u_3(0, r) = u_{30} e^{-\frac{1}{2}(r/r_{03})^2} = u_{30} e^{-S}.
\tag{6.18}
$$

where we have defined

$$
S \equiv \frac{1}{2}(r/r_{03})^2, \quad \text{and} \quad m \equiv (r_{03}/r_{02})^2.
\tag{6.19}
$$

If I_{20} denotes the peak irradiance of the signal beam, we have

$$
I_2 = I_{20} e^{-2mS}.
\tag{6.20}
$$

With the focusing parameter of the signal beam defined as

$$
\xi_2 \equiv \frac{\ell}{2z_{02}} = \frac{\ell\lambda_2}{4\pi r_{02}^2 n_2}
\tag{6.21}
$$

we have

$$
\kappa_1^2 = \frac{I_2}{I_b} = \frac{I_{20}}{I_b} e^{-2mS} = \frac{P_2}{P_b} \xi_2 e^{-2mS}
\tag{6.22}
$$

where

$$
P_2 = \pi r_{02}^2 I_{20} \quad \text{and} \quad P_b \equiv \frac{n_1 n_3 c \varepsilon_0 \lambda_1 \lambda_2 \lambda_3}{32\pi^2 d_{\text{eff}}^2 \ell}.
\tag{6.23}
$$

The single pass gain in signal power ΔP_2 is then calculated, as in Ref. [8] using the relation

$$
\begin{aligned}
\Delta P_2 &= -2\omega_2 \, \mathcal{I}m \int d\mathbf{r} E_2^* \, P^{\omega_2} \\
&= -8d_{\text{eff}}\varepsilon_0\omega_2 \, \mathcal{I}m \int d\mathbf{r} E_2^* \, E_1^* E_3 \\
&= -8d_{\text{eff}}\varepsilon_0\omega_2 A_0^3 r_{03}^2 \ell \, \mathcal{I}m \int dx_1 dy_1 dz_1 u_2^* \, u_1^* u_3 e^{i\sigma z_1} \quad (6.24)
\end{aligned}
$$

where $\mathcal{I}m$ stands for the imaginary part of the integral and normalized coordinates defined in Eqns. 4.15 are used, with $r_0 = r_{03}$. Using the expressions for u_1, u_2, u_3 and κ_1 from Eqns. 6.3, 6.17 and 6.18 and 6.22, we obtain (after some algebra, and taking the imaginary part)

$$
\begin{aligned}
\Delta P_2 &= 8d_{\text{eff}}\varepsilon_0\omega_2 A_0^3 r_{03}^2 \ell |u_{30}|^2 w_{20} \sqrt{\frac{n_3\lambda_3}{n_1\lambda_1} \frac{P_2}{P_b}} \xi_2 \\
&\quad \times \frac{2\pi}{\beta} \int_0^\infty dS e^{-2S(m+1)} \int_0^1 dz_1 \sin(\beta z_1) \cos(\beta z_1) \\
&= 2\frac{\lambda_3}{\lambda_2} \frac{P_2 P_3(0)}{P_b} \xi_2 \int_0^\infty dS e^{-2S(m+1)} \text{sinc}^2(\beta/\pi). \quad (6.25)
\end{aligned}
$$

To derive Eqn. 6.25, the following relationships were used:

$$
A_0^2 |u_{30}|^2 \pi r_{03}^2 = \frac{P_3}{2n_3 c\varepsilon_0}
$$

$$
A_0 w_{20} = \sqrt{\frac{P_2}{2n_2 c\varepsilon_0 \pi r_{02}^2}} = \sqrt{\frac{2P_2\xi_2}{c\varepsilon_0 \ell \lambda_2}}.
$$

The oscillation condition for the signal beam is then written as

$$
\epsilon_2 = \frac{\Delta P_2}{P_2} = 2\frac{\lambda_3}{\lambda_2} \frac{P_3(0)}{P_b} \xi_2 \int_0^\infty dS e^{-2S(m+1)} \text{sinc}^2(\beta/\pi) \quad (6.26)
$$

with

$$
\begin{aligned}
\beta^2 &= \frac{P_2}{P_b}\xi_2 e^{-2Sm} + \frac{\sigma^2}{4} \\
&= \psi e^{-2Sm} + \frac{\sigma^2}{4} \quad (6.27)
\end{aligned}
$$

where we have defined

$$
\psi \equiv \frac{P_2}{P_b}\xi_2 \quad (6.28)
$$

and used Eqns. 6.27 and 6.22. Using Eqn. 6.26 we find that the oscillating signal power P_2 (contained in β) is dependent on the incident pump power

$P_3(0)$ through the relation

$$P_3(0) = \epsilon_2 \frac{\lambda_2}{\lambda_3} \frac{P_b}{2\xi_2 \int_0^\infty dS e^{-2S(m+1)} \mathrm{sinc}^2(\beta/\pi)}. \qquad (6.29)$$

At the onset of oscillation, i.e., at threshold, the signal power is small, i.e., $P_2 \approx 0$ and $\beta \approx \sigma$, thus providing the threshold value of the pump power P_{3t} for the collimated case to be

$$
\begin{aligned}
P_{3t} &= \epsilon_2 \frac{\lambda_2}{\lambda_3} \frac{P_b}{2\xi_2 \int_0^\infty dS e^{-2S(m+1)} \mathrm{sinc}^2 \dfrac{\sigma}{2\pi}}. \\
&= \epsilon_2 \frac{\lambda_2}{\lambda_3} \frac{P_b(m+1)}{\xi_2 \, \mathrm{sinc}^2 \dfrac{\sigma}{2\pi}} \\
&= \epsilon_2 \frac{n_1 n_2 n_3 c \varepsilon_0 \lambda_1 \lambda_2}{16 \pi d_{\mathrm{eff}}^2 \ell^2} \left(\frac{w_{02}^2 + w_{03}^2}{\mathrm{sinc}^2 \dfrac{\sigma}{2\pi}} \right) \qquad (6.30)
\end{aligned}
$$

where the beam waists w_{02} and w_{03} are related to r_{02} and r_{03} by

$$w_{02} = \sqrt{2} r_{02} \quad \text{and} \quad w_{03} = \sqrt{2} r_{03}. \qquad (6.31)$$

If R denotes the reflectivity of the output coupling mirror of the resonator at the signal wavelength, then for low values of other losses,

$$\epsilon_2 \approx 1 - R. \qquad (6.32)$$

The external power conversion efficiency for the signal beam is then

$$\eta_2 \equiv \frac{(1-R)P_2}{P_3(0)} \approx \frac{\epsilon_2 P_2}{P_3(0)}. \qquad (6.33)$$

Denoting by N_P the number of times the incident pump power $P_3(0)$ is above the threshold P_{3t}, we have from Eqns. 6.29 and 6.30

$$
\begin{aligned}
N_P \equiv \frac{P_3(0)}{P_{3t}} &= \frac{\mathrm{sinc}^2 \dfrac{\sigma}{2\pi}}{2(m+1) \displaystyle\int_0^\infty dS e^{-2S(m+1)} \mathrm{sinc}^2(\beta/\pi)} \\
&= \frac{\mathrm{sinc}^2 \dfrac{\sigma}{2\pi}}{2(m+1) \displaystyle\int_0^\infty dS e^{-2S(m+1)} \mathrm{sinc}^2 \left(\dfrac{\sqrt{\psi e^{-2Sm} + \dfrac{\sigma^2}{4}}}{\pi} \right)}.
\end{aligned}
$$

$$(6.34)$$

Using the middle equation of Eqn. 6.30 and Eqn. 6.33

$$
\begin{aligned}
\psi \equiv \xi_2 \frac{P_2}{P_b} &= \xi_2 \frac{P_2}{P_3(0)} \frac{P_3(0)}{P_{3t}} \frac{P_{3t}}{P_b} \\
&= \xi_2 \frac{P_2}{P_3(0)} \frac{P_3(0)}{P_{3t}} \epsilon_2 \frac{\lambda_2}{\lambda_3} \frac{m+1}{\xi_2 \operatorname{sinc}^2 \dfrac{\sigma}{2\pi}} \\
&= \eta_2 N_P \frac{\lambda_2}{\lambda_3} \frac{(m+1)}{\operatorname{sinc}^2 \dfrac{\sigma}{2\pi}}.
\end{aligned}
\tag{6.35}
$$

Thus

$$
\eta_2 = \frac{\lambda_3}{\lambda_2} \frac{\psi \operatorname{sinc}^2 \dfrac{\sigma}{2\pi}}{N_P (m+1)}.
\tag{6.36}
$$

For given values of σ and m, the variable ψ can be obtained as a function of N_P using Eqn. 6.34. The equation has real valued solutions only for $N_P \geq 1$. With ψ thus known as a function of N_P, the external conversion efficiency η_2 as a function of N_P is found from Eqn. 6.36. The results of these calculations, given in the next subsection, provide the maximum values of the SRO conversion efficiencies obtainable from collimated Gaussian beams in the presence of phase mismatch. Beam walk-off and losses from absorption, scattering or additional reflection of the different beams can only lower the values of the conversion efficiency.

6.1.3 Results for the case of collimated Gaussian beams including phase mismatch

Values of the external signal power conversion efficiency η_2 are plotted in Figs. 6.1 for N_P from 1 to 10, i.e., pump power ranging from the threshold to 10 times the threshold value, for two values of δ, the 'quantum defect' parameter, which was defined (in Eqn. 4.90) as the signal to pump wavelength ratio, and for various values of m, with phase mismatch parameter σ equal to zero. As δ increases from 1.1 to 1.5, the peak achievable efficiency decreases from 90% to 67%, showing that the peak achievable efficiency is equal to the inverse of δ. For a given value of δ, as the pump to signal area ratio m increases, the efficiency goes down.

Figures 6.2 show the dependence of η_2 on N_p for two values of the wavelength ratio δ, for σ ranging from 0 to 5 and with m equal to 0.1 (or smaller). As expected, with increasing values of phase mismatch parameter σ, the signal conversion efficiency η_2 is decreased.

6.1.4 SRO with focused Gaussian beams

Equation 6.30 shows that to lower the pump power threshold of an OPO it is necessary to reduce the size of the pump and the signal beams, i.e., to focus

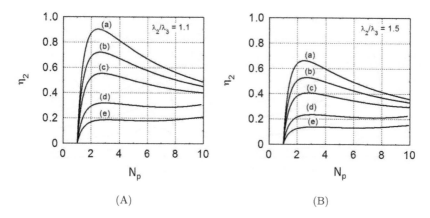

FIGURE 6.1: The external signal conversion efficiency plotted as a function of N_p for $\sigma = 0$ and for different values of m : (a) 0.1 (or smaller), (b) 1, (c) 2, (d) 5 and (e) 10. $\delta = \lambda_2/\lambda_3$ is equal to 1.1 for (A) and to 1.5 (B).

them tightly. However, for the tight focusing cases (i.e., with ξ_2 or ξ_3 close to and larger than 1), the collimated beam approximation is not valid and the diffraction terms in the coupled wave equations cannot be ignored. The solution of the equations including the diffraction as well as the absorption and walk-off terms can be obtained numerically using computer programs similar to those developed to analyze the focused SFG, SHG or DFG cases, which were discussed in the last chapter.

We assume here that both the pump beam incident on the nonlinear crystal and the resonating signal beam have circular Gaussian cross-sections and that both are focused at the center of the nonlinear crystal - these assumptions being made only to reduce the number of independent variables in the analysis. Redefining normalized coordinates x_1 and y_1 in terms of the pump beam radius r_{03} as

$$x_1 \equiv \frac{x}{r_{03}} \qquad y_1 \equiv \frac{y}{r_{03}} \tag{6.37}$$

we write the normalized pump irradiance at the entrance face of the nonlinear crystal in terms of the normalized coordinates x_1, y_1 and z_1 as

$$u_3(x_1, y_1, 0) = \frac{u_{30}}{1 - i\xi_3(1 - \mu)} e^{-\frac{1}{2}\frac{x_1^2 + y_1^2}{1 - i\xi_3(1 - \mu)}} \tag{6.38}$$

and the normalized amplitude of the signal beam through the crystal as

$$u_2(x_1, y_1, z_1) = \frac{u_{20}}{1 + i\xi_2[2z_1 - (1 - \mu)]} e^{-\frac{m}{2}\frac{x_1^2 + y_1^2}{1 + i\xi_2[2z_1 - (1 - \mu)]}} \tag{6.39}$$

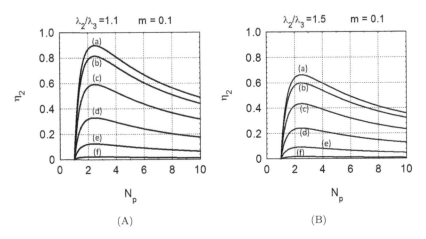

FIGURE 6.2: The external signal conversion efficiency plotted as a function of N_p for $m = 0.1$ for different values of σ : (a) 0, (b) 1, (c) 2, (d) 3, (e) 4, and (f) 5. $\delta = \lambda_2/\lambda_3$ is equal to 1.1 for (A) and to 1.5 for (B)

where $m = (r_{03}/r_{02})^2$. Furthermore, the amplitude u_{20} is assumed here to be real.

The parameter μ in Eqn. 6.39 is defined as in Ref. [9] to be

$$\mu \equiv \frac{\ell - 2f}{\ell} \tag{6.40}$$

where f denotes the distance of the focus of the signal beam from the entrance face (which is at $z_1 = 0$). For focusing at the center of the crystal, $f = \ell/2$ so that $\mu = 0$. In certain resonators, if the beams are focused at the exit face of the nonlinear medium, then $f = \ell$ and we will have $\mu = -1$.

To determine the pump power threshold and signal efficiency for the focused case a further normalization of the already normalized field amplitudes is useful. Three new dimensionless variables denoted by U_1, U_2 and U_3 are defined as

$$U_1 \equiv \frac{u_1}{u_{30}}, \quad U_2 \equiv \frac{u_2}{u_{20}}, \quad \text{and,} \quad U_3 \equiv \frac{u_3}{u_{30}} \tag{6.41}$$

and the coupled equations for the idler and pump beams, obtained from Eqns. 4.19 are then written in terms of the new variables as

$$\frac{\partial U_1}{\partial z_1} = i\xi_3 \nabla_{T_1}^2 U_1 + \sqrt{2k_3 l \xi_3} \tan \rho_1 \frac{\partial U_1}{\partial x_1} - \frac{a_1}{2} U_1 + i\frac{\lambda_3 n_3}{\lambda_1 n_1} u_{20} \kappa U_3 U_2^* e^{i\sigma z_1}$$

$$\frac{\partial U_3}{\partial z_1} = i\frac{k_1}{k_3}\xi_1 \nabla_{T_1}^2 U_3 + \sqrt{2k_3 l \xi_3} \tan \rho_3 \frac{\partial U_3}{\partial x_1} - \frac{a_3}{2} U_1 + i u_{20} \kappa U_1 U_2 e^{-i\sigma z_1}$$

$$\tag{6.42}$$

where κ (defined earlier in Chapter 4, Eqn. 4.137) is given by

$$\kappa = \frac{4\pi A_0 d_{\text{eff}} l}{\lambda_3 n_3}.$$

(6.43)

Since the peak irradiance of the signal beam is given by

$$I_{20} = 2n_2 c \varepsilon_0 (A_0 u_{20})^2 = \frac{P_2}{\pi r_{02}^2} = \frac{4n_2 P_2 \xi_2}{l \lambda_2}$$

(6.44)

we have

$$\begin{aligned}
\kappa^2 u_{20}^2 &= \frac{16\pi^2 d_{\text{eff}}^2 l^2}{\lambda_3^2 n_3^2} \frac{4n_2 P_2 \xi_2}{2n_2 c \varepsilon_0 l \lambda_2} \\
&= \frac{P_2}{P_c} \xi_2 \\
&= K_3 \xi_2
\end{aligned}$$

(6.45)

where

$$K_3 = \frac{P_2}{P_c} \quad \text{and} \quad P_c \equiv \frac{\lambda_2 \lambda_3^2 n_3^2 c \varepsilon_0}{32\pi^2 d^2 l}.$$

(6.46)

With the walk-off parameters for the idler and the pump beams defined as

$$B_1 = \frac{\sqrt{l k_3}}{2} \tan\rho_1 \quad \text{and} \quad B_3 = \frac{\sqrt{l k_3}}{2} \tan\rho_3$$

(6.47)

the Eqns. 6.42 for the idler and the pump beams can be rewritten as

$$\begin{aligned}
\frac{\partial U_1}{\partial z_1} &= i\xi_3 \nabla_{T_1}^2 U_1 + 2\sqrt{2\xi_3} B_1 \frac{\partial U_1}{\partial x_1} - \frac{a_1}{2} U_1 + i\frac{\lambda_3 n_3}{\lambda_1 n_1} \sqrt{K_3 \xi_2} U_3 U_2^* e^{i\sigma z_1} \\
\frac{\partial U_3}{\partial z_1} &= i\frac{k_1}{k_3} \xi_1 \nabla_{T_1}^2 U_3 + 2\sqrt{2\xi_3} B_3 \frac{\partial U_3}{\partial x_1} - \frac{a_3}{2} U_1 + i\sqrt{K_3 \xi_2} U_1 U_2 e^{-i\sigma z_1}.
\end{aligned}$$

(6.48)

Assuming the incident pump beam and the resonated signal beam are both focused at the center of the crystal, the initial conditions for the variables U_1 and U_3 at $z_1 = 0$ are given by

$$\begin{aligned}
U_1(0) &= 0 \\
U_3(x_1, y_1, 0) &= \frac{1}{1 - i\xi_3} e^{-\frac{1}{2} \frac{x_1^2 + y_1^2}{1 - i\xi_3}}
\end{aligned}$$

(6.49)

and the expression for U_2 is given by

$$U_2(x_1, y_1, z_1) = \frac{1}{1 + i\xi_2 \{2z_1 - 1\}} e^{-\frac{m}{2} \frac{x_1^2 + y_1^2}{1 + i\xi_2 \{2z_1 - 1\}}}.$$

(6.50)

The Eqns. 6.48 can be solved numerically using the split step method with ξ_2, ξ_3, σ, B_1, B_3, a_1, a_2, a_3, $\dfrac{\lambda_3 n_3}{\lambda_1 n_1}$ and K_3 as input parameters to the computer code.

Using Eqn. 6.24 the signal beam power gain is given by

$$
\begin{aligned}
\Delta P_2 &= -8d_{\text{eff}}\varepsilon_0 \omega_2 A_0^3 r_{03}^2 \ell \ \mathcal{I}m \int dx_1 dy_1 dz_1 U_2^* \ U_1^* U_3 e^{i\sigma z_1} \\
&= 8d_{\text{eff}}\varepsilon_0 \omega_2 A_0^3 r_{03}^2 \ell |u_{30}|^2 u_{20} \kappa_3 \mathcal{F}(K_3)
\end{aligned}
\tag{6.51}
$$

where

$$
\begin{aligned}
\mathcal{F}(K_3) &\equiv -\frac{1}{\kappa_3} \ \mathcal{I}m \int dx_1 dy_1 dz_1 U_2^* \ U_1^* U_3 e^{i\sigma z_1}, \\
\kappa_3 &\equiv \kappa u_{20} = \sqrt{K_3 \xi_2}.
\end{aligned}
\tag{6.52}
$$

From Eqn. 6.51 it can be shown that

$$
\begin{aligned}
\Delta P_2 &= \frac{2}{\pi} \frac{\lambda_3}{\lambda_2} \frac{P_2}{P_c} \xi_2 P_3(0) \mathcal{F}(K_3) \\
&= \epsilon_2 P_2
\end{aligned}
\tag{6.53}
$$

where the last equation holds for oscillation of the signal beam.

Further defining a function $g(K_3)$

$$
g(K_3) \equiv \frac{\pi}{2} \frac{\lambda_2}{\lambda_3} \frac{1}{\xi_2 \mathcal{F}(K_3)}
\tag{6.54}
$$

we obtain the oscillation condition

$$
P_3(0) = \epsilon_2 P_c g(K_3).
\tag{6.55}
$$

At threshold, K_3 goes to 0, thereby providing the threshold value of the pump power as

$$
P_{3t} = \epsilon_2 P_c g(0).
\tag{6.56}
$$

Since N_P denotes the number of times the incident pump power is above the threshold power, i.e., the ratio of $P_3(0)$ to P_{3t}, we have

$$
N_P \equiv \frac{P_3(0)}{P_{3t}} = \frac{g(K_3)}{g(0)}
\tag{6.57}
$$

and the external signal conversion efficiency is

$$
\begin{aligned}
\eta_2 &= \frac{\epsilon_2 P_2}{P_3(0)} \\
&= \frac{\epsilon_2 P_c}{P_3(0)} \frac{P_2}{P_c} \\
&= \frac{K_3}{g(K_3)}.
\end{aligned}
\tag{6.58}
$$

Since $g(K_3)$ can be numerically evaluated from Eqn. 6.54 for all values of K_3 including K_3 equal to zero, the threshold and efficiency of the SRO in the focused case can be obtained from Eqns. 6.56 and 6.58 for different values of ξ_2, ξ_3 and σ.

The function $g(K_3)$ defined in Eqn. 6.54 contains the dependence of the pump power threshold and external signal power conversion efficiency of a singly resonant OPO on various parameters. These include, the focusing parameters ξ_2 and ξ_3 of the signal and the pump beams, the phase mismatch σ, the absorption coefficients and beam walk-off angles. Eqn. 6.56 shows that the pump power threshold is proportional to the value of $g(0)$, which is the limit of $g(K_3)$ as K_3 goes to 0 and Eqn. 6.58 shows that the external signal conversion efficiency is inversely proportional to $g(K_3)$.

A computer program was used to calculate $g(K_3)$ by solving Eqns. 6.48 using the split-step technique (using FFT for the linear propagation and finite difference technique for the nonlinear propagation parts). This approach is further described in Chapter 7. Using the values of the normalized field amplitudes, the triple integral in Eqn. 6.52 was numerically evaluated. Some results of the calculation are shown below.

6.1.5 Results of optimization of the focusing parameters in an SRO

Assuming no beam walk-offs and no absorption, the value of $g(K3)$ was determined as a function of ξ_2, ξ_3, σ, δ and K_3. The values of $g(K3)$ optimized (minimized) with respect to σ is denoted by $g_m(K_3)$, the values of $g_m(K_3)$ minimized with respect to ξ_2 are denoted by $g_{mm}(K_3)$, with the values of ξ_2 at which the minimum occurs being denoted by ξ_{2m}; and finally the values of $g_{mm}(K_3)$ minimized with respect to ξ_1 are denoted by $g_{mmm}(K_3)$.

The values of $g_{mm}(K_3)$ obtained at low values of K_3 (less than 0.0001) are denoted by $g_{mm}(0)$ and are plotted in Fig. 6.3 against the ξ_3 for three values of the wavelength ratio $\delta = \lambda_2/\lambda_3$: 1.1, 1.5 and 2.0. It is seen that the minimum of $g_{mm}(0)$ occurs at ξ_3 near 3.0. Corresponding values of ξ_{2m} plotted as a function of ξ_3 are shown in Fig. 6.4 for the case of $\delta = 1.5$, and it is seen that at ξ_3 around 3, the value of ξ_{2m} is also approximately 3, as predicted by the Boyd and Kleinman analysis.

The values of $g_{mmm}(0)$, optimized with respect to σ, ξ_2 and ξ_3 are plotted in Fig. 6.5 as a function of the wavelength ratio δ. It is seen that at wavelength ratio close to 1, i.e., with small δ, the values of $g_{mmm}(0)$ are high. As the signal to pump wavelength ratio increases, say from 1.05 to 2, $g_{mmm}(0)$, and with it the pump power threshold, decreases by 2 orders of magnitude.

Finally, optimized values $g_{mm}(K_3)$ are determined as a function of K_3 for a range of K_3 values, including the very small values of K_3 (say below 0.0001). From the values at low K_3, the limiting values $g_{mm}(0)$ are found. At any value of K_3, Eqn. 6.57 is used to find the value of N_p and Eqn. 6.58 is used to find optimized value of the external signal conversion efficiency η_{2mm}. From such

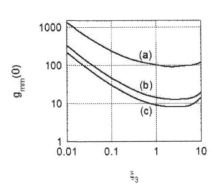

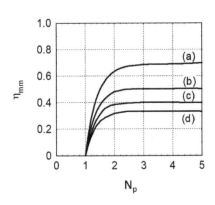

FIGURE 6.3: The parameter $g_{mm}(0)$ plotted against ξ_3 for different values of the wavelength ratio $\delta = \lambda_2/\lambda_3$: (a) $\delta = 1.1$, (b) $\delta = 1.5$, and (c) $\delta = 2.0$

FIGURE 6.4: The parameter ξ_{2m} plotted against ξ_3 for $\delta = 1.5$.

a table of K_3, N_p and η_{2mm}, the values of η_{2mm} can be plotted as a function of N_p, and such plots are shown in Fig. 6.6 for four values of δ. Optimized external signal conversion efficiency η_{2mm} is found to saturate to a value of $1/\delta$ when N_p exceeds about 2.5.

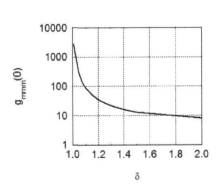

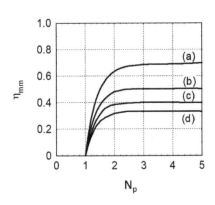

FIGURE 6.5: The parameter $g_{mmm}(0)$ plotted against the wavelength ratio $\delta = \lambda_2/\lambda_3$.

FIGURE 6.6: Optimized external signal conversion efficiency η_{mm} plotted against N_p, the number of times the pump power is above the pump power threshold, for different values of δ: (a) $\delta = 1.5$, (b) $\delta = 2.0$, (c) $\delta = 2.5$ and (d) $\delta = 3.0$

Bibliography

[1] J. A. Giordmaine and R. C. Miller, Tunable Coherent Oscillation in LiNbO$_3$ at Optical Frequencies, *Phys. Rev. Lett.* **14**, 973, 1965.

[2] A. Yariv, *Quantum Electronics*, John Wiley and Sons, Inc., 1978.

[3] Y. R. Shen, *The Principles of Nonlinear Optics*, John Wiley and Sons, New York, 1984.

[4] R. L. Sutherland, *Handbook of Nonlinear Optics*, Marcel Dekkar, Inc., 1996.

[5] P. E. Powers, *Fundamentals of Nonlinear Optics*, CRC Press, Boca Roaton, 2011.

[6] A. E. Siegman, Nonlinear Optical Effects: An Optical Power Limiter, *Appl. Opt.* **1**, 739–744, 1962.

[7] J. E. Bjorkholm, Some Effects of Spatially Nonluniform Pumping in Pulsed Optical Parametric Oscilaltors, *IEEE J. Quantum Electron.* **7**, 109, 1971.

[8] G. D. Boyd and A. Ashkin, Theory of Parametric Oscillator Threshold with Single-Mode Optical Masers and Observation of Amplification in LiNbO$_3$, *Phys. Rev.* **146**, 187, 1966.

[9] G. D. Boyd and D. A. Kleinman, Parametric Interaction of Focused Gaussian Light Beams, *J. Appl. Phys.* **39**, 3597, 1968.

[10] S. E. Harris, Tunable Optical Parametric Oscillators, *Proc. IEEE* **57**, 2096, 1969.

[11] L. B. Kreuzer, Single and Multimode Oscillation of the Singly Resonant Optical Parametric Oscillator, *Proc. Joint Conf. Lasers and Opto Electronics*, University of Southampton, Southampton, England, 52–63, 1969.

[12] J. A. Armstrong, N. Bloembergen, J. Ducuing and P. S. Pershan, Interactions between Light Waves in a Nonlinear Dielectric, *Phys. Rev.* **127**, 1918, 1962.

7

Numerical Beam Propagation Methods

7.1 Introduction

Analytical or semi-analytical solutions of the beam coupling equations, such as the ABDP expressions [1] given in terms of Jacobian elliptic functions or the Boyd and Kleinman solutions [2] in terms of double integrals, provide insight, allow quick analysis and can serve as checks for the accuracy of the full numerical solutions of the equations. However, in many cases of interest, simplifying assumptions cannot be made and analytic solutions cannot be found. In these cases numerical beam propagation methods need to be used to find the solutions.

Much work has been done to find efficient methods of numerically solving the problem of electromagnetic field propagation through air, fibers or optical systems. This chapter contains an overview of two transform methods used to solve the problem of propagation of light through linear media for radially symmetric and arbitrarily shaped light distributions. For propagation in *non-linear* media, a splitting method is used to separate the linear and nonlinear propagation operations. This chapter ends with examples of numerical beam propagation methods applied to two examples, one for the linear propagation case and the other for the case of propagation in a nonlinear medium.

The general equation describing the propagation of light of electric field strength A and traveling in the z direction is given by the expression

$$\frac{\partial A}{\partial z} = \widehat{P} A + \widehat{NL} \tag{7.1}$$

where $\widehat{P}$ is a linear propagation operator and $\widehat{NL}$ is a nonlinear operator on the field A. $\widehat{NL}$ may be a function of A and/or dependent on additional coupled fields. The electric field can be expressed by

$$A = A_0(r,t)e^{i\phi(r)} \tag{7.2}$$

or an arbitrary spatial distribution

$$A = A_0(x,y,t)e^{i\phi(x,y)}. \tag{7.3}$$

The amplitude A_0 and phase ϕ of the field A can have radial symmetry (Eqn.

237

7.2) or arbitrary transverse distributions (Eqn. 7.3). The phase of the field is assumed to be time dependent.

Assuming the refractive index of the medium of light propagation to be given by n_0, the irradiance is given by

$$I = 2n_0\varepsilon_0 c|A|^2. \tag{7.4}$$

Typical experimental outputs such as measurements of power or energy consist of the respective areal and temporal integrations of the irradiance.

Additional assumptions used in this chapter are:

1. Slowly varying envelope/paraxial approximation is made where

$$\frac{\partial^2 A}{\partial z^2} \ll k\frac{\partial A}{\partial z} \tag{7.5}$$

$$\frac{\partial^2 A}{\partial z^2} \approx 0. \tag{7.6}$$

2. The electro-magnetic field(s) are composed of single frequency, quasi-monochromatic light. The treatment here is not completely valid for ultrafast pulses of large bandwidths (i.e. valid for pulsewidths > 100 fs).

3. Vector notation is not used here for the electric fields as the light beams are assumed to be linearly polarized and only one transverse component of the field is considered.

7.2   Propagation in Linear Media

In this section, light propagation between parallel planes within a linear medium is described, with the $\widehat{NL}$ operator in Eqn. 7.1 set to zero. The electric field distribution of light is assumed to be known in one plane, say the plane denoted by P_1 in Fig. 7.1 located at $z = z_1$, and the goal is to efficiently and accurately calculate the field distribution in the the plane denoted by P_2 at $z = z_2$. While a variety of numerical techniques can be used to solve this problem, Sections 7.2.1 and 7.2.2 describe the use of transform methods for beam propagation calculations.

7.2.1   Hankel Transform Method

In an isotropic medium, where the refractive index is independent of propagation direction, the Hankel transform can be used to describe propagation of radially symmetric electric fields. This method can also be used for description

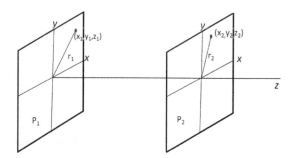

FIGURE 7.1: Propagation of light between parallel planes.

of propagation in an anisotropic medium when the refractive index is constant along the propagation direction, such as along a crystallographic axis.

We assume that the light distribution in plane P_1 is radially symmetric, and is described by a function $f(r_1, z_1)$ where r_1 is the radial coordinate of a point in the z_1 plane shown in Fig. 7.1. Due to the orthogonality relation of Bessel functions

$$\int_0^\infty J_0(mr) J_0(m'r) r \, dr = \frac{\delta(m - m')}{m} \tag{7.7}$$

any function $f(r)$ can be expressed as an integral over a function $g(\rho)$, where

$$f(r) = 2\pi \int_0^\infty \rho \, g(\rho) \, J_0(2\pi \rho r) \, d\rho \tag{7.8}$$

and

$$g(\rho) = 2\pi \int_0^\infty r \, f(r) \, J_0(2\pi \rho r) \, dr. \tag{7.9}$$

Equations 7.9 and 7.8 are known as the Hankel transform (HT) and its inverse Hankel transform (IHT) [3] with r the radial spatial coordinate and ρ the corresponding spatial frequency. The field distribution $f(r, z)$ at any plane can therefore be written as

$$f(r, z) = 2\pi \int_0^\infty \rho \, g(\rho, z) \, J_0(2\pi \rho r) \, d\rho. \tag{7.10}$$

and $f(r, z)$ must also satisfy the wave equation

$$\frac{1}{2ik} \nabla_T^2 f(r, z) + \frac{\partial f(r, z)}{\partial z} = 0. \tag{7.11}$$

Inserting Eqn. 7.10 in Eqn. 7.11 and using the relation

$$\nabla_T^2 = \frac{1}{r} \frac{\partial}{\partial r} \left(r \frac{\partial}{\partial r} \right) \tag{7.12}$$

we obtain

$$2ik\frac{\partial g(\rho, z)}{\partial z} = (2\pi\rho)^2 g(\rho, z) \qquad (7.13)$$

which can be solved to provide

$$g(\rho, z_2) = g(\rho, z_1)e^{-i\pi\rho^2\lambda\Delta z/n_0} \qquad (7.14)$$

where $\Delta z = z_2 - z_1$. The field distribution at the plane z_2 in terms of the radial coordinate r_2 is now obtained by inserting Eqn. 7.14 in the transform expression for the field at z_2

$$f(r_2, z_2) = 2\pi \int_0^\infty g(\rho, z_2) J_0(2\pi\rho r_2)\rho d\rho \qquad (7.15)$$

and using the transform expansion for $g(\rho, z_1)$ from Eqn. 7.9. The integral over ρ in Eqn. 7.15 can be easily evaluated [4] (Eqn. 6.631-4 page 717) to provide the field at plane z_2 in terms of the field at the plane z_1

$$f(r_2, z_2) = \frac{k}{i\Delta z} \int_0^\infty \left[f(r_1, z_1) \exp\left(\frac{ik}{2\Delta z}(r_1^2 + r_2^2)\right) \right] J_0\left(\frac{k r_1 r_2}{\Delta z}\right) r_1 dr_1. \qquad (7.16)$$

Equation 7.16 is also known as the Fresnel diffraction integral [3] for a radially symmetric field.

For a field distribution known at N points in plane z_1, the distribution at N points in plane z_2 can be obtained by directly evaluating the integral in Eqn. 7.16 through N^2 complex multiplication operations. This integral can also be evaluated through a much fewer number of steps ($2N \log(2N)$ complex operations) by using the discrete Hankel transform technique with $f(r_2, z_2)$ being the transform of the function within the square brackets in the integrand. However, when a large number of steps in the z direction are required, such as when using the split-step method as discussed in Sect. 7.3.1, it is computationally advantageous to use the following approach:

1. Given a distribution $f(r, z_1)$ with r going from 0 to r_{max}, obtain the transform $g(\rho, z_1)$ for ρ ranging from 0 to ρ_{max} (using Eqn. 7.9).

2. Multiply $g(\rho, z_1)$ by the factor $e^{-i\pi\rho^2\lambda\Delta z/n_0}$ to obtain $g(\rho, z_2)$ (as in Eqn. 7.14).

3. Calculate the transform of $g(\rho, z_2)$ to obtain $f(r, z_2)$ (using Eqn. 7.15).

In this scheme the Hankel transform needs to be performed repeatedly, making it important to have an efficient computational procedure for its evaluation. Much work has been done in this area from early work by Siegman on the quasi-fast Hankel transform [5] to the quasi-discrete Hankel transform approach of Li Yu [6]. The procedure developed by Li Yu and extended by others [7, 8] is presented here with a focus on the approach by Kai-Ming You et al. [8].

We describe here the method outlined by Kai-Ming You et al. [8] for efficient evaluation of Hankel transform. Given the number of radial points N at which the field is known in plane P_1 (from $r = 0$ to r_{max}), several parameters and arrays depending on N are first defined:

1. $J(N)$, the Nth zero of the zero order Bessel function $J_0(x)$. For $N = 1$ to 20, the values of $J(N)$ are given in reference [9] (page 409), with $J(1) = 2.4048$ and $J(20) = 62.04847$. For $N > 20$, the value of $J(N)$ is given by the formula from Ref. [9] (page 371)

$$J(N) = b + \frac{1}{8b} - \frac{4 \times 31}{3(8b)^3} + \frac{32 \times 3779}{15(8b)^5} - \frac{64 \times 6277237}{105(8b)^7} \quad (7.17)$$

where

$$b = \left(N - \frac{1}{4}\right)\pi. \quad (7.18)$$

2. $\alpha(N)$, the Nth zero of the first order Bessel function $J_1(x)$. For $N = 1$ to 20, the values of $\alpha(N)$ are given in reference [9] (page 409), with $\alpha(1) = 3.83171$ and $\alpha(20) = 63.61136$. For $N > 20$, the value of $\alpha(N)$ is given by the formula from reference [9] (page 371)

$$\alpha(N) = b_1 - \frac{3}{8b_1} + \frac{4}{(8b_1)^3} - \frac{7546}{(8b_1)^5} + \frac{3568000}{(8b_1)^7} \quad (7.19)$$

where

$$b_1 = \left(N + \frac{1}{4}\right)\pi. \quad (7.20)$$

With $n = 1$ to N, an array α is formed, with its nth element α_n having value equal to $\alpha(n)$.

3. k, the integer closest to $N/4$.

4. S, defined as

$$S = \frac{2}{|J_0(\alpha_k)|}\sqrt{1 + \sum_{n=1}^{N} \frac{J_0^2\left(\frac{\alpha_k \alpha_n}{J(N+1)}\right)}{J_0^2(\alpha_n)}}. \quad (7.21)$$

5. The $N \times N$ matrix c with elements

$$c_{mn} = \frac{2J_0\left(\frac{\alpha_m \alpha_n}{S}\right)}{S\,|J_0(\alpha_n)J_0(\alpha_m)|} \quad m = 1 \text{ to } N, n = 1 \text{ to } N. \quad (7.22)$$

6. The one-dimensional ($N \times 1$) array c_0 with elements

$$c_{0n} = \frac{2}{S \mid J_0(\alpha_n) \mid} \qquad n = 1 \text{ to } N. \tag{7.23}$$

7. The quantity c_{00} defined as

$$c_{00} = \frac{2}{S}. \tag{7.24}$$

8. The maximum value $\rho_{\max}$ of the spatial frequency component ρ defined as

$$\rho_{\max} = \frac{S}{2\pi r_{\max}} \tag{7.25}$$

where $\rho_{\max}$ is same as the parameter β in reference [8].

From the known values of the above constants and arrays, $(N+1) \times 1$ sized arrays F and G are formed, with elements F_n, G_m and the quantities F_0 and G_0 defined as

$$F_n \equiv \frac{f\left(\dfrac{\alpha_n}{2\pi\rho_{\max}}\right) r_{\max}}{J_0(\alpha_n)} \qquad n = 1 \text{ to } N \tag{7.26}$$

$$G_m \equiv c_{0m} F_0 + \sum_{n=1}^{N} c_{mn} F_n \qquad m = 1 \text{ to } N \tag{7.27}$$

with

$$F_0 \equiv f_0 r_{\max} \tag{7.28}$$

$$\text{and } G_0 \equiv c_{00} F_0 + \sum_{n=1}^{N} c_{0n} F(n) \tag{7.29}$$

where f_0 is the value of the function $f(r)$ at $r = 0$.

Given the arrays defined above, a one-dimensional ($N \times 1$) array g is now constructed, with element

$$g_1 = \frac{G_0}{\rho_{\max}} \tag{7.30}$$

and for $m = 2$ to N, elements

$$g_m = \frac{G_{m-1}}{\rho_{\max} \mid J_0(\alpha_{m-1}) \mid}. \tag{7.31}$$

Given an array of numbers f, which are the values of $f(r)$ at the N points r_n from $r = 0$ to $r_{\max}$, the array of numbers g constitutes the discrete Hankel transform (DHT) of f, at the spatial frequency coordinates $\rho(m) = \dfrac{\alpha_m}{2\pi r_{\max}}$.

From the known values of g_m at the known values of ρ_m, the continuous function $g(\rho)$ of a continuous variable ρ can be easily formed using an interpolation program (such as cubic spline).

We write the above procedure in a step-by-step form for clarity (and for ease of implementation in a computer program):

1. For a given value of N (the number of radial points) and $r_{\max}$, find the $N \times 1$ arrays c_0 and α, and the $N \times N$ array c and the constants S, $J(N+1)$ and $\rho_{\max}$.

2. Form an array of numbers p_n given by

$$p_n = \frac{\alpha_n}{2\pi\rho_{\max}} \qquad n = 1 \text{ to } N. \tag{7.32}$$

3. With the known initial distribution as the discrete function $f(r_n)$ of the radial variable $r_n = (n-1)r_{\max}/(N-1)$ defined for $n = 1$ to N, find the values of $f(p_n)$ through interpolation.

4. Multiply each term of the array $f(p_n)$ by $r_{\max}$ and divide by $J_0(\alpha_n)$ to form the new array F_n. Find the value of F_0 using Eqn. 7.28.

5. For $m = 1$ to N, form a new array G defined by Eqn. 7.27 using the values of F and also find the value G_0 from Eqn. 7.29.

6. From the array G form a new array g using Eqns. 7.30 and 7.31 with $\rho_{\max}$ given by Eqn. 7.25.

The array g is the desired DHT at the spatial frequency values $\rho(m) = \alpha(m)/(2\pi r_{\max})$. A subroutine, say called $\text{Hankel}(f, r_{\max}, g)$, to which the array f and the value $r_{\max}$ are the inputs and from which the array g is obtained as the output, can easily be written with the above algorithm.

The propagation of a field distribution f_1 on the plane P_1 to the distribution denoted by f_2 on the plane P_2 through a linear region of space having refractive index n_0 can now be described by the following procedure:

1. With f_1 known as a function of r_{P1}, form the array f_{1n} at the N radial points at $r_n = (n-1)r_{\max}/(N-1)$, $n = 1$ to N.

2. Call the subroutine $\text{Hankel}(f_{1n}, r_{\max}, g_{1n})$, which returns an array g_{1n} with n elements.

3. Multiply each term of the array g_{1n} by the term $\exp\left(-i2\pi^2 \dfrac{\Delta z}{n_0}\rho_n^2\right)$, where $\rho_n = \alpha_n/(2\pi r_{\max})$, to form a new array, named g_{2n}.

4. Since the procedures for the Hankel and the inverse Hankel transforms expressed in the form of Eqns. 7.9 and 7.8 are the same, the field distribution in the plane P_2 is now obtained by calling the same subroutine Hankel($g_{2n}, \rho_{max}, f_{2n}$), with ρ_{max} given by Eqn. 7.25. Here the array g_{2n} and the number ρ_{max} are the inputs and the array f_{2n} is the output.

5. The array f_{2n} thus obtained gives the values of the field in plane P_2 at the radial points $r_n = \alpha_n/(2\pi\rho_{max})$. From this array, the values of the function f_2 at the original set of radial points $r_n = (n-1)r_{max}/(N-1)$ are obtained by interpolation.

7.2.2 Fourier Transform Method

To describe propagation of light with a transverse distribution that is not necessarily radially symmetric, a two dimensional fast Fourier transform (2DFFT) method can be used. Even with a radially symmetric incident field, the 2DFFT is useful if there is Poynting vector walk-off, for example in anisotropic media.

The use of 2DFFTTs for electromagnetic field propagation is discussed in many recent publications such as in References [10, 11, 12], and 2DFFT algorithms are built-in operations in modern day mathematical and scientific computing environments such as MATLAB or IGOR. An overview of the approach is as follows.

The two-dimensional Fourier transform, $\mathfrak{F}$, of a complex function A of two variables, x and y is given in Ref. [3] as:

$$\mathfrak{F}(A(x,y)) \equiv \tilde{A}(x,y) = \iint A(x,y)\,\exp(-i2\pi(f_x x + f_y y))\,dx\,dy. \quad (7.33)$$

Starting with the propagation operator equation, Eqn. 7.1 with $\widehat{NL} = 0$

$$\frac{\partial A}{\partial z} = \widehat{P}A. \quad (7.34)$$

and taking the Fourier transform of each side of the above we obtain

$$\frac{\partial \tilde{A}}{\partial z} = \widehat{P}\tilde{A} \quad (7.35)$$

with $\widehat{P}$ defined as

$$\widehat{P} = \frac{i}{2k}\nabla_T^2 \quad (7.36)$$

applying $\widehat{P}$ to $\tilde{A}$

$$\begin{aligned}
\frac{\partial \tilde{A}}{\partial z} &= \frac{i}{2k}\left(-4\pi^2\left(f_x^2 + f_y^2\right)\right)\tilde{A} \\
&= \frac{-i\pi\lambda}{n_0}\left(f_x^2 + f_y^2\right)\tilde{A}
\end{aligned} \quad (7.37)$$

and solving for the field at $z_2 = z_1 + \Delta z$ gives

$$\tilde{A}(z_2) = \exp\left(\frac{-i\pi\lambda\Delta z}{n_0}\left(f_x^2 + f_y^2\right)\right)\tilde{A}(z_1). \qquad (7.38)$$

Taking the inverse Fourier transform to retrieve the field in real space yields

$$
\begin{aligned}
A(z_2) &= \mathfrak{F}^{-1}\left[\exp\left(\frac{-i\pi\lambda\Delta z}{n_0}\left(f_x^2 + f_y^2\right)\right)\tilde{A}(z_1)\right] \\
&= \mathfrak{F}^{-1}\left[\exp\left(\frac{-i\pi\lambda\Delta z}{n_0}\left(f_x^2 + f_y^2\right)\right)\mathfrak{F}(A(z_1))\right].
\end{aligned}
\qquad (7.39)
$$

Thus to obtain $A(z_2)$ from the distribution $A(z_1)$, the following steps are taken:

1. The 2D discrete Fourier transform of $A(z_1)$ is performed to get $\tilde{A}(z_1)$.

2. $\tilde{A}(z_2) = \tilde{A}(z_1)\exp\left(\frac{-i\pi\lambda\Delta z}{n_0}\left(f_x^2 + f_y^2\right)\right)$ is calculated.

3. The inverse Fourier transform of the function $\tilde{A}(z_2)$ is performed to get $A(z_2)$.

Note that for propagation through an anisotropic medium where for instance the field experiences walk-off in say the x direction, $\hat{P}$ becomes

$$\hat{P} = \frac{i}{2k}\nabla_T^2 - \tan\rho\frac{\partial}{\partial x} \qquad (7.40)$$

and following a similar set of steps $A(z_2)$ can be found by

$$A(z_2) = \mathfrak{F}^{-1}\left[\exp\left(\frac{-i\pi\lambda\Delta z}{n(\theta)}\left(f_x^2 + f_y^2\right) - \frac{\pi f_x \tan\rho}{k}\right)\mathfrak{F}(A(z_1))\right]. \qquad (7.41)$$

7.3 Propagation in Nonlinear Media

7.3.1 Split Step Method

The split step method [13] provides a convenient way of solving Eqn. 7.1 when both linear and nonlinear operators are present. In this method the nonlinear medium is be split into a number of slices along the propagation direction, each of same thickness, say Δz. For propagation of light through each slice, all the nonlinearities are initially set to zero, $\widehat{NL} = 0$, and the field is propagated a distance Δz using the appropriate linear propagation method described earlier in this chapter. At the end of linear propagation through the slice the linear operator is turned off, $\hat{P} = 0$, and the nonlinear portion is solved using the values of the electric field just obtained from the linear propagation through the slice.

The updated amplitude at the end of a slice is used as the input field for propagation through the next slice using the same procedure. Repeating the steps for all the slices the field at the end of the nonlinear medium is obtained. Then depending on the experiment, the field is integrated to determine the power or energy exiting the nonlinear medium or the linear propagation method is used to propagate the beam to then next element in the experiment. An example application of the split step method is given in Sect. 7.4.2.

7.4 Application Examples

7.4.1 Phase Retrieval

To accurately model propagation of a complex electro-magnetic field, both the spatial amplitude and phase distributions of the incident beam must be known. For beams which propagate as a circularly symmetric, TEM_{00} mode, simple expressions exist for describing the phase. In practice there are many situations where the laser beams are not truly TEM_{00}. Such situations include the case of high energy pulsed lasers, and laser beams generated from nonlinear processes such as OPO or SHG. The beam inside the resonator of a high energy laser is usually multimode, or the resonator is designed such that a flat top amplitude profile is resonated in order to lower the peak fluence inside the cavity while maximizing energy extraction. While this prevents damage to the optical components inside the resonator, the resulting beam is non-Gaussian.

A simple method to determine the phase distribution of a beam based on the beam propagation analysis is described in Ref. [14] and more recently in Ref. [15]. The addition of discrete wavelet transforms (DWT) to aide in noise suppression improves the accuracy of the technique. The technique consists of measuring the irradiance distribution of the beam on two planes at two locations, say at the input face of a nonlinear crystal, and on a plane where any detector or diagnostic equipment is located. Say I_1 and I_2 denote the two dimensional irradiance distributions at the two planes which are separated by a distance z_{prop}.

To determine the spatial distribution of the phase, the following iterative procedure is used:

1. At the plane of interest, say where $I_1(x, y)$ was recorded, an initial guess of the phase is made. The phase in this initial guess is usually chosen to be zero.

2. Next, the complex electric field at this plane is defined as

$$A_1(x, y) = V_1(x, y) \, e^{i\phi_1(x,y)} \qquad (7.42)$$

where $V_1(x,y)$ is proportional to $\sqrt{I_1(x,y)}$.

3. Using Eqn. 7.39, A_1 is propagated a distance z_{prop} to find A_2.

4. Next, the distribution $A_2(x,y)$ is written as

$$A_2(x,y) = V_2(x,y)\,e^{i\phi_2(x,y)}. \qquad (7.43)$$

5. The amplitude $V_2(x,y)$ is replaced with the known distribution from $\sqrt{I_2(x,y)}$.

6. A discrete wavelet transform (DWT) is applied to the phase array $\phi_2(x,y)$ to smooth out the numerical result and decrease the influence of noise.

7. Field A_2 is then propagated a distance $-z_{prop}$ back to the start plane. As before, the amplitude and phase are separated, the amplitude is replaced with the original V_1 and a DWT is reapplied to the new phase distribution $\phi_1(x,y)$.

8. The updated A_1 is then propagated the distance z_{prop} and the process is repeated until the values of $\phi_1(x,y)$ converge.

This procedure is a predictor-corrector technique which has been extensively used in the literature, as shown in References [16] and [17]. Typically 25 to 100 round trip propagations between the two planes are needed to obtain convergence of the phase to a stable solution. To verify that convergence of the solution is attained, the relative error between the phase at the incident plane is monitored after each round trip until the difference between successive iterations falls below a user defined threshold or acceptable error.

To demonstrate this technique, a high energy beam from a q-switched Nd:YAG laser was apertured and focused by a lens. This created an Airy beam profile at focus which has a non-negligible phase profile. An infrared camera was used to capture the beam images through focus. The camera was an array of 320×256 pixels with pixel size (pitch) of 30 μm on each side. The beam focus was located at a distance of 53 cm from the focusing lens. Images for phase retrieval were chosen at distances 10 and 14 cm after the focus, with the phase retrieved at the distance of 10 cm. Beam images at these planes and the numerically retrieved phase are shown in Fig. 7.2.

With the retrieved phase and the known amplitude given by $\sqrt{I_1}$ the complex field was determined at a distance of 10 cm from the focus. This field distribution was propagated to focus and the beam irradiance was then calculated at the focal plane. Comparison of the calculated and measured beam images are shown in Fig. 7.3. A good match between the predicted and measured beam shapes was found, showing the effectiveness of the technique.

7.4.2 Second Harmonic Generation

Second harmonic generation (SHG) can be described by a pair of coupled partial differential equations. Denoting the low (pump) and high (signal) fre-

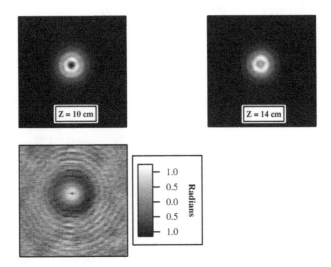

FIGURE 7.2: Top row shows the beam images used for phase retrieval at distances of $z = 10$ and 14 cm after the focusing lens. Bottom image is the numerically retrieved phase at $z = 10$ cm. All images are 4×4 mm^2.

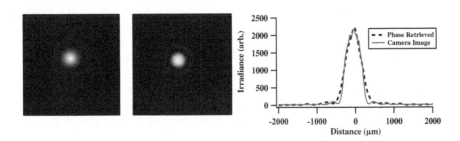

FIGURE 7.3: Beam profiles at a distance of $z = 54$ cm from the lens (at focus). Actual beam image (left) and numerical result using measured amplitude and retrieved phase at earlier plane (center) are shown. Right graph is a horizontal slice though the beam centroids for comparison.

quency beam parameters by the subscripts P and S respectively, and assuming that only the signal beam experiences walk-off, the coupled wave equations can be written as

$$\frac{\partial A_P}{\partial z} = \frac{i}{2k_P}\nabla_T^2 A_P + \frac{i2d_{\text{eff}}\omega_P}{cn_P}A_P^* A_S \exp\left(i\left(k_S - 2k_P\right)z\right) - \frac{\alpha_P}{2}A_P \quad (7.44)$$

$$\frac{\partial A_S}{\partial z} = \frac{i}{2k_S}\nabla_T^2 A_S - \tan\rho\frac{\partial A_S}{\partial x}$$

$$+ \frac{i2d_{\text{eff}}\omega_P}{cn_S}A_P A_P \exp\left(-i\left(k_S - 2k_P\right)z\right) - \frac{\alpha_S}{2}A_S. \quad (7.45)$$

The incident pump field may be known analytically or with its amplitude obtained from experimentally measured irradiance distribution and its phase determined by the method presented in Sect. 7.4.1. To determine the generated signal field, the split-step method can be applied where the right side of Eqns. 7.44 and 7.45 are separated into different operations, one for propagation and a second for absorption and nonlinearities allowing independent solution of each operator. A similar method is used in Ref. [18].

In the split step method for SHG used here, the coupled pump and signal field equations, Eqns. 7.44 and 7.45 are expressed in terms of propagation and nonlinear operators:

$$\frac{\partial A_P}{\partial z} = \widehat{P_P}A_P + \widehat{NL_P}$$

$$\frac{\partial A_S}{\partial z} = \widehat{P_S}A_S + \widehat{NL_S} \quad (7.46)$$

where the operators are

$$\widehat{P_P} = \frac{i}{2k_p}\nabla_T^2 \quad (7.47)$$

$$\widehat{P_S} = \frac{i}{2k_s}\nabla_T^2 - \tan\rho\frac{\partial}{\partial x} \quad (7.48)$$

$$\widehat{NL_P} = \frac{i2d_{\text{eff}}\omega_P}{cn_P}A_P^* A_S \exp(i\left(k_S - 2k_P\right)z) - \frac{\alpha_P}{2}A_P \quad (7.49)$$

$$\widehat{NL_S} = \frac{i2d_{\text{eff}}\omega_P}{cn_S}A_P A_P \exp(-i\left(k_S - 2k_P\right)z) - \frac{\alpha_S}{2}A_S. \quad (7.50)$$

The nonlinear medium is hypothetically divided into N slices, each of length Δz. Propagation over each Δz involves the following steps:

1. Based on the incident energy and spatial and temporal beam profiles, determine the incident pump field, $A_P(x, y, z = 0, t)$ just inside the front surface (i.e., taking into account the transmission coefficient).

2. Set the initial signal field $A_S(x, y, z = 0, t) = 0$.

3. Set $\widehat{NL} = 0$ and propagate the fields using $\widehat{P}$ for a distance Δz.

4. Set $\widehat{P} = 0$ and solve $\widehat{NL}$ for each field using finite difference techniques (discussed below) to determine A_P and A_S.

5. The A_p and A_s fields become the inputs to the next Δz slice.

6. Repeat 3, 4 and 5 until fields propagate to the end of the crystal.

7. Calculate the fields outside the crystal using the appropriate transmission coefficient.

At this point the fields are propagated to the next stage of the experiment or the output powers and energies can be determined. To determine the outputs, first the pump and signal irradiances, $I_{P,S}(x, y, z = l, t)$ just after the exit face are calculated by using Eqn. 7.4. For the case of continuous wave beams the irradiance is integrated over area to determine the total output power of each field. For pulsed beams, the irradiance is integrated both spatially and temporally to calculate the output energies.

Setting $\widehat{P} = 0$, the nonlinear portion of the coupled equations, including the linear absorption term, can be written as

$$\frac{\partial A_P}{\partial z} = \frac{i2d_{\text{eff}}\omega_P}{cn_P} A_P^* A_S \exp\left(i\left(k_S - 2k_P\right)z\right) - \frac{\alpha_P}{2} A_P \tag{7.51}$$

$$\frac{\partial A_S}{\partial z} = \frac{i2d_{\text{eff}}\omega_P}{cn_S} A_P A_P \exp\left(-i\left(k_S - 2k_P\right)z\right) - \frac{\alpha_S}{2} A_S. \tag{7.52}$$

These equations are solved using the implicit and forward finite difference method from the known values of A_P and A_S at the boundary plane $z = 0$. Equation 7.52 can be expressed in a finite difference form as

$$\frac{A_S^{z+\Delta z} - A_S^z}{\Delta z} = C_S A_P^{z+\Delta z} A_P^{z+\Delta z} - \frac{\alpha_S}{2} A_S^{z+\Delta z} \tag{7.53}$$

where

$$C_S = \frac{i2d_{\text{eff}}\omega_P}{cn_S} \exp\left(-i\left(k_S - 2k_P\right)z\right). \tag{7.54}$$

Solving for $A_S^{z+\Delta z}$ gives

$$A_S^{z+\Delta z} = \frac{A_S^z + \Delta z C_S A_P^{z+\Delta z} A_P^{z+\Delta z}}{1 + \Delta z \dfrac{\alpha_S}{2}} \tag{7.55}$$

which shows that the signal field at location $z + \Delta z$ is a function of the pump field at this location as well as the signal field at location z. Similarly, the finite difference expression for the pump field at the step $z + \Delta z$ is

$$A_P^{z+\Delta z} = \frac{A_P^z + \Delta z C_P A_P^{*,z} A_S^{z+\Delta z}}{1 + \Delta z \dfrac{\alpha_P}{2}} \tag{7.56}$$

where

$$C_P = \frac{i2d_{\text{eff}}\omega_P}{cn_P} \exp\left(i\left(k_S - 2k_P\right)z\right). \tag{7.57}$$

Using Eqns. 7.55 and 7.56, the values of A_S and A_P at the exit of the crystal can be calculated.

Bibliography

[1] J. A. Armstrong, N. Bloembergen, J. Ducuing, and P. S. Pershan, Interactions between light waves in a nonlinear dielectric, *Phys. Rev.* **127**, 1918, 1962.

[2] G. D. Boyd and D. A. Kleinman, Parametric interaction of focused gaussian light beams, *J. Appl. Phys.* **39**, 3597, 1968.

[3] J. W. Goodman, *Introduction to Fourier Optics*, 3rd ed., Roberts & Company, Greenwood Village, CO, 2005.

[4] I. S. Gradshteyn and I. M. Ryzhik, *Table of Integrals, Series and Products*, Academic Press, New York, 1980.

[5] A. E. Siegman, Quasi fast Hankel transmform, *Opt. Lett.* **1**, 1, 1977.

[6] L. Yu, M. Huang, M. Chen, W. Chen, W. Huang, and Z. Zhu, Quasi-discrete Hankel transform, *Opt. Lett.*, **23**, 409, 1998.

[7] M. Guizar-Sicairos and J. C. Gutirrez-Vega, Computation of quasi-discrete Hankel transforms of integer order for propagating optical wave fields, *J. Opt. Soc. Am. A* **21**, 53, 2004.

[8] Y. Kai-Ming, W. Shuang-Chun, C. Lie-Zun, W. You-Wen, and H. Yong-Hua, A quasi-discrete Hankel transform for nonlinear beam propagation, *Chinese Phys. B* **18**, 3893, 2009.

[9] M. Abramowitz and I. Stegun, *Handbook of Mathematical Functions*, Dover, New York, 1965.

[10] G. Lawrence, *GLAD Theory Manual Ver 5.6*, http://www.aor.com/anonymous/pub/theory.pdf, 2010.

[11] J. D. Schmidt, *Numerical Simulation of Optical Wave Propagation with Examples in MATLAB*, SPIE Press, Washington, 2010.

[12] D. Voelz, *Computational Fourier Optics: A MATLAB Tutorial*, SPIE Press, Washington, 2011.

[13] J. M. Burzler, S. Hughes and B. S. Wherrett, Split-step fourier methods applied to model nonlinear refractive effects in optically thick media, *App. Phy. B* **62**, 389, 1996.

[14] R. W. Gerchberg and W. O. Saxton, A practical algorithm for the determination of phase from image and diffraction plane pictures, Optik **35**, 237 (1972).

[15] A. R. Pandey, J. W. Haus, P. E. Powers, and P. P. Yaney, Optical field measurements for accurate modeling of nonlinear parametric interactions, *J. Opt. Soc. Am. B* **26**, 218, 2009.

[16] V. Y. Ivanov, V. P. Sivokon, and M. A. Vorontsov, Phase retrieval from a set of intensity measurements: theory and experiment, *J. Opt. Soc. Amer. A.* **9**, 1515, 1992.

[17] S. Matsuoka and K. Yamakawa, Wave-front measurements of terawatt-class ultrashort laser pulses by the Fresnel phase-retrieval method, *J. Opt. Soc. of Amer. B.* **17**, 663, 2000.

[18] M. A. Yates, C. L. Tsangaris, P. Kinsler, and G. H. C. New, Modelling of angular effects in nonlinear optical processes, *Opt. Comm.* **257**, 164, 2006.

A

Computer Codes for SFG Efficiency

A.1 The MATLAB codes for the collimated Gaussian beam case

In MATLAB environment, the following programs are used to find the sum-frequency generation power and energy conversion efficiencies (η_P and η_E, respectively) for collimated beams with Gaussian spatial and temporal distributions.

A.1.1 The file sfg_PE.m

```
%sfg_PE.m
clear all;
tic
K10 = 0.0001;
K1f = 5;
mk1 = 50;

dk1 = (K1f - K10)/mk1;

for ik = 1:mk1+1;

K1(ik) = K10 +(ik-1)*dk1;

%For eta_P, turn on the y(ik) in the next two lines
(by removing the % signs, if any)
%y(ik) = getysfg_p(K1(ik));
%y(ik) = 2*y(ik)/K1(ik);

%For eta_P, turn off the 2 lines for y(ik) below
(by inserting or retaining the % signs)

%For eta_E, turn off the y(ik) line above
%For eta_E, turn on the y(ik) lines below
```

253

```
y(ik) = getysfg_e(K1(ik));
y(ik) = 2*y(ik)/K1(ik)/sqrt(pi);

end

A = [K1',y'];

%save etaP_out_4  A -ASCII
save etaE_out_1  A -ASCII

toc
```

A.1.2   The file getysfg_p.m for the power conversion efficiency

```
function y = getysfg_p(x)

del = 1.1;
rho = 1.0;
p1 = 5;
sig = 0;

del1 = del/(del-1);
del2 = del*del1;

fr1 = @(r1) exp(-(r1.^2));
fr2 = @(r1) exp(-((rho^2)*(r1.^2)));

c0 = @(r1) p1*x^2*fr1(r1).*fr2(r1);
c1= @(r1) del*p1*x*fr2(r1) + del1*x*fr1(r1) + 0.25*sig^2;
c2 = @(r1) del2*p1*(x^2)*fr1(r1).*fr2(r1);
d1 = @(r1) sqrt(c1(r1).^2 - 4*c2(r1));
b1 = @(r1).5*(c1(r1)-d1(r1));
b2 = @(r1).5*(c1(r1)+d1(r1));
b22 = @(r1) sqrt(b2(r1));

gam2 = @(r1) (b1(r1)/b2(r1));
s1 = @(r1) ellipj(b22(r1),gam2(r1));

integrand = @(r1) r1.*b1(r1).*s1(r1).*s1(r1);

total = quad(integrand,0,3);

y = total/((1+p1/rho/rho));
```

A.1.3 The file getysfg_e.m for the energy conversion efficiency

```
function y = getysfg_e(x)

del = 1.1;
rho = 1.0;
p1 = 9.999;
sig = 0;

del1 = del/(del-1);
del2 = del*del1;

fr1 = @(r1,t1) exp(-(r1.^2)).*exp(-(t1.^2));
fr2 = @(r1,t1) exp(-((rho^2).*(r1.^2))).*exp(-(t1.^2));

c0 = @(r1,t1) p1*x^2*fr1(r1,t1).*fr2(r1,t1);
c1= @(r1,t1) del*p1*x*fr2(r1,t1) + del1*x*fr1(r1,t1) + 0.25*sig^2;
c2 = @(r1,t1) del2*p1*(x^2)*fr1(r1,t1).*fr2(r1,t1);
d1 = @(r1,t1) sqrt(c1(r1,t1).^2 - 4*c2(r1,t1));

b1 = @(r1,t1).5*(c1(r1,t1)-d1(r1,t1));
b2 = @(r1,t1).5*(c1(r1,t1)+d1(r1,t1));
b22 = @(r1,t1) sqrt(b2(r1,t1));

gam2 = @(r1,t1) (b1(r1,t1)/b2(r1,t1));
s1 = @(r1,t1) ellipj(b22(r1,t1),gam2(r1,t1));

integrand = @(r1,t1) r1.*b1(r1,t1).*s1(r1,t1).*s1(r1,t1);
total = dblquad(integrand,0,5.,0,5.,.001,@quad);

y = 2*total/((1+p1/rho/rho));

%the factor of 2 in y comes from the t integration which is
%done here from 0 to infinity but should be done from -inf to inf.
```

A.2 The Fortran Code For The Focused Beam Case of SHG

The Fortran code sfg_hmm.f finds the function h_{3mm} (h_3 optimized with respect to the phase mismatch parameter σ and the the focusing parameter ξ_2) for the sum frequency generation case, as a function of the normalized pump power K_2 and the focusing parameter ξ_1. The input parameters are read from a text file (sfg_hmm_in.txt) and the output is written to another file, the name of which is specified in a text file (sfg_hmax_fileout.txt). The three files sfg_hmm.f, sfg_hmm_in.txt and sfg_hmm_fileout.txt need to stay in the same folder when sfg_hmm.f is compiled and run.

A.2.1 The file sfg_hmm.f

```
c sfg_hmm.f

implicit real*4 (a-h,o-z)
real*4 n12,n13
real*4 data(24)
CHARACTER*25 fileout1,fileout2,fileout3

common/list1/n12,n13,del,p1,q1,aK1
common/list2/c10,c1f,c20,c2f,sig0,sigf
common/list3/mc1,mc2,msig
common/list4/rho1,rho2,rho3,alp1,alp2,alp3

open(7,file='sfg_hmax_in.txt',status='old')

do i = 1,24
read(7,*)data(i)
enddo

n12 = data(1)
n13 = data(2)
del = data(3)

c10 = data(4)
c1f = data(5)
mc1 = data(6)

c20 = data(7)
c2f = data(8)
mc2 = data(9)

sig0 = data(10)
sigf = data(11)
msig = data(12)

p1 = data(13)
q1 = data(14)
```

```
aK1 = data(15)

alp1 = data(16)
alp2 = data(17)
alp3 = data(18)

rho1 = data(19)
rho2 = data(20)
rho3 = data(21)

mx = data(22)
mz = data(23)
mt = data(24)

close(7)

open(8,file='sfg_hmax_fileout.txt',status='old')

read(8,*)fileout1

close(8)

open(unit=9,file=fileout1,status='new')

call prop_FFT_W0(mx,mz,mt)

stop
end

subroutine prop_fft_W0(mx,mz,mt)

implicit real*4 (a-h,o-z)
real*4 n12,n13

complex*8 u1(mx,mx),u2(mx,mx),u3(mx,mx),i,s1,h1

real*4 h(210), hm(210), sigmax(210)
real*4 sum1(mt)

common/list1/n12,n13,del,p1,q1,aK1
common/list2/c10,c1f,c20,c2f,sig0,sigf
common/list3/mc1,mc2,msig
common/list4/rho1,rho2,rho3,alp1,alp2,alp3

i = (0.0,1.0)
```

```
        pi = 4.0*atan(1.00)

akk3 = sqrt(aK1)
akk3 = akk3*del/sqrt(del-1.0)

dc2 = (c2f-c20)/mc2
dc1 = (c1f-c10)/mc1
dsig = (sigf-sig0)/msig

write(6,202)'c2','hmm-sfg','c1m','sigmm'
write(9,202)'c2','hmm-sfg','c1m','sigmm'

202 format(5X,A3,10x,A8,8x,A4,10x,A6)
do ic2 = 1,mc2 + 1
c2 = c20 +(ic2-1)*dc2

do ic1 = 1,mc1 + 1
c1 = c10 +(ic1-1)*dc1

do isig = 1,msig+1
sig = sig0 + (isig-1)*dsig

mxm = mx/2+1

z21 = c1
x0 = 4.*sqrt(1.+z21*z21)

p = 2.*x0/mx
q = 0.5/x0

dz1 = 1./mz
dx = p

a1 = c1/cos(rho1)/cos(rho1)
a2 = c1*n12*(del-1.)/cos(rho2)/cos(rho2)
a3 = c1*n13*(del-1.0)/del/cos(rho3)/cos(rho3)

b1 = tan(rho1)*sqrt(2.*c1*q1)
b2 = tan(rho2)*sqrt(2.*c1*q1)
b3 = tan(rho3)*sqrt(2.*c1*q1)

rho = sqrt(c2/c1/n12/(del - 1.0))

call Gauss(u1,u2,u3,c1,c2,p1,rho,mx)
```

```
do iz = 1,mz

isign = 1

c FFT1 converts 2D array to 1D array for FFT

call FFT1(u1,mx,p,isign)
call FFT1(u2,mx,p,isign)
call FFT1(u3,mx,p,isign)

call FFTPROPWO(u1,dz1,mx,a1,b1,p)
call FFTPROPWO(u2,dz1,mx,a2,b2,p)
call FFTPROPWO(u3,dz1,mx,a3,b3,p)

isign = -1

call FFT1(u1,mx,q,isign)
call FFT1(u2,mx,q,isign)
call FFT1(u3,mx,q,isign)
c end of linear propagation through dist dz1

uu = u1(mxm,mxm)*conjg(u1(mxm,mxm))

call NL_SFG_1(u1,u2,u3,mx,akk3,sig,dz1,iz)

enddo

mxm = mx/2+1

uu = real(u3(mxm,mxm))
sum1 = 0.0
sum2 = 0.0
sum3 = 0.0
do ix = 1,mx
do iy = 1,mx

sum1 = sum1 + u1(ix,iy)*conjg(u1(ix,iy))
sum2 = sum2 + u2(ix,iy)*conjg(u2(ix,iy))
sum3 = sum3 + u3(ix,iy)*conjg(u3(ix,iy))

enddo
enddo

 sum1 = sum1*dx*dx
 sum2 = sum2*dx*dx
```

```
  sum3 = sum3*dx*dx

h(isig) = c2*sum3/pi/aK1/p1

enddo
c end of the sigma loop

call fmax(msig,h,hmax,isigmax)

hm(ic1) = hmax

sigmax(ic1) = sig0+(isigmax-1)*dsig
c write(6,*)c2,c1,hm(ic1),sigmax(ic1)

enddo
c end of c1 loop

call fmax(mc1,hm,hmm,ic1m)

c1m = c10 + (ic1m-1)*dc1
sigmm = sigmax(ic1m)

del1=del*del/(del-1.0)

write(6,*)c2,hmm,c1m,sigmm
write(9,*)c2,hmm,c1m,sigmm

enddo
c end of c2 loop
write(9,*)
write(9,*)
write(9,*)
write(9,*) ' mx = ',mx, 'mz = ',mz
write(9,*)'sig0 = ',sig0,'sigf = ',sigf, 'msig = ',msig
write(9,*)'c10 = ',c10,'c1f = ',c1f, 'mc1 = ',mc1
write(9,*)'del = ',del, 'rho = ',rho
write(9,*)'aK1 = ',aK1, 'akk3 = ',akk3, 'p1 = ',p1

close(unit=9)

102 continue
return
  end

subroutine Gauss(u1,u2,u3,c1,c2,p1,rho,mx)
```

```
implicit real*4 (a-h,o-z)
real*4 n12,n13

complex*8 i,u1(mx,mx),u2(mx,mx),u3(mx,mx)


 i = (0.0, 1.0)
 pi = 4.0*atan(1.00)


z21 = c1
x0 = 4.*sqrt(1.+z21*z21)
p = 2.*x0/mx

CCCCCCCCCCCCCCCCCCCCCCCCCCCCCCCCCCCCCCCCCCCCCCCCCCC CCCCCCCCCCCCCCCCCCC
c         Initial Conditions
          do i1 = 1, mx
    do   j1 = 1,mx
x1 = (i1-1)*p - x0
   y1 = (j1-1)*p - x0
r12 = (x1*x1) + (y1*y1)
rr12 = r12*rho*rho

u1(i1,j1)=cexp(-r12*0.5/(1.0-i*c1))
u1(i1,j1)=u1(i1,j1)/(1.0-i*c1)

u2(i1,j1)=cexp(-rr12*0.5/(1.0-i*c2))
u2(i1,j1)=sqrt(p1)*u2(i1,j1)/(1.0-i*c2)

u3(i1,j1) = (0.0,0.0)

  enddo
  enddo
60 continue
cccccccCCCCCCCCCCCCCCCCCCCCCCCCCCCCCCCCCCCCCCCCCCCCCCCCCCCCCCCCCCCCCCCCC
mxm = mx/2+1

return
end


subroutine FFT1(u,mx1,dx,isign)

implicit real*4 (a-h,o-z)
dimension data(2*mx1*mx1)
```

```fortran
complex*8 v(mx1,mx1),i,u1(mx1,mx1),u(mx1,mx1)
integer nn(2)

 i = (0.0, 1.0)
 pi = 4.0*atan(1.00)

c 2D array u to 1D array 'data'
ccccccccccccccccccccccccccccccccc
 ic=0
       do j1=1,mx1
       do i1=1,mx1
        ic=ic+1
        data(ic)=real(u(i1,j1))
        ic=ic+1
        data(ic)=aimag(u(i1,j1))

enddo
       enddo
CCCCCCCCCCCCCCCCCCCCCCCCCCCCCCCCCCCCCCCC
c FFT
ccccccccccccccccccccccccccccccccccccccc
nn(1) = 1*mx1
nn1 = 2*mx1*mx1
nn(2) = mx1
ndim1 = 2

call Fourn(data,nn,ndim1,isign)
ccccccccccccccccccccccccccccccccccccccc

c 1D array data to 2D array u
cccccccccccccccccccccccccccccccc
118 continue

ic1 = 0
       do icq=1,ic,2
          j1=int(icq/(2*mx1))+1
          i1=(icq-(j1-1)*mx1*2)/2+1
          u(i1,j1)=cmplx(data(icq),data(icq+1))*(dx**2)

       enddo

return
end
```

```
subroutine FFTPROPWO(u,dz1,mx1,a,b,dx)
    implicit real*4 (a-h,o-z)
complex*8 v(mx1,mx1),i,u1(mx1,mx1),u(mx1,mx1)

real*8 rhox(mx1),rhoy(mx1)

 i = (0.0, 1.0)
 pi = 4.0*atan(1.00)

do j1 = 1,mx1/2
     rhox(j1) = (j1-1)/(mx1*dx)
     rhoy(j1) = (j1-1)/(mx1*dx)
   enddo

do i1 = (mx1/2)+1,mx1
rhox(i1) = (i1-mx1-1)/(mx1*dx)
rhoy(i1) = (i1-mx1-1)/(mx1*dx)
   enddo


do j1 = 1,mx1
          do j2 = 1,mx1

  r12 = rhox(j1)**2 + rhoy(j2)**2

 g = -4.*pi*pi*a*r12 + 2.*pi*b*rhox(j1)

  u(j1,j2) = u(j1,j2)*cexp(i*g*dz1)
enddo
enddo

return
end


Function aj0(x)
Implicit real*4 (a-z)

If (x.gt.3.0) go to 30

A1 = 2.2499997
A2 = 1.2656208
A3 = 0.3163866
A4 = 0.0444479
A5 = 0.0039444
```

```
A6 = 0.0002100

 xx = (x/3.0)**2

aj0 = 1.0  -   A1*xx + A2*xx*xx  - A3*(xx**3)
aj0 = aj0 + A4*(xx**4) -   A5*(xx**5) + A6*(xx**6)

go to 60
30 continue

B0 = 0.79788456
B1 = 0.00000077
B2 = 0.00552740
B3 = 0.00009512
B4 = 0.00137237
B5 = 0.00072805
B6 = 0.00014476

C0 = 0.78539816
C1 = 0.04166397
C2 = 0.00003954
C3 = 0.00262573
C4 = 0.00054125
C5 = 0.00029333
C6 = 0.00013558

Xx1 = 3.0/x

F0 = B0 -   B1*xx1 - B2*xx1*xx1 - B3*(xx1**3)
F0 = F0 + B4*(xx1**4) - B5*(xx1**5)+ B6*(xx1**6)

T0 = x - C0 - C1*xx1 - C2*(xx1**2) + C3*(xx1**3)
T0 = T0 - C4*(xx1**4) - C5*(xx1**5) + C6*(xx1**6)

aj0 = cos(T0)*F0/sqrt(x)
60 continue
Return
end

complex function f00(x,z21)
implicit real*4 (a-z)

complex*8 i

i = (0.0,1.0)
```

```
f00 = cexp(-x*x*0.5/(1.0+i*z21))
f00 = f00/(1.0+i*z21)
return
end


subroutine fmax(n,x,xmax,j)
implicit real*4 (a-z)
integer n,j
real*4 x(n)

xmax = x(1)
j = 1
if(n.eq.1)go to 7
do 6 i = 2,n
if(x(i).le.xmax) go to 6
xmax = x(i)
j = i
6 continue
7 return
end


      SUBROUTINE fourn(data,nn,ndim,isign)

         The source-code for this subroutine is available in
         Numerical Recipes, The Art of Scientific Computing,
         W.H. Press, B.P. Flannery, S.A. Teukolsky and W.T.
         Vetterling, Cambridge University Press, Cambridge, 1986

         return
         END


subroutine NL_SFG_1(u1,u2,u3,mx,akk3,sig,dz1,iz)
implicit real*4 (a-h,o-z)
real*4 n12,n13

complex*8 i,u1(mx,mx),u2(mx,mx),u3(mx,mx),d,d1
complex*8 u1c,u2c
c real*4 x(mxa),y(mxa)


common/list1/n12,n13,del,p1,q1,aK1
common/list2/c10,c1f,c20,c2f,sig0,sigf
```

```
common/list3/mc1,mc2,msig
common/list4/rho1,rho2,rho3,alp1,alp2,alp3

i = (0.0,1.0)

akk1 = akk3*(del-1.0)/del/cos(rho1)/cos(rho1)

akk2 = akk3/del/cos(rho2)/cos(rho2)

z11 =  iz*dz1

do iy = 1,mx
  do ix = 1,mx

    u1c = conjg(u1(ix,iy))
    u2c = conjg(u2(ix,iy))

d = cexp(-i*sig*z11)
d1 = cexp(i*sig*z11)

      u1(ix,iy) = u1(ix,iy) + i*akk1*dz1*u2c*u3(ix,iy)*d1
      u1(ix,iy) = u1(ix,iy) - 0.5*alp1*u1(ix,iy)*dz1

       u2(ix,iy) = u2(ix,iy) + i*akk2*dz1*u1c*u3(ix,iy)*d1
      u2(ix,iy) = u2(ix,iy) - 0.5*alp2*u2(ix,iy)*dz1

       u3(ix,iy) = u3(ix,iy) + i*akk3*dz1*u1(ix,iy)*u2(ix,iy)*d
      u3(ix,iy) = u3(ix,iy) - 0.5*alp3*u3(ix,iy)*dz1

    enddo
  enddo
  return
  end
```

A.2.2 sfgh_{mm}_in.txt

The file sfgh_{mm}_in.txt should contain just a row of numbers for the input variables, as shown below as an example:

1.0 1.0 1.5 2 3.5 150 1 3 2 -2 -5 30 .1 000.0001 .0001 0.0 0.0 0.0 0.0 0.0 0.0 32 25 1

A.2.3   sfg_hmm_fileout.txt

The file sfg_hmm_fileout.txt should contain just the name of a desired output file, for example

```
sfg_hmm_out1
```

B

Computer Codes for SHG Efficiency

B.1 MATLAB code for SHG efficiency of collimated Gaussian beams

In MATLAB environment, the following programs are used to find the second harmonic generation conversion efficiency for collimated beams with Gaussian spatial and temporal distributions.

B.1.1 shg_IPE.m

The program shg_IPE.m calls files getyshg_I.m, getyshg_P.m and the file getyshg_E.m to calculate the irradiance, power and energy conversion efficiencies of collimated Gaussian beams as a function of the intensity parameter K_1 of the pump beam for the phase matched ($\sigma = 0$) case.

```
%shg_IPE.m
clear all;
tic
K10 = 0.00001;
K1f = 50;
mk1 = 200;

alk1 = log10(K10);
alk2 = log10(K1f);

dk1 = (K1f - K10)/mk1;

for ik = 1:mk1+1;

K1(ik) = K10 +(ik-1)*dk1;
%ak1(ik) = alk1 + (ik - 1)*dk1;
%K1(ik) = 10^ak1(ik);

%y(ik) = getyshg_I(K1(ik));
%y(ik) = getyshg_p(K1(ik));
```

```
y(ik) = getyshg_E(K1(ik));
end
A = [K1',y'];
%save shg_I_1  A -ASCII
%save shg_p_1  A -ASCII
save shg_E_1  A -ASCII
toc
```

B.1.2   The file getyshg_I.m for the irradiance conversion efficiency

```
%filename getyshg_I.m
function y = getyshg_I(x)

y=tanh(sqrt(x))*tanh(sqrt(x));
```

B.1.3   The file getyshg_P.m for the power conversion efficiency

```
%filename getyshg_P.m
function y = getyshg_P(x)

sig = 0;

fr1 = @(r1) exp(-(r1.^2));

c1 = @(r1) x*fr1(r1);

c2 = @(r1) tanh(sqrt(c1(r1)));

integrand = @(r1) r1.* fr1(r1).*c2(r1).*c2(r1);

total = quad(integrand,0,4);

y = 2*total;
```

B.1.4   The file getyshg_E.m for the energy conversion efficiency

```
%filename getyshg_E.m
function y = getyshg_E(x)

fr1 = @(r1,t1) exp(-(r1.^2)).*exp(-(t1.^2));
```

```
c1 = @(r1,t1) x*fr1(r1,t1);

c2 = @(r1,t1) tanh(sqrt(c1(r1,t1)));

integrand = @(r1,t1) r1.* fr1(r1,t1).*c2(r1,t1).*c2(r1,t1);

total = dblquad(integrand,0,4,0,4);

y = 4*total/sqrt(pi);
```

B.2 The Fortran Code For The Focused Beam Case of SHG

The Fortran code shg_hmax.f finds the function h_{sm} (h_s optimized with respect to the phase mismatch parameter σ) for the second harmonic generation case, as a function of the normalized pump power K_2 and the pump focusing parameter ξ_p. The input parameters are read from a text file (shg_hmax_in.txt) and the output is written to another file, the name of which is specified in a text file (shg_hmax_fileout.txt). The three files shg_hmax.f, shg_hmax_in.txt and shg_hmax_fileout.txt need to stay in the same folder when shg_hmax.f is compiled and run.

B.2.1 The file sfg_hmax.f

```
c       SHG_hmax.f

        implicit real*8 (a-h,o-z)
real*8 kappa,k2,K20,K2f,mu
real*8 data(18)
CHARACTER*25 fileout1,fileout2,fileout3

CHARACTER*20 fileout

common/list1/ kappa,B1,B2,mu,AA1,AA2,ak
common/list2/K20,K2f,mk2
common/list3/cp0,cpf,mcp
common/list4/sig0,sigf,msig
common/list5/dx,dy

open(7,file='shg_hmax_in.txt',status='old')
```

```
do i = 1,18
read(7,*)data(i)
enddo

K20 = data(1)
K2f = data(2)
mk2 = data(3)

sig0 = data(4)
sigf = data(5)
msig = data(6)

cp0 = data(7)
cpf = data(8)
mcp = data(9)

B1 = data(10)
B2 = data(11)

AA1 = data(12)
AA2 = data(13)

mz = data(14)
mx = data(15)
my = data(16)

mu = data(17)
ak = data(18)

close(7)

open(8,file='shg_hmax_fileout.txt',status='old')

read(8,*)fileout1

close(8)

open(unit=9,file=fileout1,status='new')

call prop_FFT_W0(mz,mx,my)

stop
end
```

```
subroutine prop_fft_W0(mz,mx,my)

      implicit real*8 (a-h,o-z)
real*8 kappa,k2,K20,K2f,mu
complex*16 u1(mx,my),u2(mx,my),i,s1,h1
real*8 h(5000), hm(5000), sigmax(5000)
real*4 aK2, ahmm, acpm, asigmm

common/list1/ kappa,B1,B2,mu,AA1,AA2,ak
common/list2/K20,K2f,mk2
common/list3/cp0,cpf,mcp
common/list4/sig0,sigf,msig
common/list5/dx,dy

      i = (0.0,1.0)
      pi = 4.0*datan(1.0d0)

B = B2

dsig = (sigf-sig0)/msig
dcp = (cpf-cp0)/mcp
dk2 = (K2f-K20)/mk2

write(6,202)'K2','c_p','h_m','sigm'
write(9,202)'K2','c_p','h_m','sigm'
202 format(6X,A3,8x,A8,8x,A4,10x,A6)

do ik2 = 1,mk2

K2 = K20 + (ik2-1)*dk2

do icp = 1,mcp

cp = cp0 + (icp-1)*dcp

kappa = dsqrt(k2*cp)

do isig = 1,msig+1

sig = sig0 + (isig-1)*dsig

z21 = 2.*cp*mu
x0 = 3.5*dsqrt(1.+z21*z21)

a1 = cp
```

```
a2 = ak*cp

b1 = 0.
b2 = B*dsqrt(8.*cp)
amu = mu

px = (2.*x0 + b2)/mx
qx = 1./mx/px

py = (2.*x0)/my
qy = 1./my/py

dz1 = 1./mz
dx = px
dy = py


call Gauss(u1,u2,amu,cp,mx,my,x0)

do iz = 1,mz

isign = 1

call FFT1(u1,mx,my,px,py,isign)
call FFT1(u2,mx,my,px,py,isign)

call FFTPROPW0(u1,dz1,mx,my,a1,b1)
call FFTPROPW0(u2,dz1,mx,my,a2,b2)

isign = -1

call FFT1(u1,mx,my,qx,qy,isign)
call FFT1(u2,mx,my,qx,qy,isign)

uu = u1(mxm,mxm)*dconjg(u1(mxm,mxm))

call NL_SHG_1(u1,u2,mx,my,dz1,sig,iz)

enddo

sum = 0.0
do ix = 1,mx
do iy = 1,my
sum = sum + u2(ix,iy)*conjg(u2(ix,iy))
enddo
```

```
enddo

 sum = sum*dx*dy

h(isig) = 2.*sum/pi/k2

enddo
c endo of the sigma loop

call fmax(msig,h,hmax,isigmax)

hm(icp) = hmax
sigmax(icp) = sig0+(isigmax-1)*dsig
write(6,101)k2,cp,hm(icp),sigmax(icp)
write(9,101)k2,cp,hm(icp),sigmax(icp)
enddo
c end of cp loop
write(6,*)' '
write(9,*)' '

enddo
c end of K2 loop
write(9,*)
write(9,*)
write(9,*) ' mx,my = ',mx, 'mz = ',mz
write(9,*)'sig0 = ',sig0,'sigf = ',sigf, 'msig = ',msig
write(9,*)'cp0 = ',cp0,'cpf = ',cpf, 'mcp = ',mcp
write(9,*) 'rho = ',rho

close(unit=9)

101 format(e14.5,2x,f10.4,2x,e14.6,2x,e14.6,2x,e14.5)
return
   end

subroutine Gauss(u1,u2,zif,cp,mx,my,x0)
    implicit real*8 (a-h,o-z)
complex*16 i,u1(mx,my),u2(mx,my)
common/list5/dx,dy

 i = (0.0, 1.0)
 pi = 4.0*datan(1.0d0)

cccccccccccccccccccccccccccccccccccccccccccccccccc ccccccccccccccccccc
c          Initial Conditions
```

```
          do i1 = 1, mx
      do  j1 = 1,my
x1 = (i1-1)*dx - x0
    y1 = (j1-1)*dy - x0
r12 = (x1*x1) + (y1*y1)
u1(i1,j1)=cdexp(-r12*0.5/(1.0-2.*i*cp*zif))
u1(i1,j1)=u1(i1,j1)/(1.0-2.*i*cp*zif)
u2(i1,j1) = (0.0,0.0)
   enddo
   enddo
60 continue
cccccccCCCCCCCCCCCCCCCCCCCCCCCCCCCCCCCCCCCCCCCCCCCCCCCCCCCCCCCCCCC
mxm = mx/2+1

c write(6,*)'u2_peak',u1(mxm,mxm)
return
end

subroutine FFT1(u,mx1,my1,px,py,isign)
    implicit real*8 (a-h,o-z)
dimension data(2*mx1*my1)
complex*16 v(mx1,my1),i,u1(mx1,my1),u(mx1,my1)
integer nn(2)
common/list5/dx,dy

 i = (0.0, 1.0)
 pi = 4.0*datan(1.0d0)

c 2D array u to 1D array 'data'
cccccccccccccccccccccccccccccccccc
 ic=0
       do j1=1,my1
       do i1=1,mx1
        ic=ic+1
        data(ic)=real(u(i1,j1))
        ic=ic+1
        data(ic)=aimag(u(i1,j1))

enddo
      enddo
CCCCCCCCCCCCCCCCCCCCCCCCCCCCCCCCCCCCCCCCCCCC
c do id = 1,2*my1*mx1
c write(6,*)id,data(id)
c enddo
```

```
c FFT
ccccccccccccccccccccccccccccccccccccccccc
nn(1) = 1*mx1
nn1 = 2*mx1*my1
nn(2) = my1
ndim1 = 2
c isign = 1

call Fourn(data,nn,ndim1,isign)
ccccccccccccccccccccccccccccccccccccccccc

c 1D array data to 2D array u
ccccccccccccccccccccccccccccccccccc
118 continue

ic1 = 0
      do icq=1,ic,2
        j1=int(icq/(2*mx1))+1
        i1=(icq-(j1-1)*mx1*2)/2+1
        u(i1,j1)=dcmplx(data(icq),data(icq+1))*(px*py)
c        u(i1,j1)=dcmplx(data(icq),data(icq+1))

      enddo

return
end

subroutine FFTPROPWO(u,dz1,mx1,my1,a1,b1)
    implicit real*8 (a-h,o-z)
complex*16 v(mx1,my1),i,u1(mx1,my1),u(mx1,my1)
common/list5/dx,dy
real*8 rhox(mx1),rhoy(my1)


 i = (0.0, 1.0)
 pi = 4.0*datan(1.0d0)

do ix = 1,mx1/2
      rhox(ix) = (ix-1)/(mx1*dx)
enddo

do ix = (mx1/2)+1,mx1
rhox(ix) = (ix-mx1-1)/(mx1*dx)
enddo
```

```
do iy = 1,my1/2
     rhoy(iy) = (iy-1)/(my1*dy)
        enddo

do iy = (my1/2)+1,my1
rhoy(iy) = (iy-my1-1)/(my1*dy)
enddo

do i1 = 1,mx1
          do j2 = 1,my1

  r12 = rhox(i1)**2 + rhoy(j2)**2

  g = -4.*pi*pi*a1*r12 - 2.*pi*b1*rhox(i1)

  u(i1,j2) = u(i1,j2)*cdexp(i*g*dz1)
enddo
enddo

return
end


Function aj0(x)
Implicit real*8 (a-z)

If (x.gt.3.0) go to 30

A1 = 2.2499997
A2 = 1.2656208
A3 = 0.3163866
A4 = 0.0444479
A5 = 0.0039444
A6 = 0.0002100

 xx = (x/3.0)**2

aj0 = 1.0  -  A1*xx + A2*xx*xx  - A3*(xx**3)
aj0 = aj0 + A4*(xx**4) -  A5*(xx**5) + A6*(xx**6)

go to 60
30 continue

B0 = 0.79788456
```

```
B1 = 0.00000077
B2 = 0.00552740
B3 = 0.00009512
B4 = 0.00137237
B5 = 0.00072805
B6 = 0.00014476

C0 = 0.78539816
C1 = 0.04166397
C2 = 0.00003954
C3 = 0.00262573
C4 = 0.00054125
C5 = 0.00029333
C6 = 0.00013558

Xx1 = 3.0/x

F0 = B0 -  B1*xx1 - B2*xx1*xx1 - B3*(xx1**3)
F0 = F0 + B4*(xx1**4) - B5*(xx1**5)+ B6*(xx1**6)

T0 = x - C0 - C1*xx1 - C2*(xx1**2) + C3*(xx1**3)
T0 = T0 - C4*(xx1**4) - C5*(xx1**5) + C6*(xx1**6)

aj0 = dcos(T0)*F0/dsqrt(x)
60 continue
Return
end

complex function f00(x,z21)
implicit real*8 (a-z)

complex*16 i

i = (0.0,1.0)
f00 = cdexp(-x*x*0.5/(1.0+i*z21))
f00 = f00/(1.0+i*z21)
return
end


subroutine fmax(n,x,xmax,j)
implicit real*8 (a-z)
integer n,j
real*8 x(n)
```

```
xmax = x(1)
j = 1
if(n.eq.1)go to 7
do 6 i = 2,n
if(x(i).le.xmax) go to 6
xmax = x(i)
j = i
6 continue
7 return
end


        SUBROUTINE fourn(data,nn,ndim,isign)

        The source-code for this subroutine is available in
        Numerical Recipes, The Art of Scientific Computing,
        W.H. Press, B.P. Flannery, S.A. Teukolsky and W.T.
        Vetterling, Cambridge University Press, Cambridge, 1986

        return
        END

subroutine NL_SHG_1(u1,u2,mx,my,dz1,sig,iz)
implicit real*8 (a-h,o-z)
real*8 kappa,ka,k2,mu
complex*16 i,u1(mx,my),u2(mx,my),d,d1,u11,u1c


common/list1/ kappa,B1,B2,mu,AA1,AA2,ak

i = (0.0,1.0)
ka = kappa
z11 =  (iz)*dz1

d = cdexp(-i*sig*z11)

d1 = dcos(sig*z11) + i*dsin(sig*z11)

do iy = 1,my
  do ix = 1,mx
    u1c = conjg(u1(ix,iy))
u11 = u1(ix,iy)

    u1(ix,iy) = u1(ix,iy) + 0.5*i*ka*dz1*u1c*u2(ix,iy)*d/ak
     u1(ix,iy) = u1(ix,iy)*(1.0 - 0.5*AA1*dz1)
```

```
u11 = u1(ix,iy)
      u2(ix,iy) = u2(ix,iy) + i*ka*dz1*u11*u11*d1
    u2(ix,iy) = u2(ix,iy)*(1. - 0.5*AA2*dz1)
   enddo
enddo
mxm = mx/2 + 1
uu = u1(mxm,mxm)*dconjg(u1(mxm,mxm))
return
end
```

B.2.2 shg_hmax_in.txt

The text file shg_hmax_in.txt should contain only a column of numbers, for example, the column below given as an example

```
0.0001
1.
2
0
5.
50
2
4
20
0.0
0.0
0.
0.
100
32
32
0.5
0.5
```

B.2.3 shg_hmax_fileout.txt

The text file shg_hmax_fileout.txt should contain only the name of the output file, for example

```
shg_hmax_out1
```

C

The Fortran Source Code for QPM-SHG Efficiency

C.1 qpmshg.f

The Fortran code qpmshg.f finds the function h_s for the quasi-phase matched second harmonic generation case, as a function of the parameter δ_1, for given values of the pump focusing parameter ξ_p and the normalized pump power denoted by K_2. The input parameters are read from a text file (qpmshg_in.txt) and the output is written to another file, the name of which is specified in a text file (qpmshg_fileout.txt). The three files qpmshg.f, qpmshg_in.txt and qpmshg_fileout.txt need to stay in the same folder when qpmshg.f is compiled and run.

```
c qpmshg_h.f

      implicit real*8 (a-h,o-z)
real*8 kappa,K2,mu,data(12)
CHARACTER*20 fileout
CHARACTER*35 fileout1,fileout2,fileout3

common/list1/kappa,mu,AA1,AA2,del
common/list2/K2,cp,del0,delf,mdel
common/list3/dx,dy

open(7,file='qpmshg_in.txt',status='old')

do i = 1,12
read(7,*)data(i)
enddo

K2 = data(1)
cp = data(2)

del0 = data(3)
delf = data(4)
```

```
mdel = data(5)

AA1 = data(6)
AA2 = data(7)

mz = data(8)
mx = data(9)
my = data(10)

mu = data(11)
del = data(12)
close(7)

open(8,file='qpmshg_fileout.txt',status='old')

read(8,*)fileout1

close(8)

open(unit=9,file=fileout1,status='new')

call prop_FFT_W0(mz,mx)

stop
end

subroutine prop_fft_W0(mz,mx)
      implicit real*8 (a-h,o-z)
real*8 kappa,K2,mu,data(11)
CHARACTER*20 fileout
complex*16 u1(mx,mx),u2(mx,mx),i,s1,h1,s2

common/list1/kappa,mu,AA1,AA2,del
common/list2/K2,cp,del0,delf,mdel
common/list3/dx,dy

      i = (0.0,1.0)
      pi = 4.0*datan(1.0d0)

   write(6,202)'delta_1','h'
   write(9,202)'delta_1','h'
202    format(10X,A3,13x,A8)

ddel = (delf-del0)/mdel
do idel = 1,mdel+1
```

```
del1 = del0 + (idel-1)*ddel

sig = 1.*pi*mz*(1.+del1)
kappa = dsqrt(K2*cp)
z21 = cp*(1.-mu)
x0 = 4.*dsqrt(1.+z21*z21)

p = 2.*x0/mx
q = 0.5/x0

dz1 = 1./mz
dx = p

alc1 = dz1

a1 = cp
a2 = del*cp

b1 = 0.
b2 = 0.

amu = mu

call Gauss(u1,u2,amu,cp,mx,x0)

do iz = 1,mz
isign = 1

c FFT1 converts 2D array to 1D array for FFT

call FFT1(u1,mx,p,isign)
call FFT1(u2,mx,p,isign)

call FFTPROPWO(u1,alc1,mx,a1,b1,p)
call FFTPROPWO(u2,alc1,mx,a2,b2,p)

isign = -1
c dx = 1.0/(mx*dx)

call FFT1(u1,mx,q,isign)
call FFT1(u2,mx,q,isign)

c end of linear propagation through dist dz1
```

```
uu = u1(mxm,mxm)*dconjg(u1(mxm,mxm))
c write(6,*)(iz)*dz1,uu

ajsign = 1.0

mz2 = 10
dz2 = dz1/mz2

do iz2 = 1,mz2

ccccccccccccccccccccccccccccccccccc

call NL_SHG_1(u1,u2,mx,dz1,dz2,sig,iz,iz2)

ccccccccccccccccccccccccccccccccccc

z121 = (iz-1)*dz1+(iz2-1)*dz2

mxm = mx/2+1
uu = cdabs(u2(mxm,mxm))

c write(6,*)z121,uu*uu/K2
c write(9,*)z121,uu*uu/K2

enddo
c end of iz2 loop

enddo
c end of iz loop

sum = 0.0
      do ix = 1,mx
  do iy = 1,mx
    sum = sum + u2(ix,iy)*conjg(u2(ix,iy))
  enddo
  enddo

sum = sum*dx*dx
h = 2.*sum/pi/K2

write(9,*)del1,h
write(6,*)del1,h

enddo
c endo of the delta1 loop
```

```fortran
      write(9,*)' '
      write(9,*)' '
      write(9,*)'K2 = ',K2, 'cp = ',cp
      write(6,*)'K2 = ',K2, 'cp = ',cp
      write(9,*)'mz =',mz
      write(6,*)'mz =',mz

      return
        end


      subroutine Gauss(u1,u2,amu,cp,mx,x0)
      implicit real*8 (a-h,o-z)
      complex*16 i,u1(mx,mx),u2(mx,mx)

       i = (0.0, 1.0)
       pi = 4.0*datan(1.0d0)

          p = 2.*x0/mx
       zif = 1.-amu
CCCCCCCCCCCCCCCCCCCCCCCCCCCCCCCCCCCCCCCCCCCCCCCCCCCCCC CCCCCCCCCCCCCCCCCCC
c         Initial Conditions
             do i1 = 1, mx
        do  j1 = 1,mx
      x1 = (i1-1)*p - x0
         y1 = (j1-1)*p - x0
      r12 = (x1*x1) + (y1*y1)
      u1(i1,j1)=cdexp(-r12*0.5/(1.0-1.*i*cp*zif))
      u1(i1,j1)=u1(i1,j1)/(1.0-1.*i*cp*zif)
      u2(i1,j1) = (0.0,0.0)
        enddo
        enddo
60 continue
ccccccCCCCCCCCCCCCCCCCCCCCCCCCCCCCCCCCCCCCCCCCCCCCCCCCCCCCCCCCCCCCCCCCC
      return
      end

      subroutine FFT1(u,mx1,dx,isign)
      implicit real*8 (a-h,o-z)
      dimension data(2*mx1*mx1)
      complex*16 v(mx1,mx1),i,u1(mx1,mx1),u(mx1,mx1)
      integer nn(2)

       i = (0.0, 1.0)
```

```
 pi = 4.0*datan(1.0d0)

c 2D array u to 1D array 'data'
cccccccccccccccccccccccccccccccc
 ic=0
       do j1=1,mx1
       do i1=1,mx1
        ic=ic+1
        data(ic)=real(u(i1,j1))
        ic=ic+1
        data(ic)=aimag(u(i1,j1))

enddo
       enddo

cccccccccccccccccccccccccccccccccccccccccc
c FFT
cccccccccccccccccccccccccccccccccccccccccc
nn(1) = 1*mx1
nn1 = 2*mx1*mx1
nn(2) = mx1
ndim1 = 2
c isign = 1

call Fourn(data,nn,ndim1,isign)
cccccccccccccccccccccccccccccccccccccccccc

c 1D array data to 2D array u
cccccccccccccccccccccccccccccccccccccccc
118 continue

ic1 = 0
       do icq=1,ic,2
         j1=int(icq/(2*mx1))+1
         i1=(icq-(j1-1)*mx1*2)/2+1
         u(i1,j1)=dcmplx(data(icq),data(icq+1))*(dx**2)
c        u(i1,j1)=dcmplx(data(icq),data(icq+1))

       enddo

return
end


subroutine FFTPROPW0(u,dz1,mx1,a1,b1,dx)
```

```fortran
      implicit real*8 (a-h,o-z)
      complex*16 v(mx1,mx1),i,u1(mx1,mx1),u(mx1,mx1)
      real*8 rhox(mx1),rhoy(mx1)

       i = (0.0, 1.0)
      pi = 4.0*datan(1.0d0)

      do j1 = 1,mx1/2
           rhox(j1) = (j1-1)/(mx1*dx)
           rhoy(j1) = (j1-1)/(mx1*dx)
      enddo

      do i1 = (mx1/2)+1,mx1
      rhox(i1) = (i1-mx1-1)/(mx1*dx)
      rhoy(i1) = (i1-mx1-1)/(mx1*dx)
      enddo

      do j1 = 1,mx1
                do j2 = 1,mx1
         r12 = rhox(j1)**2 + rhoy(j2)**2
             g = -4.*pi*pi*a1*r12 - 2.*pi*b1*rhox(j1)
        u(j1,j2) = u(j1,j2)*cdexp(i*g*dz1)
      enddo

      enddo

      return
      end


      Function aj0(x)
      Implicit real*8 (a-z)

      If (x.gt.3.0) go to 30

      A1 = 2.2499997
      A2 = 1.2656208
      A3 = 0.3163866
      A4 = 0.0444479
      A5 = 0.0039444
      A6 = 0.0002100

       xx = (x/3.0)**2

      aj0 = 1.0  -  A1*xx + A2*xx*xx  - A3*(xx**3)
```

```
aj0 = aj0 + A4*(xx**4) -   A5*(xx**5) + A6*(xx**6)

go to 60
30 continue

B0 = 0.79788456
B1 = 0.00000077
B2 = 0.00552740
B3 = 0.00009512
B4 = 0.00137237
B5 = 0.00072805
B6 = 0.00014476

C0 = 0.78539816
C1 = 0.04166397
C2 = 0.00003954
C3 = 0.00262573
C4 = 0.00054125
C5 = 0.00029333
C6 = 0.00013558

Xx1 = 3.0/x

F0 = B0 -   B1*xx1 - B2*xx1*xx1 - B3*(xx1**3)
F0 = F0 + B4*(xx1**4) - B5*(xx1**5)+ B6*(xx1**6)
T0 = x - C0 - C1*xx1 - C2*(xx1**2) + C3*(xx1**3)
T0 = T0 - C4*(xx1**4) - C5*(xx1**5) + C6*(xx1**6)

aj0 = dcos(T0)*F0/dsqrt(x)
60 continue
Return
end

complex function f00(x,z21)
implicit real*8 (a-z)
complex*16 i

i = (0.0,1.0)
f00 = cdexp(-x*x*0.5/(1.0+i*z21))
f00 = f00/(1.0+i*z21)
return
end


subroutine fmax(n,x,xmax,j)
```

```
implicit real*8 (a-z)
integer n,j
real*8 x(n)

xmax = x(1)
j = 1
if(n.eq.1)go to 7
do 6 i = 2,n
if(x(i).le.xmax) go to 6
xmax = x(i)
j = i
6 continue
7 return
end
```

```
         SUBROUTINE fourn(data,nn,ndim,isign)

         The source-code for this subroutine is available in
         Numerical Recipes, The Art of Scientific Computing,
         W.H. Press, B.P. Flannery, S.A. Teukolsky and W.T.
         Vetterling, Cambridge University Press, Cambridge, 1986

         return
         END
```

```
subroutine NL_SHG_1(u1,u2,mx,dz1,dz2,sig,iz,iz2)

implicit real*8 (a-h,o-z)
real*8 kappa,K2,mu,ka
complex*16 u1(mx,mx),u2(mx,mx),i,s1,h1
complex*16 d,d1,u1c,u11

common/list1/ kappa,mu,AA1,AA2,del
common/list2/K2,cp,del0,delf,mdel
common/list3/dx,dy

i = (0.0,1.0)

ka = kappa
z11 =   iz*dz1
z12 = z11+iz2*dz2

sign = (-1.0)**(iz)
ka = ka*sign
```

```
d = cdexp(-i*sig*z12)
d1 = cdexp(i*sig*z12)

do iy = 1,mx
  do ix = 1,mx
    u1c = conjg(u1(ix,iy))
      u1(ix,iy) = u1(ix,iy) + 0.5*i*ka*dz2*u1c*u2(ix,iy)*d/del
u11 = u1(ix,iy)
      u2(ix,iy) = u2(ix,iy) + i*ka*dz2*u11*u11*d1
    enddo
enddo
mxm = mx/2 + 1
uu = u1(mxm,mxm)*dconjg(u1(mxm,mxm))
return
end
```

C.2   qpmshg_in.txt

The text file qpmshg_in.txt should contain only a column of numbers, for example, the column below given as an example

```
5
3
-.02
.04
150
0
0
400
32
32
0
0.5
```

C.3   qpmshg_fileout.txt

The text file qpmshg_fileout.txt should contain only the name of the output
file, for example

qpmshg_out1

D

The Fortran Source Code for OPO Threshold and Efficiency

D.1 OPO.f

The Fortran code OPO.f used to calculate the function g_{mm} is presented below. For a given value of the ξ_p, the confocal parameter of the pump beam, the program maximizes the function h with respect to the confocal parameter of the signal beam ξ_s and the phase mismatch parameter σ. The ranges of ξ_p and σ over which the maximum of h occurs are set by trial and error.

```
c OPO.f

      implicit real*8 (a-h,o-z)
      real*8 kappa2,k2,mu
      CHARACTER*20 fileout

      common/list1/ kappa2,alpp,alpi,alps,cp,cs,etas,del
      common/list2/ cp0,cpf,cs0,csf,sig0,sigf
      common/list3/ mcp,mcs,msig

del = 1.1

write(6,*)'enter kappa2'
read(5,*)kappa2

write(6,*)'enter cp0,cpf,mcp'
read(5,*)cp0,cpf,mcp

write(6,*)'enter cs0,csf,mcs'
read(5,*)cs0,csf,mcs

write(6,*)'enter sig0,sigf,msig'
read(5,*)sig0,sigf,msig
```

```
write(6,*)'Enter Output File Name'
read(5,*)fileout

B = 0.0

open(unit=7,file=fileout,status='new')

mu = 0.5
zif = mu
mz = 100
mx = 1*32
call prop_FFT_WO(mz,mx)

stop
end


subroutine prop_fft_WO(mz,mx)

    implicit real*8 (a-h,o-z)
real*8 kappa2,k2

complex*16 up(mx,mx),ui(mx,mx),us(mx,mx),i,s1,h1
real*8 h2(1000),h2m(1000)

common/list1/ kappa2,alpp,alpi,alps,cp,cs,etas,del
common/list2/ cp0,cpf,cs0,csf,sig0,sigf
common/list3/ mcp,mcs,msig

    i = (0.0,1.0)

     pi = 4.0*datan(1.0d0)

alpp = 0.0
alps = 0.0
alpi = 0.0

dcp = (cpf-cp0)/mcp

dcs = (csf-cs0)/mcs

write(6,202)'cp','csm','sigmax','h2mm','gmm'
write(7,202)'cp','csm','sigmax','h2mm','gmm'
202 format(7X,A3,9x,A4,6x,A7,9x,A5,11x,A4)
```

```fortran
do icp = 1,mcp+1

do ics = 1,mcs+1

cp = cp0 + (icp - 1)*dcp
cs = cs0 + (ics - 1)*dcs

kappa = dsqrt(k2*cp)

dsig = (sigf-sig0)/msig

do isig = 1,msig+1

sig = sig0 + (isig-1)*dsig

mxm = mx/2+1

z21 = cp
x0 = 4.*dsqrt(1.+z21*z21)

p = 2.*x0/mx
q = 0.5/x0

dz1 = 1./mz
dx = p

ap = cs/del
ai = cs/(del-1.)

call Gauss(up,ui,mx,x0)

vs = 0.0

do iz = 1,mz

isign = 1

c FFT1 converts 2D array to 1D array for FFT

call FFT1(up,mx,p,isign)
call FFT1(ui,mx,p,isign)

call FFTPROPWO(up,dz1,mx,ap,p)
call FFTPROPWO(ui,dz1,mx,ai,p)
```

```
isign = -1

call FFT1(up,mx,q,-1)
call FFT1(ui,mx,q,-1)

c end of linear propagation through dist dz1

call NL_OPO_1(up,ui,sum1,mx,dz1,sig,iz,x0)

vs = vs + sum1
enddo

h2(isig) = vs*dz1

enddo

call fmax(msig,h2,h2max,isigmax)

sigmax = sig0+(isigmax-1)*dsig

   h2m(ics) = h2max

enddo

call fmax(mcs,h2m,h2mm,icsm)

csm = cs0 +(icsm-1)*dcs

gmm = kappa2/h2mm

write(7,101)cp,csm,sigmax,h2mm,gmm
write(6,101)cp,csm,sigmax,h2mm,gmm

enddo
write(7,*)' '
write(6,*)' '

write(7,*)'K2 = ',k2
write(7,*)'cp0,cpf',cp0,cpf

write(7,*)'cs0,csf,mcs',cs0,csf,mcs
write(7,*)'sig0,sigf,msig',sig0,sigf,msig


101 format(3f12.4,2x,2e15.4,2x,e12.4,2x,e12.4,2x,e13.4)
```

```
102 continue
return
   end


subroutine Gauss(up,ui,mx,x0)

    implicit real*8 (a-h,o-z)
complex*16 i,up(mx,mx),ui(mx,mx)
real*8 kappa2,k2

common/list1/ kappa2,alpp,alpi,alps,cp,cs,etas,del

 i = (0.0, 1.0)
 pi = 4.0*datan(1.0d0)

p = 2.*x0/mx

b1 = del*cp/cs


CCCCCCCCCCCCCCCCCCCCCCCCCCCCCCCCCCCCCCCCCCCCCCCCCCC CCCCCCCCCCCCCCCCCCC
c         Initial Conditions
          do i1 = 1, mx
   do   j1 = 1,mx
x1 = (i1-1)*p - x0
   y1 = (j1-1)*p - x0
r12 = (x1*x1) + (y1*y1)
up(i1,j1)=cdexp(-r12*b1*0.5/(1.0-i*cp))
up(i1,j1)=up(i1,j1)/(1.0-i*cp)
ui(i1,j1) = (0.0,0.0)
   enddo
   enddo
60 continue
ccccccCCCCCCCCCCCCCCCCCCCCCCCCCCCCCCCCCCCCCCCCCCCCCCCCCCCCCCCCCCCCCCCCC
mxm = mx/2+1

c write(6,*)'u2_peak',u1(mxm,mxm)
return
end


subroutine FFT1(u,mx1,dx,isign)
    implicit real*8 (a-h,o-z)
c parameter (mxa = 4*128)
```

```
dimension data(2*mx1*mx1)
complex*16 i,u(mx1,mx1)
c real*4 x(mxa),y(mxa)
c real*8 fx(mxa,mxa),fy(mxa,mxa)
integer nn(2)

 i = (0.0, 1.0)
 pi = 4.0*datan(1.0d0)

c 2D array u to 1D array 'data'
cccccccccccccccccccccccccccccccc
 ic=0
       do j1=1,mx1
       do i1=1,mx1
        ic=ic+1
        data(ic)=real(u(i1,j1))
        ic=ic+1
        data(ic)=aimag(u(i1,j1))

enddo
      enddo
CCCCCCCCCCCCCCCCCCCCCCCCCCCCCCCCCCCCCCCC


c FFT
ccccccccccccccccccccccccccccccccccccccc
nn(1) = 1*mx1
nn1 = 2*mx1*mx1
nn(2) = mx1
ndim1 = 2
c isign = 1

call Fourn(data,nn,ndim1,isign)
ccccccccccccccccccccccccccccccccccccccc

c 1D array data to 2D array u
cccccccccccccccccccccccccccccccc
118 continue

ic1 = 0
       do icq=1,ic,2
          j1=int(icq/(2*mx1))+1
          i1=(icq-(j1-1)*mx1*2)/2+1
          u(i1,j1)=dcmplx(data(icq),data(icq+1))*(dx**2)
c         u(i1,j1)=dcmplx(data(icq),data(icq+1))
```

```
          enddo

      return
      end


      subroutine FFTPROPWO(u,dz1,mx1,a1,dx)
          implicit real*8 (a-h,o-z)
c parameter (mxa = 4*128)
c dimension data(2*mx1*mx1)
      complex*16 i,u(mx1,mx1)

c real*4 x(mxa),y(mxa)
      real*8 rhox(mx1),rhoy(mx1)

c write(6,*)'a1 = ',a1
c write(6,*)'dx = ',dx

      b1 = 0.0

       i = (0.0, 1.0)
       pi = 4.0*datan(1.0d0)

      do j1 = 1,mx1/2
          rhox(j1) = (j1-1)/(mx1*dx)
          rhoy(j1) = (j1-1)/(mx1*dx)
        enddo

      do i1 = (mx1/2)+1,mx1
      rhox(i1) = (i1-mx1-1)/(mx1*dx)
      rhoy(i1) = (i1-mx1-1)/(mx1*dx)
        enddo


      do j1 = 1,mx1
              do j2 = 1,mx1

       r12 = rhox(j1)**2 + rhoy(j2)**2
c       r12 = r12*2.*pi*pi

       g = -4.*pi*pi*a1*r12 - 2.*pi*b1*rhox(j1)

c write(6,*)j1,j2,rhox(j1),rhoy(j2)
         u(j1,j2) = u(j1,j2)*cdexp(i*g*dz1)
```

```
enddo
enddo

return
end


Function aj0(x)
Implicit real*8 (a-z)

If (x.gt.3.0) go to 30

A1 = 2.2499997
A2 = 1.2656208
A3 = 0.3163866
A4 = 0.0444479
A5 = 0.0039444
A6 = 0.0002100

 xx = (x/3.0)**2

aj0 = 1.0  -  A1*xx + A2*xx*xx  - A3*(xx**3)
aj0 = aj0 + A4*(xx**4) -  A5*(xx**5) + A6*(xx**6)

go to 60
30 continue

B0 = 0.79788456
B1 = 0.00000077
B2 = 0.00552740
B3 = 0.00009512
B4 = 0.00137237
B5 = 0.00072805
B6 = 0.00014476

C0 = 0.78539816
C1 = 0.04166397
C2 = 0.00003954
C3 = 0.00262573
C4 = 0.00054125
C5 = 0.00029333
C6 = 0.00013558

Xx1 = 3.0/x
```

```fortran
F0 = B0 -  B1*xx1 - B2*xx1*xx1 - B3*(xx1**3)
F0 = F0 + B4*(xx1**4) - B5*(xx1**5)+ B6*(xx1**6)


T0 = x - C0 - C1*xx1 - C2*(xx1**2) + C3*(xx1**3)
T0 = T0 - C4*(xx1**4) - C5*(xx1**5) + C6*(xx1**6)

aj0 = dcos(T0)*F0/dsqrt(x)
60 continue
Return
end

complex function f00(x,z21)
implicit real*8 (a-z)

complex*16 i

i = (0.0,1.0)
f00 = cdexp(-x*x*0.5/(1.0+i*z21))
f00 = f00/(1.0+i*z21)
return
end


subroutine fmax(n,x,xmax,j)
implicit real*8 (a-z)
integer n,j
real*8 x(n)

xmax = x(1)
j = 1
if(n.eq.1)go to 7
do 6 i = 2,n
if(x(i).le.xmax) go to 6
xmax = x(i)
j = i
6 continue
7 return
end

        SUBROUTINE fourn(data,nn,ndim,isign)

        The source-code for this subroutine is available in
        Numerical Recipes, The Art of Scientific Computing,
        W.H. Press, B.P. Flannery, S.A. Teukolsky and W.T.
        Vetterling, Cambridge University Press, Cambridge, 1986
```

```
      return
      END

subroutine NL_OPO_1(up,ui,sum1,mx,dz1,sig,iz,x0)
implicit real*8 (a-h,o-z)
real*8 kappa2,ka,k2
complex*16 i,up(mx,mx),ui(mx,mx),us(mx,mx)
complex*16 e1,e2,u11,u1c,ds,sum,uic,uss,usc,upp,upc

common/list1/ kappa2,alpp,alpi,alps,cp,cs,etas,del

      i = (0.0,1.0)
       pi = 4.0*datan(1.0d0)

p = 2.*x0/mx
ka = kappa2
z11 =   (iz)*dz1

ds = (1.-i*cs) + 2.*i*cs*z11

b1 = 1.0

e1 = cdexp(-i*sig*z11)
e2 = dconjg(e1)

sum = 0.0
do iy = 1,mx
  do ix = 1,mx

x1 = (ix-1)*p - x0
   y1 = (iy-1)*p - x0
r12 = (x1*x1) + (y1*y1)
us(ix,iy)=(cdexp(-r12*0.5*b1/ds))/ds
usc = dconjg(us(ix,iy))

up(ix,iy)=up(ix,iy)+i*ka*cs*dz1*ui(ix,iy)*us(ix,iy)*e1*0.

upp = up(ix,iy)

ui(ix,iy)=ui(ix,iy)+i*dz1*up(ix,iy)*dconjg(us(ix,iy))

    enddo
 enddo
```

```fortran
sum = (0.0,0.0)
do iy = 1,mx
  do ix = 1,mx

upp = up(ix,iy)
upc = dconjg(upp)
uss = us(ix,iy)

uic = dconjg(ui(ix,iy))
usc = dconjg(us(ix,iy))

sum = sum + upp*usc*e2*(1.-1./del)*dconjg(ui(ix,iy))
  enddo
enddo
sum = sum*p*p*e2

sum1 = - 2.*ka*cp*dimag(sum)/pi
mxm = mx/2+1
uus = us(mxm,mxm)*dconjg(us(mxm,mxm))
return
end
```

Index